THERMOELASTIC DEFORMATIONS

SOLID MECHANICS AND ITS APPLICATIONS
Volume 48

Aims and Scope of the Series

The fundamental questions arising in mechanics are: *Why?*, *How?*, and *How much?* The aim of this series is to provide lucid accounts written by authoritative researchers giving vision and insight in answering these questions on the subject of mechanics as it relates to solids.

The scope of the series covers the entire spectrum of solid mechanics. Thus it includes the foundation of mechanics; variational formulations; computational mechanics; statics, kinematics and dynamics of rigid and elastic bodies; vibrations of solids and structures; dynamical systems and chaos; the theories of elasticity, plasticity and viscoelasticity; composite materials; rods, beams, shells and membranes; structural control and stability; soils, rocks and geomechanics; fracture; tribology; experimental mechanics; biomechanics and machine design.

The median level of presentation is the first year graduate student. Some texts are monographs defining the current state of the field; others are accessible to final year undergraduates; but essentially the emphasis is on readability and clarity.

Thermoelastic Deformations

by

D. IEŞAN

Department of Mathematics,
University of Iasi,
Rumania

and

A. SCALIA

Department of Mathematics,
University of Catania,
Italy

SPRINGER-SCIENCE+BUSINESS MEDIA, B.V.

A C.I.P. Catalogue record for this book is available from the Library of Congress.

ISBN 978-90-481-4752-6 ISBN 978-94-017-3517-9 (eBook)
DOI 10.1007/978-94-017-3517-9

Printed on acid-free paper

Contents

Introduction

The theory of thermoelasticity studies the interaction between thermal and mechanical fields in elastic bodies. This theory is of interest both for the mathematical and technical point of view. Intense interest has been shown recently in this field owing to the great practical importance of dynamical effects in aeronautics, nuclear reactors, and its potential importance in cryogenic applications. This work is concerned mainly with basic problems of the theory of thermoelasticity. Thermoelasticity of polar materials and the theories of thermoelasticity with finite wave speeds are not considered here. The reader interested in these subjects will find a full account in the works of Nowacki [280], Chandrasekharaiah [60] and Ignaczak [195]. Our purpose in this work is to present a systematic treatment of some results established in the theory of thermoelasticity. On the whole, the subject matter is directed towards recent developments.

Chapter 1 is concerned mainly with the development of the fundamental equations of the theory of thermoelasticity. The kinematics and primitive concepts associated with the basic principles are developed and emphasized only to the extent that they are needed in our treatment of the subject. Chapter 2 is devoted to a study of linear thermoelastic deformations for prestressed bodies. We have attempted to isolate those conceptual and mathematical difficulties which arise over and above those inherent in the problems concerned with unstressed bodies. The greatest of these arises from the dependence of the coefficients of the field equations on the functions describing the primary deformation. Chapter 3 deals with the dynamic theory of thermoelastic bodies without initial stress. Special attention is given to the study of steady vibrations by the method of the potential theory, transient waves and certain results on the propagation of singular surfaces. Chapter 4 is concerned with the linear theory of thermoelastostatics. This chapter includes a number of results pertaining to the boundary value problems of the equilibrium theory and the solutions of some special problems. Chapter 5 is devoted to the nonlinear theory of thermoelasticity. The applications included are problems considered relevant to the purpose of the text. We are cognizant of certain omissions which might properly belong in an article of this type. However, our selection of topics has been based upon limitation of space coupled with a desire to give precedence to basic aspects and applications of the theory. An account of the historical developments as well as references to other various contributions may be found in some of the works cited.

Most of the material is self-contained and the only prerequisite is a course in the theory of elasticity. We make no claim to completeness. Neither the contents, nor the list of works cited are exhaustive. Nevertheless, it is hoped that the present work gives a unified, accessible treatment of a part of the contributions that have been made to the subject.

1 The Foundations of the Theory of Thermoelasticity

1.1 Notation. The Basic Laws of Thermomechanics

We consider a body that at time t_0 occupies the region B_0 of Euclidean three-dimensional space E^3. Throughout this book, unless specified to the contrary, B_0 will denote a bounded regular region in the sense of Kellogg . The configuration of the body at time t_0 is taken as the reference configuration. The motion of the body is referred to the reference configuration and a fixed system of rectangular Cartesian axes. Let B_0 be at rest relative to the considered system of axes. We identify a typical particle X of the body with its position $\boldsymbol{X}$ in the reference configuration. The coordinates of a typical particle X in B_0 are X_A $(A = 1, 2, 3)$. The coordinates of this particle in the position $\boldsymbol{x}$ at time t are denoted by x_i. We have

$$\boldsymbol{x} = \boldsymbol{x}(\boldsymbol{X}, t) , \qquad (\boldsymbol{X}, t) \in \bar{B}_0 \times I , \tag{1.1.1}$$

where I is a given interval of time. We assume the continuous differentiability of $\boldsymbol{x}$ with respect to the variables X_A and t as many times as required and

$$\det \left(\frac{\partial x_i}{\partial X_A}\right) > 0 \qquad \text{on } \bar{B}_0 \times I . \tag{1.1.2}$$

Clearly, a motion of B_0 is a one-parameter family of mappings $\boldsymbol{x}^t : B_0 \to E^3$, $t \in I$. The image $\boldsymbol{x}^t(M)$ of any set of points $M \subset B_0$ is called the configuration of M at the time t. The configuration of the body at the time t is denoted by B and is called present configuration. We shall employ the usual summation and differentiation conventions: lower and upper case Latin subscripts (unless otherwise specified) are understood to range over the integers (1, 2, 3) whereas Greek subscripts are confined to the range (1, 2), summation over repeated subscripts is implied and subscripts preceded by a comma denote partial differentiation with respect to the corresponding Cartesian coordinate. Throughout this work, a superposed dot denotes the material derivative with respect to the time. Letters in boldface stand for tensors of an order $p \geq 1$, and if $\boldsymbol{v}$ has the order p, we write $v_{ij...k}$ (p subscripts) for the rectangular Cartesian components of $\boldsymbol{v}$. Let (t_0, t_1) denote a given interval of time. We say that g is of class $C^{M,N}$ on $B_0 \times (t_0, t_1)$ if g is continuous on $B_0 \times (t_0, t_1)$ and the functions $\dfrac{\partial^m}{\partial X_A \partial X_B ... \partial X_K}\left(\dfrac{\partial^n g}{\partial t^n}\right)$, $m \in \{0, 1, ..., M\}$, $n \in \{0, 1, ..., N\}$, $m + n \leq max(M, N)$, exist and are continuous on $B_0 \times (t_0, t_1)$. We write C^N for $C^{N,N}$. We consider an arbitrary material region $\mathcal{P}$ in the continuum, bounded by a surface $\partial\mathcal{P}$ at time t, and we suppose that P is the corresponding region in the reference configuration, bounded by the surface ∂P. Let the outward unit normal at

∂P be N_A, and let the outward unit normal at $\partial\mathcal{P}$ be n_i, referred to the rectangular frame of reference.

We are now in position to postulate the conservation laws (or balance principles) employed in classical continuum mechanics. We restrict our attention to motions in which mass elements are conserved. Thus, the *law of conservation of mass* is

$$\int_{\mathcal{P}} \varrho dv = \int_{\mathrm{P}} \varrho_0 dV, \tag{1.1.3}$$

for every part $\mathcal{P}$ and every time, where ϱ is the mass density in the present configuration, ϱ_0 is the mass density in the reference configuration, dv is the element of volume in the present configuration and dV is the element of volume in the reference configuration. We assume that the mass density is a positive function of class C^1 on $B_0 \times (t_0, t_1)$. The equation (1.1.3) leads by usual arguments to the following local form of the equation of conservation of mass

$$\varrho = \varrho_0 J, \tag{1.1.4}$$

where

$$J = \det\,(x_{i,A}). \tag{1.1.5}$$

With reference to the present configuration we write *the conservation of linear momentum*, for every regular subregion $\mathcal{P}$ of B and every time t, in the form

$$\frac{d}{dt}\int_{\mathcal{P}} \varrho \dot{\boldsymbol{x}} dv = \int_{\mathcal{P}} \varrho \boldsymbol{f} dv + \int_{\partial\mathcal{P}} \boldsymbol{t} da, \tag{1.1.6}$$

where $\boldsymbol{f}$ is the body force per unit mass, $\boldsymbol{t}$ is the stress vector measured per unit area of $\partial\mathcal{P}$ and da is the element of area in the present configuration. We assume that $\boldsymbol{f}$ is a continuous vector field on $B_0 \times (t_0, t_1)$ and $\boldsymbol{t}$ is continuous on $\bar{B}_0 \times [t_0, t_1)$ and of class $C^{1,0}$ on $B_0 \times (t_0, t_1)$. In view of (1.1.4), the equation (1.1.6) becomes

$$\int_{\mathcal{P}} \varrho \boldsymbol{f} dv + \int_{\partial\mathcal{P}} \boldsymbol{t} da = \int_{\mathcal{P}} \varrho \ddot{\boldsymbol{x}} dv. \tag{1.1.7}$$

If the components of stress across the coordinate planes are t_{ji}, it follows from (1.1.7) that (see, for example, Green and Zerna [152], p.70, Gurtin [158], p.45, Eringen [123], p.101) the well-known Cauchy's formula

$$t_i = t_{ji} n_j, \tag{1.1.8}$$

holds. The tensor t_{ji} is the Cauchy stress tensor. Using (1.1.5), we can reduce the equation (1.1.7) to

$$t_{ji,j} + \varrho f_i = \varrho \ddot{x}_i \qquad \text{on } B \times (t_0, t_1). \tag{1.1.9}$$

We postulate *the conservation of moment of momentum,* for every $\mathcal{P} \subset B$ and every time t, in the form

$$\frac{d}{dt}\int_{\mathcal{P}} \varrho \boldsymbol{x} \times \dot{\boldsymbol{x}} dv = \int_{\mathcal{P}} \varrho \boldsymbol{x} \times \boldsymbol{f} dv + \int_{\partial\mathcal{P}} \boldsymbol{x} \times \boldsymbol{t} da. \tag{1.1.10}$$

In view of (1.1.4), (1.1.8) and (1.1.9), from (1.1.10) we obtain the following local form of the conservation law of moment of momentum

$$t_{ij} = t_{ji}. \tag{1.1.11}$$

We postulate *the conservation of energy (the first law of thermodynamics)* for every regular subregion $\mathcal{P}$ of B and every time t, in the form (cf. Coleman and Noll [83], Green and Adkins [148], p.259),

$$\frac{d}{dt}\int_{\mathcal{P}} \varrho(\frac{1}{2}\dot{\boldsymbol{x}}^2 + \varepsilon)dv = \int_{\mathcal{P}} \varrho(\boldsymbol{f}\cdot\boldsymbol{v} + s)dv + \int_{\partial\mathcal{P}} (\boldsymbol{t}\cdot\boldsymbol{v} + h)da, \tag{1.1.12}$$

where ε is the internal energy per unit mass, s is the heat supply per unit mass and unit time, $\boldsymbol{v} = \dot{\boldsymbol{x}}$, and h is the heat flux across the surface $\partial\mathcal{P}$ measured per unit area of $\partial\mathcal{P}$, per unit time. We assume that ε is of class $C^{0,1}$, h is of class $C^{1,0}$, and s is continuous on $B_0 \times (t_0, t_1)$. By (1.1.8),

$$\int_{\partial\mathcal{P}} t_i \dot{x}_i da = \int_{\mathcal{P}} (t_{ji,j} v_i + t_{ji} v_{i,j})dv, \tag{1.1.13}$$

where $v_i = \dot{x}_i$. In view of (1.1.4), (1.1.9) and (1.1.13), the equation (1.1.12) becomes

$$\int_{\mathcal{P}} \varrho\dot{\varepsilon} dv = \int_{\mathcal{P}} (t_{ji} v_{i,j} + \varrho s)dv + \int_{\partial\mathcal{P}} h da. \tag{1.1.14}$$

We apply this equation to an arbitrary tetrahedron bounded by coordinate planes through the point $\mathbf{x}$ and by a plane whose unit normal is n_k, to obtain

$$h = q_i n_i, \tag{1.1.15}$$

where q_i is the heat flux across the x_i-plane at $\boldsymbol{x}$, per unit area of this plane and per unit time. Then, using (1.1.15) in (1.1.14) and applying the equation to an arbitrary domain, we find

$$\varrho\dot{\varepsilon} = t_{ji} v_{i,j} + q_{i,i} + \varrho s \qquad \text{on } B \times (t_0, t_1). \tag{1.1.16}$$

The *second law of thermodynamics* is the assertion that (cf. Truesdell and Noll [354], p.295)

$$\frac{d}{dt}\int_{\mathcal{P}} \varrho\eta dv - \int_{\mathcal{P}} \frac{1}{T}\varrho s dv - \int_{\partial\mathcal{P}} \frac{1}{T} h da \geq 0, \tag{1.1.17}$$

for every $\mathcal{P} \subset B$ and every time t. Here η is the entropy per unit mass, and T is the absolute temperature, which is assumed to be positive. We assume that η is of class $C^{0,1}$ and T is of class $C^{2,1}$ on $B_0 \times (t_0, t_1)$. In view of (1.1.15), the second law of thermodynamics is equivalent to

$$\varrho T\dot{\eta} - \varrho s - q_{i,i} + \frac{1}{T} q_i T_{,i} \geq 0 . \tag{1.1.18}$$

1.2 Alternative Forms of the Basic Laws

The work of the previous section is sufficiently general to be applied to any continuum, whether solid or fluid. When the continuum has a reference configuration through which it passes at time t_0, it is convenient to have an alternative form of the theory in which the stresses and heat flux are measured with respect to this configuration. We recall that $\mathcal{P}$ is the image of the regular region $P \subset B_0$.

The law of conservation of linear momentum (1.1.16) can be written in the form

$$\int_P \varrho_0 \ddot{x}_i dV = \int_P \varrho_0 f_i dV + \int_{\partial P} T_i dA , \tag{1.2.1}$$

where dV and dA are elements of volume and area in the reference configuration, and $\boldsymbol{T}$ is the stress vector associated with the surface $\partial\mathcal{P}$, but measured per unit area of the surface ∂P. We apply the equation (1.2.1) to a domain which in the reference state was a tetrahedron bounded by coordinate planes through the point $\boldsymbol{X}$ and by a plane whose unit normal is N_A, to obtain

$$T_i = T_{Ki} N_K , \tag{1.2.2}$$

where T_{Ki} is the first Piola -Kirchhoff stress tensor (see, e.g. Truesdell and Noll [354], p.124, Eringen [123], p.109). Then, using (1.2.2) in (1.2.1) and applying the equation to an arbitrary region from B_0, we get

$$T_{Ki,K} + \varrho_0 f_i = \varrho_0 \ddot{x}_i \qquad \text{on } B_0 \times (t_0, t_1) . \tag{1.2.3}$$

The law of conservation of moment of momentum (1.1.10) becomes

$$\int_P \varrho_0 \boldsymbol{x} \times \ddot{\boldsymbol{x}} dV = \int_P \varrho_0 \boldsymbol{x} \times \boldsymbol{f} dV + \int_{\partial P} \boldsymbol{x} \times \boldsymbol{T} dA , \tag{1.2.4}$$

for every regular region P $\subset B_0$ and every time. In view of (1.2.2) and (1.2.3), the law of conservation (1.2.4) reduces to the local form

$$x_{i,K} T_{Kj} = x_{j,K} T_{Ki} . \tag{1.2.5}$$

The first law of thermodynamics (1.1.12) is replaced by

$$\int_P \varrho_0(\dot{x}_i\ddot{x}_i + \dot{\varepsilon})dV = \int_P \varrho_0(f_i\dot{x}_i + s)dV + \int_{\partial P}(T_i\dot{x}_i + H)dA\,, \tag{1.2.6}$$

where H is the heat flux across the surface $\partial\mathcal{P}$, measured per unit area of ∂P, per unit time. If we follow an argument similar to that used to derive (1.2.2), then from (1.2.6) we obtain

$$H = Q_A N_A\,, \tag{1.2.7}$$

where Q_A is the heat flux across the surface that was originally the coordinate plane perpendicular to the X_A-axis through the point $\boldsymbol{X}$, measured per unit area of this plane and per unit time. In view of (1.2.2), (1.2.3) and (1.2.7), the equation (1.2.6) leads to the local equation of energy

$$\varrho_0\dot{\varepsilon} = T_{Ai}\dot{x}_{i,A} + \varrho_0 s + Q_{A,A} \qquad \text{on } B_0 \times (t_0, t_1)\,. \tag{1.2.8}$$

If we define the second Piola -Kirchhoff stress tensor T_{AB} by

$$T_{Ai} = x_{i,B}T_{AB}\,, \tag{1.2.9}$$

then the local balance of energy (1.2.8) becomes

$$\varrho_0\dot{\varepsilon} = T_{KL}\dot{x}_{i,K}x_{i,L} + \varrho_0 s + Q_{L,L}\,. \tag{1.2.10}$$

It follows from (1.2.5) and (1.2.9) that

$$T_{KL} = T_{LK}\,. \tag{1.2.11}$$

We denote by E_{KL} the Lagrangian strain tensor, defined by

$$2E_{KL} = x_{i,K}x_{i,L} - \delta_{KL}\,, \tag{1.2.12}$$

where δ_{KL} is Kroneker's delta. The right Cauchy -Green tensor C_{AB} and the left Cauchy -Green tensor b_{ij} are defined, respectively, by (see, e.g. Eringen [123], p.11)

$$C_{AB} = x_{i,A}x_{i,B}\,, \qquad b_{ij} = x_{i,K}x_{j,K}\,. \tag{1.2.13}$$

In view of (1.2.11) and (1.2.12), the equation of energy (1.2.10) becomes

$$\varrho_0\dot{\varepsilon} = T_{KL}\dot{E}_{KL} + \varrho_0 s + Q_{L,L}\,. \tag{1.2.14}$$

We record the following relations between the stress tensors (see, for example, Truesdell and Noll [354], p.124, Eringen [123], p.109)

$$Jt_{ij} = x_{i,K}T_{Kj} = x_{i,A}x_{j,B}T_{AB}\,. \tag{1.2.15}$$

We also note that (cf. Green and Adkins [148], p.260, Eringen [123], p.163)

$$Jq_i = x_{i,K}Q_K\,. \tag{1.2.16}$$

The entropy production inequality (1.1.17) becomes

$$\int_P \varrho_0\dot{\eta}dV - \int_P \frac{1}{T}\varrho_0 s dV - \int_{\partial P} \frac{1}{T}HdA \geq 0\,, \tag{1.2.17}$$

for every regular region P $\subset B_0$ and every time $t \in (t_0, t_1)$. If we use (1.2.7) in (1.2.9), then we obtain the local form of the entropy production inequality

$$\varrho_0 T\dot{\eta} - \varrho_0 s - Q_{A,A} + \frac{1}{T}Q_A T_{,A} \geq 0\,. \tag{1.2.18}$$

Introducing the Helmholtz free-energy per unit of initial volume,

$$\psi = \varrho_0(\varepsilon - T\eta)\,, \tag{1.2.19}$$

and combining (1.2.14) and (1.2.18) we arrive at the following local dissipation inequality

$$T_{KL}\dot{E}_{KL} - \varrho_0\eta\dot{T} - \dot{\psi} + \frac{1}{T}Q_K T_{,K} \geq 0\,. \tag{1.2.20}$$

1.3 Constitutive Equations. Consequences of the Second Law

A thermoelastic material is defined as one for which the following constitutive equations hold

$$\begin{aligned} \psi &= \tilde{\psi}\left(x_{i,A}, T, T_{,B}, X_M\right), \\ T_{KL} &= \tilde{T}_{KL}\left(x_{i,A}, T, T_{,B}, X_M\right), \\ \eta &= \tilde{\eta}\left(x_{i,A}, T, T_{,B}, X_M\right), \\ Q_L &= \tilde{Q}_L\left(x_{i,A}, T, T_{,B}, X_M\right). \end{aligned} \tag{1.3.1}$$

We assume that the response functions $\tilde{\psi}$, $\tilde{T}_{KL}$, $\tilde{\eta}$, and $\tilde{Q}_L$ are of class C^2 on their domain $\mathcal{C}$ which is the set of all $(x_{i,A}$, T, $T_{,B}$, $X_M)$ with det $(x_{i,A}) \neq 0$, $T > 0$, $\boldsymbol{X} \in B_0$. In the case of homogeneous bodies, $\tilde{\psi}$, $\tilde{T}_{KL}$, $\tilde{\eta}$ and $\tilde{Q}_L$ do not depend explicitly on X_M. The constitutive equations must satisfy invariance conditions under superposed rigid-body motions. Hence

$$\tilde{\psi}\left(x_{i,A}, T, T_{,B}, X_M\right) = \tilde{\psi}\left(Q_{ir}x_{r,A}, T, T_{,B}, X_M\right), \tag{1.3.2}$$

for all proper orthogonal tensors Q_{ir}. The functionals $\tilde{T}_{KL}$, $\tilde{\eta}$ and $\tilde{Q}_L$ must satisfy the same invariance conditions as $\tilde{\psi}$. In view of (1.1.2) we may write $x_{i,A}$ in the polar form

$$x_{i,A} = R_{iK} M_{KA} \,,$$

where M_{KA} is a positive definite symmetric tensor and R_{iK} is a rotation tensor,

$$R_{iK} R_{iL} = \delta_{KL} \,, \qquad R_{iK} R_{jK} = \delta_{ij} \,, \qquad \det\,(R_{iK}) = 1 \,.$$

We denote by δ_{KL}, δ_{ij} and δ_{iK} the Kronecker's delta. We may take the special value $R_{jK}\delta_{Kr}$ for Q_{rj} in (1.3.2) so that

$$\tilde{\psi}\,(x_{i,A}, T, T_{,B}, X_N) = \tilde{\psi}\,(M_{KL}, T, T_{,B}, X_N) \,.$$

Recalling (1.2.12), we have $2E_{KL} = M_{KP}M_{PL} - \delta_{KL}$. Since M_{KL} is a positive definite symmetric tensor, a single-valued function of M_{KL} can be replaced by a single-valued function of E_{AB}, so that we can replace $\tilde{\psi}$ by a different function,

$$\psi = \hat{\psi}\,(E_{KL}, T, T_{,N}, X_M) \,. \tag{1.3.3}$$

We may verify that $\hat{\psi}$ satisfies the condition (1.3.2) for arbitrary proper orthogonal values of Q_{ij}. In view of (1.3.3), the dissipation inequality (1.2.20) can be written in the form

$$(T_{KL} - \frac{\partial \hat{\psi}}{\partial E_{KL}})\dot{E}_{KL} - (\varrho_0 \eta + \frac{\partial \hat{\psi}}{\partial T})\dot{T} - \frac{\partial \hat{\psi}}{\partial T_{,K}}\dot{T}_{,K} + \frac{1}{T} Q_K T_{,K} \geq 0 \,. \tag{1.3.4}$$

We assume that ψ in (1.3.4) is arranged as a symmetric function of E_{KL}. For a given deformation and temperature, the inequality (1.3.4) is valid for all arbitrary values of $\dot{T}$, $\dot{T}_{,K}$ and $\dot{E}_{KL}$, subject to $E_{KL} = E_{LK}$. This fact implies that (see Coleman and Noll [83], Carlson [46], p.306),

$$T_{KL} = \frac{\partial \hat{\psi}}{\partial E_{KL}} \,, \qquad \varrho_0 \eta = -\frac{\partial \hat{\psi}}{\partial T} \,, \qquad \frac{\partial \hat{\psi}}{\partial T_{,K}} = 0 \,,$$

and

$$Q_K T_{,K} \geq 0 \,. \tag{1.3.5}$$

Thus, we conclude that the constitutive equations of thermoelastic bodies are given by

$$\begin{aligned} &\psi = \hat{\psi}\,(E_{KL}, T, X_M) \,, \\ &T_{KL} = \frac{\partial \hat{\psi}}{\partial E_{KL}} \,, \qquad \varrho_0 \eta = -\frac{\partial \hat{\psi}}{\partial T} \,, \\ &Q_L = \hat{Q}_L\,(E_{KN}, T, T_{,M}, X_A) \,. \end{aligned} \tag{1.3.6}$$

The inequality (1.3.5) implies that

$$\hat{Q}_M(E_{KL}, T, 0, X_A) = 0\,. \tag{1.3.7}$$

This relation has been established by Pipkin and Rivlin [301] who described it as showing the "non-existence of a piezocaloric effect."

In view of (1.2.19) and (1.3.6), the energy equation (1.2.14) reduces to

$$\varrho_0 T\dot{\eta} = Q_{K,K} + \varrho_0 s \qquad \text{on} \quad B_0 \times (t_0, t_1)\,. \tag{1.3.8}$$

1.4 Isotropic Bodies. Incompressible Media

When the body is *homogeneous and isotropic*, the free energy has the form

$$\psi = \hat{\psi}(I_1, I_2, I_3, T)\,, \tag{1.4.1}$$

where I_1, I_2 and I_3 are the principal invariants of the strain tensor C_{KL}

$$I_1 = C_{LL}\,, \quad I_2 = \frac{1}{2}\left(C_{LL}C_{KK} - C_{KL}C_{KL}\right)\,, \quad I_3 = J^2. \tag{1.4.2}$$

From (1.2.12), (1.2.13) and (1.4.2) we obtain

$$\frac{\partial I_1}{\partial E_{KL}} = 2\delta_{KL}\,, \quad \frac{\partial I_2}{\partial E_{KL}} = 2\Lambda_{KL}\,, \quad \frac{\partial I_3}{\partial E_{KL}} = 2I_3 C_{KL}^{(-1)}, \tag{1.4.3}$$

where

$$\Lambda_{KL} = I_1\delta_{KL} - C_{KL}\,, \qquad C_{KL}^{(-1)} = X_{K,i}X_{L,i}\,. \tag{1.4.4}$$

It follows from (1.3.6) and (1.4.3) that

$$T_{KL} = \gamma_0\delta_{KL} + \gamma_1 C_{KL} + \gamma_2 C_{KL}^{(-1)}, \tag{1.4.5}$$

where

$$\gamma_0 = 2\left(\frac{\partial\hat{\psi}}{\partial I_1} + I_1\frac{\partial\hat{\psi}}{\partial I_2}\right), \quad \gamma_1 = -2\frac{\partial\hat{\psi}}{\partial I_2}, \quad \gamma_2 = 2I_3\frac{\partial\hat{\psi}}{\partial I_3}\,. \tag{1.4.6}$$

We introduce the notations

$$\Phi = \frac{2}{\sqrt{I_3}}\frac{\partial\hat{\psi}}{\partial I_1}, \quad \Psi = \frac{2}{\sqrt{I_3}}\frac{\partial\hat{\psi}}{\partial I_2}, \quad p = 2\sqrt{I_3}\frac{\partial\hat{\psi}}{\partial I_3}\,. \tag{1.4.7}$$

Then, we get

$$T_{KL} = \sqrt{I_3}\left(\Phi\delta_{KL} + \Psi\Lambda_{KL} + pC_{KL}^{(-1)}\right). \tag{1.4.8}$$

It follows from (1.2.13), (1.2.15) and (1.4.5) that

$$t_{ij} = \Phi b_{ij} + \Psi B_{ij} + p\delta_{ij}\,, \tag{1.4.9}$$

where

$$B_{ij} = x_{i,K}x_{j,L}\Lambda_{KL} = I_1 b_{ij} - b_{is}b_{sj}\,. \tag{1.4.10}$$

For an isotropic body we have (Pipkin and Rivlin [301], Green and Adkins [148], p.269, Eringen and Suhubi [124])

$$\hat{Q}_L = (K_1\delta_{LM} + K_2E_{LM} + K_3E_{LN}E_{NM})T_{,M}\,, \tag{1.4.11}$$

where K_i are functions of the invariants

$$I_1,\ I_2,\ I_3,\ I_4 = T_{,K}T_{,K},\ I_5 = E_{KL}T_{,K}T_{,L},\ I_6 = E_{BC}E_{CA}T_{,B}T_{,A},\ T. \tag{1.4.12}$$

If the continuum is incompressible then

$$I_3 = 1\,. \tag{1.4.13}$$

The inequality (1.3.4) is now subject to (1.4.13). It follows from (1.4.13) and (1.4.3) that $C^{(-1)}_{KL}\dot{E}_{KL} = 0$ so that, introducing the scalar function p as a Lagrangian multiplier, we obtain

$$T_{KL} = pC^{(-1)}_{KL} + \frac{\partial\psi}{\partial E_{KL}}\,.$$

For bodies which are homogeneous, isotropic, and incompressible, we have

$$\psi = \hat{\psi}\,(I_1, I_2, T)\,, \tag{1.4.14}$$

and the equations (1.4.8) are replaced by

$$T_{KL} = pC^{(-1)}_{KL} + \Phi\delta_{KL} + \Psi\Lambda_{KL}\,, \tag{1.4.15}$$

where Λ_{KL} is given by (1.4.4), p is an unknown function, and

$$\Phi = 2\frac{\partial\hat{\psi}}{\partial I_1}\,, \qquad \Psi = 2\frac{\partial\hat{\psi}}{\partial I_2}\,. \tag{1.4.16}$$

The function p will be found from the equations of motion and the boundary conditions.

1.5 Boundary-Initial-Value Problems

The complete system of field equations of nonlinear thermoelasticity consists of the equations of motion (1.2.3), the energy equation (1.3.8), the constitutive equations

(1.3.6), and the geometrical equations (1.2.12). The response functions $\hat{Q}_K$ are subjected to the restriction (1.3.5). We consider the following boundary conditions

$$\begin{aligned} x_i = \hat{x}_i \quad \text{on}\,\bar{\Sigma}_1 \times [t_0, t_1), \qquad T_{Ai}N_A = \hat{T}_i \quad \text{on}\;\Sigma_2 \times [t_0, t_1), \\ T = \hat{T} \quad \text{on}\;\bar{\Sigma}_3 \times [t_0, t_1)\,, \qquad Q_K N_K = \hat{Q} \quad \text{on}\;\Sigma_4 \times [t_0, t_1), \end{aligned} \tag{1.5.1}$$

where Σ_i $(i = 1,2,3,4)$ are subsurfaces of ∂B_0 such that $\bar{\Sigma}_1 \cup \Sigma_2 = \bar{\Sigma}_3 \cup \Sigma_4 = \partial B_0$, $\Sigma_1 \cap \Sigma_2 = \Sigma_3 \cap \Sigma_4 = \emptyset$, and $\hat{x}_i$, $\hat{T}_i$, $\hat{T}$ and $\hat{Q}$ are prescribed functions. We assume that $\hat{x}_i$ are continuous on $\bar{\Sigma}_1 \times [t_0, t_1)$, $\hat{T}_i$ are piecewise regular on $\Sigma_2 \times [t_0, t_1)$ and continuous in time, $\hat{T}$ is continuous on $\Sigma_3 \times [t_0, t_1]$ and $\hat{Q}$ is piecewise regular on $\Sigma_4 \times [t_0, t_1)$ and continuous in time. To the above equations and boundary conditions we add the initial conditions

$$\boldsymbol{x}(\boldsymbol{X},0) = \boldsymbol{a}(\boldsymbol{X})\,, \quad \dot{\boldsymbol{x}}(\boldsymbol{X},0) = \boldsymbol{b}(\boldsymbol{X})\,, \quad \eta(\boldsymbol{X},0) = \eta_0(\boldsymbol{X})\,, \quad \boldsymbol{X} \in B_0, \tag{1.5.2}$$

where $\boldsymbol{a}$, $\boldsymbol{b}$ and η_0 are prescribed continuous functions on B_0. If the solution $\boldsymbol{x} = \boldsymbol{x}(\boldsymbol{X},t)$, $T = T(\boldsymbol{X},t)$, $(\boldsymbol{X},t) \in B_0 \times (t_0, t_1)$ is determined then the density ϱ is given by (1.1.4). Sometimes *the convection condition* on the boundary is considered. This condition has the form

$$Q_K N_K = h(T - T_e) \qquad \text{on} \quad \partial B_0 \times [t_0, t_1)\,,$$

where T_e is the temperature of surrounding medium and $h(\geq 0)$ is the heat transfer coefficient.

It is sometimes useful to have the above relations expressed in terms of quantities referred to the present configuration. These quantities are the Cauchy stress tensor and the heat flux q_i. Now, the equations of motion (1.2.3) are replaced by (1.1.9), and the energy equation (1.3.8) becomes

$$\varrho T \dot{\eta} = q_{i,i} + \varrho s \qquad \text{on} \quad B \times (t_0, t_1). \tag{1.5.3}$$

The boundary conditions corresponding to (1.5.1) are

$$\begin{aligned} x_i = \tilde{x}_i \quad \text{on}\;\bar{S}_1 \times [t_0, t_1), \quad t_{ji}n_j = \tilde{t}_i \quad \text{on}\;S_2 \times [t_0, t_1), \\ T = \tilde{T} \quad \text{on}\;\bar{S}_3 \times [t_0, t_1), \quad q_i n_i = \tilde{Q} \quad \text{on}\;S_4 \times [t_0, t_1), \end{aligned} \tag{1.5.4}$$

where $\tilde{x}_i$, $\tilde{t}_i$, $\tilde{T}$ and $\tilde{q}$ are prescribed functions, and S_r is the image of $\Sigma_r (r = 1,2,3,4)$ by motion.

1.6 Curvilinear Coordinates

The region B_0 of Euclidean three-dimensional space can be described by a general curvilinear set of coordinates $(\theta^1, \theta^2, \theta^3)$. Thus we have

$$X_K = X_K(\theta^1, \theta^2, \theta^3)\,, \tag{1.6.1}$$

where the functions (1.1.6) are assumed to be of class C^3. Moreover, we suppose that the determinant of the transformation is not zero. It follows from (1.1.1) and (1.1.6) that

$$x_i = x_i(\theta^1, \theta^2, \theta^3, t)\,. \tag{1.6.2}$$

We denote by $\boldsymbol{E}_K$ the unit vectors along the fixed axes of the Cartesian coordinate system. We define the base vectors $\boldsymbol{g}_i, \boldsymbol{g}^i$ and metric tensors g_{ij}, g^{ij} by

$$\begin{aligned} \boldsymbol{g}_i &= \frac{\partial X_K}{\partial \theta^i}\boldsymbol{E}_K\,, \qquad & \boldsymbol{g}^i &= \frac{\partial \theta^i}{\partial X_K}\boldsymbol{E}_K\,, \\ g_{ij} &= \frac{\partial X_K}{\partial \theta^i}\frac{\partial X_K}{\partial \theta^j}\,, \qquad & g^{ij} &= \frac{\partial \theta^i}{\partial X_K}\frac{\partial \theta^j}{\partial X_K}\,. \end{aligned} \tag{1.6.3}$$

Moreover, we define the base vectors $\boldsymbol{G}_i, \boldsymbol{G}^i$ and the metric tensors G_{ij}, G^{ij} associated to the present configuration by

$$\begin{aligned} \boldsymbol{G}_i &= \frac{\partial x_s}{\partial \theta^i}\boldsymbol{E}_s\,, \qquad & \boldsymbol{G}^i &= \frac{\partial \theta^i}{\partial x_s}\boldsymbol{E}_s\,, \\ G_{ij} &= \frac{\partial x_s}{\partial \theta^i}\frac{\partial x_s}{\partial \theta^j}\,, \qquad & G^{ij} &= \frac{\partial \theta^i}{\partial x_s}\frac{\partial \theta^j}{\partial x_s}\,, \end{aligned} \tag{1.6.4}$$

where $\boldsymbol{E}_i = \delta_{iK}\boldsymbol{E}_K$. Clearly,

$$g^{is}g_{sj} = \delta^i_j\,, \quad G^{is}G_{sj} = \delta^i_j\,, \quad \boldsymbol{g}^i = g^{is}\boldsymbol{g}_s\,, \quad \boldsymbol{G}^i = G^{is}\boldsymbol{G}_s\,. \tag{1.6.5}$$

If we denote

$$g = \det\,(g_{ij})\,, \qquad G = \det\,(G_{ij})\,, \tag{1.6.6}$$

then

$$J = \sqrt{G/g}. \tag{1.6.7}$$

The strain tensor γ_{ij} is defined by

$$2\gamma_{ij} = G_{ij} - g_{ij}. \tag{1.6.8}$$

The displacement vector $\boldsymbol{u} = \boldsymbol{x} - \boldsymbol{X}$ can be expressed in terms of the above base vectors in the form

$$\boldsymbol{u} = u^i\boldsymbol{g}_i = u_i\boldsymbol{g}^i = U^i\boldsymbol{G}_i = U_i\boldsymbol{G}^i\,. \tag{1.6.9}$$

We note that

$$\frac{\partial \boldsymbol{u}}{\partial \theta^i} = u_{r|i}\boldsymbol{g}^r = U_{r\|i}\boldsymbol{G}^r, \tag{1.6.10}$$

where we have introduced the following covariant derivatives

$$u_{r|i} = \frac{\partial u_r}{\partial \theta^i} - {}_0\Gamma^s_{ri}u_s\,, \qquad U_{r\|i} = \frac{\partial U_r}{\partial \theta^i} - \Gamma^s_{ri}U_s\,, \tag{1.6.11}$$

with

$$\begin{aligned} {}_0\Gamma^s_{ij} &= g^{rs}{}_0\Gamma_{ijs}\,, \qquad & {}_0\Gamma_{ijs} &= \frac{\partial^2 X_K}{\partial\theta^i\partial\theta^j}\frac{\partial X_K}{\partial\theta^s}\,, \\ \Gamma^r_{ij} &= G^{rs}\Gamma_{ijs}\,, \qquad & \Gamma_{ijs} &= \frac{\partial^2 x_r}{\partial\theta^i\partial\theta^j}\frac{\partial x_r}{\partial\theta^s}\,. \end{aligned}$$

It follows from (1.6.3), (1.6.4), (1.6.8) and (1.6.10) that

$$\begin{aligned} 2\gamma_{ij} &= u_{i|j} + u_{j|i} + u^r|_i u_{r|j} = \\ &= U_{i\|j} + U_{j\|i} - U^r\|_i U_{r\|j}\,. \end{aligned} \tag{1.6.12}$$

We consider the following mixed strain tensor

$$\gamma^i_j = g^{ir}\gamma_{rj} = \frac{1}{2}(g^{ir}G_{rj} - \delta^i_j)\,.$$

Clearly,

$$\begin{aligned} \gamma^i_j + \lambda\delta^i_j &= \frac{1}{2}g^{ir}G_{rj} + (\lambda - \frac{1}{2})\delta^i_j = \frac{1}{2}\left[g^{ir}G_{rj} + \right. \\ &\left. + (2\lambda - 1)\right]\delta^i_j = \frac{1}{2}(g^{ir}G_{rj} + \mu\delta^i_j)\,, \end{aligned}$$

where $\mu = 2\lambda - 1$. Thus,

$$\begin{aligned} \det\,(g^{ir}G_{rj} + \mu\delta^i_j) &= \det q\left[2\gamma^i_j + (\mu+1)\delta^i_j\right] = \\ &= \mu^3 + I_1\mu^2 + I_2\mu + I_3\,. \end{aligned}$$

The invariants I_1, I_2 and I_3 of the mixed tensor γ^i_j coincide with the invariants (1.4.2) and can be expressed in the form

$$\begin{aligned} I_1 &= 3 + 2\gamma^r_r = g^{rs}G_{rs}\,, \\ I_2 &= 3 + 4\gamma^r_r + 2(\gamma^r_r\gamma^s_s - \gamma^s_r\gamma^r_s) = \frac{1}{2}(I_1^2 - g^{rm}g^{sn}G_{rs}G_{mn})\,, \\ I_3 &= \det\,(\delta^r_s + 2\gamma^r_s) = \det\,(g^{ri}G_{is}) = G/g\,. \end{aligned} \tag{1.6.13}$$

We now consider the law of balance of linear momentum. At the point M of B we construct a tetrahedron whose edges are formed by the coordinate curves MM_i

where $\boldsymbol{MM}_1 = \boldsymbol{G}_1 d\theta^1, \boldsymbol{MM}_2 = \boldsymbol{G}_2 d\theta^2, \boldsymbol{MM}_3 = \boldsymbol{G}_3 d\theta^3$. We denote by dS_i the area of the surface θ^i = constant of the tetrahedron, and by dS the area of surface $M_1M_2M_3$. Let $\boldsymbol{n} = n_i \boldsymbol{G}^i$ be the unit normal to this surface. Then (see Green and Zerna [152], p.59)

$$dS_i = n_i \sqrt{G^{ii}} dS \,. \tag{1.6.14}$$

By applying the equation (1.1.7) to the tetrahedron $MM_1M_2M_3$ we find that

$$\boldsymbol{t} = \sum_{i=1}^{3} \boldsymbol{t}_i n_i \sqrt{G^{ii}} \,, \tag{1.6.15}$$

where $\mathbf{t}_i$ is the stress vector associated with the surface θ^i = constant. In view of (1.6.15) we can introduce the contravariant and mixed stress tensors τ^{ij} and τ^i_j by

$$\boldsymbol{t}_i \sqrt{G^{ii}} = \tau^{ij} \boldsymbol{G}_j = \tau^i_j \boldsymbol{G}^j \,. \tag{1.6.16}$$

We also introduce the covariant stress tensor

$$\tau_{ij} = G_{ir} \tau^r_j \,.$$

It follows from (1.6.15) and (1.6.16) that

$$\boldsymbol{t}_i = \frac{1}{\sqrt{G}} \boldsymbol{T}_i n_i = \tau^{ij} n_i \boldsymbol{G}_j, \tag{1.6.17}$$

where

$$\boldsymbol{T}_i = \boldsymbol{t}_i \sqrt{G G^{ii}} = \sqrt{G} \tau^{ij} \boldsymbol{G}_j \,. \tag{1.6.18}$$

We note that

$$\frac{1}{\sqrt{G_{ss}}} \boldsymbol{G}_s \,,$$

are unit vectors along the coordinate curves. We define the physical components of the stress tensor by

$$\sigma_{ij} = \sqrt{\frac{G_{jj}}{G^{ii}}} \tau^{ij} \,. \tag{1.6.19}$$

Clearly,

$$\boldsymbol{t}_i = \sum_{j=1}^{3} \frac{\boldsymbol{G}_j}{\sqrt{G_{jj}}} \sigma_{ij} \,. \tag{1.6.20}$$

It follows from (1.1.7), (1.6.17) and the divergence theorem that

$$\frac{\partial \boldsymbol{T}_i}{\partial \theta^i} + \varrho \boldsymbol{f} \sqrt{G} = \varrho \ddot{\boldsymbol{x}} \sqrt{G} \,. \tag{1.6.21}$$

The law of conservation of moment of momentum (1.1.10) reduces to

$$\tau^{ij} = \tau^{ji}. \tag{1.6.22}$$

In view of (1.6.17), the equations of motion (1.6.21) can be expressed in the form

$$\tau^{ij}\|_i + \varrho f^j = \varrho a^j, \tag{1.6.23}$$

where

$$\boldsymbol{f} = f^i \boldsymbol{G}_i, \qquad \ddot{\boldsymbol{x}} = a^i \boldsymbol{G}_i. \tag{1.6.24}$$

The stress vector $\boldsymbol{t}$ is associated to a surface in the present configuration and measured per unit area of this surface.

Let us consider the element of area

$$dA_1 = |\boldsymbol{G}_2 d\theta^2 \times \boldsymbol{G}_3 d\theta^3| = \sqrt{GG^{11}}d\theta^2 d\theta^3.$$

The force acting on dA_1 is given by

$$\boldsymbol{t}_1\sqrt{GG^{11}}d\theta^2 d\theta^3 = \boldsymbol{T}_1 d\theta^2 d\theta^3.$$

In general, the force acting on the element of area $dA_i = \sqrt{GG^{ii}}d\theta^j d\theta^k$ is given by $\boldsymbol{T}_i d\theta^j d\theta^k$($i$ not summed; $i,\ j,\ k$ unequal). This force can be expressed in the form

$$\boldsymbol{T}_i d\theta^j d\theta^k = {}_0\boldsymbol{t}_i\sqrt{gg^{ii}}d\theta^j d\theta^k,$$

where ${}_0\boldsymbol{t}_i$ is the stress vector acting on the θ^i-surface in the present configuration but measured per unit area of the corresponding surface in the reference configuration. If we put

$${}_0\boldsymbol{t}_i\sqrt{g^{ii}} = p^{ij}\boldsymbol{G}_j,$$

then we have

$$\boldsymbol{T}_i = {}_0\boldsymbol{t}_i\sqrt{gg^{ii}} = \sqrt{g}p^{ij}\boldsymbol{G}_j. \tag{1.6.25}$$

It follows from (1.6.18) and (1.6.25) that

$$p^{ij} = \tau^{ij}\sqrt{I_3}. \tag{1.6.26}$$

Let S be the image of the oriented surface S_0 by motion. We denote by $\boldsymbol{N}$ the unit normal to S_0 and assume that

$$\boldsymbol{N} = {}_0n_i\boldsymbol{g}^i = {}_0n^i\boldsymbol{g}_i. \tag{1.6.27}$$

Then the stress vector across the surface S and measured per unit area of S_0 is given by

$$_0\boldsymbol{t} = {_0n_i} p^{ij} \boldsymbol{G}_j = \frac{1}{\sqrt{g}} {_0n_i} \boldsymbol{T}_i. \tag{1.6.28}$$

Clearly,

$$\boldsymbol{G}_i = \boldsymbol{g}_i + \frac{\partial \boldsymbol{u}}{\partial \theta^i} = \left(\delta_i^s + u^s|_i\right) \boldsymbol{g}_s,$$

so that the equations (1.6.18) and (1.6.25) can be expressed in the form

$$\boldsymbol{T}_i = \sqrt{G}\pi^{ij}\mathbf{g}_j = \sqrt{g}S^{ij}\mathbf{g}_j, \tag{1.6.29}$$

where

$$\pi^{ij} = \tau^{is}\left(\delta_s^j + u^j|_s\right), \qquad S^{ij} = p^{ir}\left(\delta_r^j + u^j|_r\right). \tag{1.6.30}$$

It follows from (1.6.18), (1.6.25) and (1.6.29) that

$$\tau^{ij}\boldsymbol{G}_j = \pi^{ij}\boldsymbol{g}_j, \qquad p^{ij}\boldsymbol{G}_j = S^{ij}\boldsymbol{g}_j.$$

In view of (1.6.21) and (1.6.29) we can express the equations of motion in terms of stress tensor S^{ij},

$$S^{ji}|_j + \varrho_0 \hat{f}^i = \varrho_0 \hat{a}^i, \tag{1.6.31}$$

where

$$\boldsymbol{f} = \hat{f}^i \boldsymbol{g}_i, \qquad \ddot{\boldsymbol{x}} = \hat{a}^i \boldsymbol{g}_i.$$

With an argument similar to that used to obtain the relation (1.6.15), from (1.1.14) we get

$$h = h_i n_i \sqrt{G^{ii}}, \tag{1.6.32}$$

where h_i is the heat flux associated to the surface θ^i = constant. We define the heat flux vector $\boldsymbol{q}$ by

$$\boldsymbol{q} = q^i \boldsymbol{G}_i = q_i \boldsymbol{G}^i, \tag{1.6.33}$$

where

$$q^i = h_i \sqrt{G^{ii}}. \tag{1.6.34}$$

The vector $\boldsymbol{q}$ is associated to a surface in the present configuration and measured per unit area of this surface and per unit time. From (1.6.32) we get

$$h = q^i n_i, \tag{1.6.35}$$

Similarly,

$$H = {_0n^i} h_i^0 \sqrt{g^{ii}}. \tag{1.6.36}$$

The heat flux across the surface S and measured per unit area of S_0 and per unit time is defined by

$$\boldsymbol{Q} = Q^i \boldsymbol{g}_i = Q_i \boldsymbol{g}^i, \tag{1.6.37}$$

where

$$Q^i = h_i^0 \sqrt{g^{ii}}\,. \tag{1.6.38}$$

The relation

$$h_i^0 \sqrt{g g^{ii}} = h_i \sqrt{G G^{ii}}\,,$$

implies that

$$Q^i = q^i \sqrt{I_3}\,. \tag{1.6.39}$$

We note also that

$$Q^i|_i = q^s\|_s \sqrt{I_3}\,. \tag{1.6.40}$$

The surface integrals in (1.1.12) become

$$\begin{aligned} \int\limits_{\partial\mathcal{P}} \boldsymbol{t}\cdot\boldsymbol{v}da &= \int\limits_{\partial\mathcal{P}} \frac{1}{\sqrt{G}}\boldsymbol{T}_i\cdot\boldsymbol{v}n_i da = \int\limits_{\mathcal{P}} \frac{1}{\sqrt{G}}\left(\frac{\partial \boldsymbol{T}_i}{\partial\theta^i}\cdot\boldsymbol{v} + \right. \\ &\left. + \boldsymbol{T}_i\cdot\frac{\partial\boldsymbol{v}}{\partial\theta^i}\right)dv\,, \qquad \int\limits_{\partial\mathcal{P}} h da = \int\limits_{\mathcal{P}} q^i\|_i dv\,. \end{aligned} \tag{1.6.41}$$

We note that

$$\dot\gamma_{ij} = \frac{1}{2}\dot G_{ij} = \frac{1}{2}(\dot{\boldsymbol{G}}_i\cdot\boldsymbol{G}_j + \boldsymbol{G}_i\cdot\dot{\boldsymbol{G}}_j) = \frac{1}{2}(\boldsymbol{G}_i\cdot\boldsymbol{v}_{,j} + \boldsymbol{G}_j\cdot\boldsymbol{v}_{,i})\,.$$

Thus, from (1.1.12), (1.6.21), (1.6.22) and (1.6.41) we obtain the following local form of energy equation

$$\varrho\dot\varepsilon = \tau^{ij}\dot\gamma_{ij} + q^i\|_i + \varrho s\,. \tag{1.6.42}$$

Similarly, from (1.2.6) we obtain

$$\varrho_0\dot\varepsilon = p^{ij}\dot\gamma_{ij} + Q^i|_i + \varrho_0 s\,. \tag{1.6.43}$$

The entropy production inequality (1.2.17) reduces to

$$\varrho_0 T\dot\eta - \varrho_0 s - Q^i|_i + \frac{1}{T}Q^i T|_i \geq 0\,. \tag{1.6.44}$$

From (1.2.19), (1.6.43) and (1.6.44) we get

$$-\varrho_0\eta\dot T - \dot\psi + p^{ij}\dot\gamma_{ij} + \frac{1}{T}Q^i T|_i \geq 0\,. \tag{1.6.45}$$

The constitutive equations (1.3.6) are replaced by

$$\psi = \bar{\psi}(\gamma_{ij},\, T,\, \theta^s)\,,$$

$$p^{ij} = \frac{1}{2}\left(\frac{\partial\bar{\psi}}{\partial\gamma_{ij}} + \frac{\partial\bar{\psi}}{\partial\gamma_{ji}}\right), \qquad \varrho_0\eta = -\frac{\partial\bar{\psi}}{\partial T}\,, \tag{1.6.46}$$

$$Q^i = \bar{Q}^i\left(\gamma_{rs}, T, \frac{\partial T}{\partial\theta^s}, \theta^m\right).$$

Here the differentiation with respect to γ_{ij} is understood to be performed holding all other components of the strain tensor constant including γ_{ji} if $i \neq j$. In view of (1.6.26) and (1.6.30) we obtain

$$\tau^{ij} = \frac{1}{2\sqrt{I_3}}\left(\frac{\partial\bar{\psi}}{\partial\gamma_{ij}} + \frac{\partial\bar{\psi}}{\partial\gamma_{ji}}\right), \qquad S^{ij} = \pi^{ij}\sqrt{I_3} = \frac{\partial\bar{\psi}}{\partial u_{j|i}}\,. \tag{1.6.47}$$

The energy equation (1.3.8) becomes

$$\varrho_0 T\dot{\eta} = Q^i|_i + \varrho_0 s\,. \tag{1.6.48}$$

When the body is homogeneous and isotropic, the free energy has the form (1.4.1) where I_1, I_2 and I_3 are given by (1.6.13). It follows from (1.6.13) and (1.6.47) that

$$\tau^{ij} = \Phi g^{ij} + \Psi B^{ij} + pG^{ij}, \tag{1.6.49}$$

where

$$\Phi = \frac{2}{\sqrt{I_3}}\frac{\partial\bar{\psi}}{\partial I_1}\,, \quad \Psi = \frac{2}{\sqrt{I_3}}\frac{\partial\bar{\psi}}{\partial I_2}\,, \quad p = 2\sqrt{I_3}\frac{\partial\bar{\psi}}{\partial I_3}\,, \tag{1.6.50}$$

$$B^{ij} = I_1 g^{ij} - g^{im}g^{jn}G_{mn}\,.$$

In the case of an incompressible body p is an unknown function.

The constitutive equation for the heat flux vector becomes (see Green and Adkins [148], p.268)

$$Q^i = (C_1\delta^i_j + C_2\gamma^i_j + C_3\gamma^i_m\gamma^m_j)T|^j\,, \tag{1.6.51}$$

where

$$T|^j = g^{jr}T|_r = g^{jr}\frac{\partial T}{\partial\theta^r}\,,$$

and C_s are functions of the invariants I_1, I_2, I_3, $I_4 = T|_i T|^i$, $I_5 = T|^i T|_j\gamma^j_i$, $I_6 = T|^i T_{|j}\gamma^j_s\gamma^s_i$, T. The equations (1.6.51) may be expressed in the form

$$Q^i = (C^*_1\delta^i_j + C^*_2\gamma^i_j + C^*_3\gamma^i_m\gamma^m_j)T\|^j\,, \tag{1.6.52}$$

where

$$T\|^i = G^{ir}T\|_r = G^{ir}\frac{\partial T}{\partial \theta^r},$$

and C_s^* are functions of the invariants I_1, I_2, I_3, $I_4^* = T\|_i T\|^i$, $I_5^* = T\|^i T\|_j \gamma_i^j$, $I_6^* = T\|^i T\|_j \gamma_s^j \gamma_i^s$, and T.

1.7 Singular Surfaces

In this section we present basic results for the study of surfaces of discontinuity in continuum mechanics. The modern treatment of singular surfaces goes back to the works of Thomas [349], p.22, [350], p.87, Truesdell and Toupin [353], p.498, Chadwick and Powdrill [51], Chen [64], p.301, Eringen [123], p.274, Mc Carthy [249], p.453.

a) **Preliminaries.** We express the position vector of a point of Euclidean three-dimensional space E^3 in the special form

$$\boldsymbol{x} = \boldsymbol{r}_0 + \theta^3 \boldsymbol{a}_3, \tag{1.7.1}$$

where $\boldsymbol{r}_0$ is a vector field which depends on θ^1 and θ^2 only and $\boldsymbol{a}_3$ is a unit vector field which also depends only on θ^1 and θ^2. The equation $\theta^3 = 0$ determines the surface

$$\boldsymbol{x} = \boldsymbol{r}_0(\theta^1, \theta^2). \tag{1.7.2}$$

Clearly, the vector $\boldsymbol{a}_3$ is perpendicular to the surface (1.7.2). We denote

$$\boldsymbol{a}_\alpha = \frac{\partial \boldsymbol{r}_0}{\partial \theta^\alpha}. \tag{1.7.3}$$

The covariant base vectors for the space are given by

$$\boldsymbol{g}_\alpha = \boldsymbol{a}_\alpha + \theta_3 \frac{\partial \boldsymbol{a}_3}{\partial \theta^\alpha}, \quad \boldsymbol{g}_3 = \boldsymbol{a}_3.$$

We have

$$\bar{\boldsymbol{a}}_3 \cdot \bar{\boldsymbol{a}}_\alpha = 0, \quad \boldsymbol{a}_3 \cdot \boldsymbol{a}_3 = 1, \quad \boldsymbol{a}_3 \cdot \frac{\partial \boldsymbol{a}_3}{\partial \theta^\alpha} = 0. \tag{1.7.4}$$

We note that for $\theta^3 = 0$ we have $\boldsymbol{g}_i = \boldsymbol{a}_i$. The vector fields $\boldsymbol{a}_1$ and $\boldsymbol{a}_2$ are the covariant base vectors of the surface. The values of $\boldsymbol{g}^i$, g_{mn}, g^{mn}, g for $\theta^3 = 0$ are denoted by $\boldsymbol{a}^i$, a_{mn}, a^{mn}, a, respectively. Thus, we have

$$\begin{aligned} &\boldsymbol{a}_\alpha \cdot \boldsymbol{a}_\beta = a_{\alpha\beta}, \quad a_{3i} = \delta_{i3}, \qquad a_{\alpha\beta}a^{\beta\gamma} = \delta_\alpha^\gamma, \quad a^{3i} = \delta^{3i}, \\ &\boldsymbol{a}^\alpha = a^{\alpha\beta}\boldsymbol{a}_\beta, \quad \boldsymbol{a}_\alpha = a_{\alpha\beta}\boldsymbol{a}^\beta, \quad \boldsymbol{a}^\alpha \cdot \boldsymbol{a}^\beta = a^{\alpha\beta}, \quad \boldsymbol{a}^3 = \boldsymbol{a}_3. \end{aligned} \tag{1.7.5}$$

Since

$$d\boldsymbol{r}_0 = \boldsymbol{a}_\alpha d\theta^\alpha,$$

the line element of the surface is

$$ds^2 = a_{\alpha\beta} d\theta^\alpha d\theta^\beta.$$

The scalar product

$$d\boldsymbol{r}_0 \cdot d\boldsymbol{a}_3 = -b_{\alpha\beta} d\theta^\alpha d\theta^\beta \,, \tag{1.7.6}$$

is the second fundamental form of the surface. Clearly,

$$b_{\alpha\beta} = -\boldsymbol{a}_\alpha \cdot \frac{\partial \boldsymbol{a}_3}{\partial \theta^\beta} = -\boldsymbol{a}_\beta \cdot \frac{\partial \boldsymbol{a}_3}{\partial \theta^\alpha} = \boldsymbol{a}_3 \cdot \frac{\partial \boldsymbol{a}_\alpha}{\partial \theta^\beta} = \boldsymbol{a}_3 \cdot \frac{\partial \boldsymbol{a}_\beta}{\partial \theta^\alpha} \,. \tag{1.7.7}$$

We have

$$b^\alpha_\beta = a^{\alpha\lambda} b_{\beta\lambda} = a_{\beta\lambda} b^{\alpha\lambda} \,, \qquad b^{\alpha\beta} = a^{\alpha\lambda} b^\beta_\lambda \,, \qquad b_{\alpha\beta} = a_{\alpha\lambda} b^\lambda_\beta \,. \tag{1.7.8}$$

The mean curvature of the surface is given by

$$H = \frac{1}{2} b^\alpha_\alpha \,. \tag{1.7.9}$$

The Gaussian curvature of the surface is

$$K = \frac{1}{a} \det\,(b_{\alpha\beta}) \,. \tag{1.7.10}$$

We can introduce the Christoffel symbols with respect to the surface $\theta^3 = 0$. Thus we have

$$\tilde{\Gamma}^\alpha_{\beta\gamma} = \boldsymbol{a}^\alpha \cdot \frac{\partial \boldsymbol{a}_\beta}{\partial \theta^\gamma} \,, \qquad \tilde{\Gamma}^\alpha_{\beta 3} = \boldsymbol{a}^\alpha \cdot \frac{\partial \boldsymbol{a}_3}{\partial \theta^\beta} = a^{\varrho\alpha} \boldsymbol{a}_\varrho \cdot \frac{\partial \boldsymbol{a}_3}{\partial \theta^\beta} =$$

$$= -a^{\alpha\varrho} b_{\varrho\beta} = -b^\alpha_\beta \,, \qquad \tilde{\Gamma}^3_{\alpha\beta} = \boldsymbol{a}_3 \cdot \frac{\partial \boldsymbol{a}_\alpha}{\partial \theta^\beta} = -\boldsymbol{a}_\beta \cdot \frac{\partial \boldsymbol{a}^3}{\partial \theta^\alpha} = b_{\alpha\beta} \,,$$

$$\tilde{\Gamma}^3_{\alpha 3} = \boldsymbol{a}^3 \cdot \frac{\partial \boldsymbol{a}_3}{\partial \theta^\alpha} = 0 \,, \qquad \tilde{\Gamma}^i_{33} = 0 \,.$$

The derivatives of the base vectors of the surface have the expressions

$$\frac{\partial \boldsymbol{a}_\alpha}{\partial \theta^\beta} = \tilde{\Gamma}^\varrho_{\alpha\beta} \boldsymbol{a}_\varrho + b_{\alpha\beta} \boldsymbol{a}_3 \,, \qquad \frac{\partial \boldsymbol{a}^\alpha}{\partial \theta^\beta} = -\tilde{\Gamma}^\alpha_{\beta\lambda} \boldsymbol{a}^\lambda + b^\alpha_\beta \boldsymbol{a}_3 \,,$$
$$\frac{\partial \boldsymbol{a}_3}{\partial \theta^\alpha} = -b^\lambda_\alpha \boldsymbol{a}_\lambda \,. \tag{1.7.11}$$

By (1.7.11),

$$\frac{\partial \boldsymbol{a}_3}{\partial \theta^\alpha} \cdot \frac{\partial \boldsymbol{a}_3}{\partial \theta^\beta} = b_{\alpha\lambda} b^\lambda_\beta ,$$

so that the third fundamental form of the surface can be expressed as

$$d\boldsymbol{a}_3 \cdot d\boldsymbol{a}_3 = b_{\alpha\beta} b^\beta_\gamma d\theta^\alpha d\theta^\gamma . \tag{1.7.12}$$

Clearly, a vector $\boldsymbol{w}$ can be written in the forms

$$\boldsymbol{w} = w^\alpha \boldsymbol{a}_\alpha + w^3 \boldsymbol{a}_3 = w_\alpha \boldsymbol{a}^\alpha + w_3 \boldsymbol{a}_3 .$$

In view of (1.7.11), we have

$$\begin{aligned} \frac{\partial \boldsymbol{w}}{\partial \theta^\alpha} &= \left(w^\lambda_{;\alpha} - b^\lambda_\alpha w^3\right) \boldsymbol{a}_\lambda + \left(\frac{\partial w_3}{\partial \theta^\alpha} + b_{\lambda\alpha} w^\lambda\right) \mathbf{a}_3 = \\ &= (w_{\lambda;\alpha} - b_{\alpha\lambda} w_3) \boldsymbol{a}^\lambda + \left(\frac{\partial w_3}{\partial \theta^\alpha} + b^\lambda_\alpha w_\lambda\right) \mathbf{a}_3 , \end{aligned} \tag{1.7.13}$$

where

$$w_{\lambda;\alpha} = \frac{\partial w_\lambda}{\partial \theta^\alpha} - \tilde{\Gamma}^\varrho_{\lambda\alpha} w_\varrho , \qquad w^\lambda_{;\alpha} = \frac{\partial w^\lambda}{\partial \theta^\alpha} + \tilde{\Gamma}^\lambda_{\varrho\alpha} w^\varrho . \tag{1.7.14}$$

b) **Moving surfaces.** Let $\mathcal{E} = E^3 \times I$ where I is a given interval of time. We consider the hypersurface σ in $\mathcal{E}$ which admits the representation

$$x_i = \varphi_i(\theta^1, \theta^2, t) , \qquad (\theta^1, \theta^2) \in \Theta , \quad t \in I . \tag{1.7.15}$$

For a given value of $t \in I$ the equations (1.7.15) define a surface in E^3 denoted by $\sigma(t)$. The totality of surfaces $\sigma(t)$ for $t \in I$ represents a moving surface in E^3 whose configuration at time t_* is $\sigma(t_*)$.

We assume that the hypersurface σ is smooth, i.e. the functions φ_i are one-one and of class C^2 on $\Theta \times I$ and the matrix $(\partial\varphi_i/\partial\theta^\alpha)$ is, at all points of σ, of rank two. We denote by $\boldsymbol{n}$ a unit normal vector to $\sigma(t)$ so that in (1.7.1) we take

$$\boldsymbol{a}_3 = \boldsymbol{n} . \tag{1.7.16}$$

It follows from (1.7.3) and (1.7.4) that

$$\boldsymbol{a}_\alpha = \varphi_{i;\alpha} \boldsymbol{E}_i , \qquad \boldsymbol{n} = \frac{1}{a} \boldsymbol{a}_1 \times \boldsymbol{a}_2 . \tag{1.7.17}$$

Thus, if we put $\boldsymbol{n} = n_i \boldsymbol{E}_i$, we obtain

$$n_i = \frac{1}{a} e_{irs} \varphi_{r;1} \varphi_{s;2} , \tag{1.7.18}$$

where e_{irs} are the x-components of the alternating tensor. We note that an x-components of a tensor is treated as a scalar when differentiated covariantly with respect to the θ^α. By (1.7.17),

$$n_i n_i = 1\,, \qquad n_i \varphi_{i;\alpha} = 0\,. \tag{1.7.19}$$

It follows from (1.7.5), (1.7.7) and (1.7.17) that

$$a_{\alpha\beta} = \varphi_{i;\alpha}\varphi_{i;\beta}\,, \qquad b_{\alpha\beta} = \frac{\partial^2 \varphi_i}{\partial\theta^\alpha \partial\theta^\beta} n_i\,. \tag{1.7.20}$$

In view of (1.7.11), we get

$$\begin{aligned} \frac{\partial^2 \varphi_i}{\partial\theta^\alpha \partial\theta^\beta} &= b_{\alpha\beta} n_i + \tilde{\Gamma}^\lambda_{\alpha\beta}\varphi_{i;\lambda}\,, \\ n_{i;\alpha} &= -b^\lambda_\alpha \varphi_{i;\lambda} = -a^{\lambda\varrho} b_{\varrho\alpha}\varphi_{i;\lambda}\,. \end{aligned} \tag{1.7.21}$$

With the help of $(1.7.19)_2$ and

$$\varphi_{i;\alpha\beta} = \frac{\partial^2 \varphi_i}{\partial\theta^\alpha \partial\theta^\beta} - \tilde{\Gamma}^\varrho_{\alpha\beta}\frac{\partial \varphi_i}{\partial\theta^\varrho}\,,$$

we can write $(1.7.20)_2$ in the form

$$b_{\alpha\beta} = \varphi_{i;\alpha\beta} n_i. \tag{1.7.22}$$

Thus, from (1.7.21) we obtain

$$\varphi_{i;\alpha\beta} = b_{\alpha\beta} n_i\,, \qquad n_{i;\alpha} = -a^{\lambda\varrho}\varphi_{j;\varrho\alpha}\varphi_{i;\lambda} n_j\,. \tag{1.7.23}$$

If we put in (1.7.1), $\boldsymbol{x} = y_i \boldsymbol{E}_i$, then from (1.7.15) and (1.7.16) we find

$$y_i = \varphi_i + \theta^3 n_i\,. \tag{1.7.24}$$

It follows from (1.6.5) that

$$g^{rs}\frac{\partial y_i}{\partial\theta^r}\frac{\partial y_j}{\partial\theta^s} = \delta_{ij}\,,$$

so that, by (1.7.24), we get

$$g^{\alpha\beta}\left(\varphi_{i;\alpha} + \theta^3 n_{i;\alpha}\right)\left(\varphi_{j;\beta} + \theta^3 n_{j;\beta}\right) + n_i n_j = \delta_{ij}\,, \tag{1.7.25}$$

For the surface $\sigma(t)$, the relation (1.7.25) becomes

$$a^{\alpha\beta}\varphi_{i;\alpha}\varphi_{j;\beta} = \delta_{ij} - n_i n_j\,. \tag{1.7.26}$$

Let $\boldsymbol{\nu} = \nu_i \boldsymbol{E}_i$ be the velocity of a point of $\sigma(t)$. We have

$$\nu_i = \frac{\partial \varphi_i}{\partial t} . \tag{1.7.27}$$

The speed of σ in the direction of the unit normal vector on $\sigma(t)$ is

$$u_n = \nu_i n_i . \tag{1.7.28}$$

If the moving surface is represented by

$$f(x_1, x_2, x_3, t) = 0 , \tag{1.7.29}$$

where f is of class C^2 on its domain, then

$$\boldsymbol{n} = \frac{1}{|\text{grad } f|} \text{grad } f , \qquad |\text{grad } f| = \left(\frac{\partial f}{\partial x_i} \frac{\partial f}{\partial x_i} \right)^{\frac{1}{2}} . \tag{1.7.30}$$

From (1.7.29) and (1.7.27) we obtain

$$\frac{\partial f}{\partial t} + \boldsymbol{\nu} \cdot \text{grad } f = 0. \tag{1.7.31}$$

In view of (1.7.28), (1.7.30) and (1.7.31),

$$u_n = -\frac{1}{|\text{grad } f|} \frac{\partial f}{\partial t} . \tag{1.7.32}$$

Let $g : \Omega \to \mathbb{R}$ be a real scalar-valued function. We assume that $\Omega = B \times I$ where B is an open set in E^3 which, for all $t \in I$, has non-empty intersection with $\sigma(t)$. In what follows we are interested only in that part of σ contained in Ω. Thus we shall replace $\Omega \cap \sigma$ by σ and $B \cap \sigma(t)$ by $\sigma(t)$. We are concerned with the calculation of the first and second partial derivatives of the function g on σ. We assume first that $g \in C^1(\Omega)$. Then, we have on σ

$$\begin{aligned} & g = g(\varphi_i(\theta^1, \theta^2, t), t) , \\ & \frac{dg}{dt} = \frac{\partial g}{\partial t} + \frac{\partial g}{\partial x_i} \frac{\partial \varphi_i}{\partial t} , \qquad g_{;\alpha} = \frac{\partial g}{\partial x_i} \varphi_{i;\alpha} . \end{aligned} \tag{1.7.33}$$

It follows from (1.7.26) and $(1.7.33)_2$ that

$$\frac{\partial g}{\partial x_i} = a^{\alpha\beta} g_{;\alpha} \varphi_{i;\beta} + n_i \frac{\partial g}{\partial n} , \qquad \frac{\partial g}{\partial n} = \frac{\partial g}{\partial x_i} n_i . \tag{1.7.34}$$

Then, from $(1.7.33)_1$ we get

$$\frac{\partial g}{\partial t} = \frac{dg}{dt} - a^{\alpha\beta} g_{;\alpha}\varphi_{i;\beta}\nu_i - u_n \frac{\partial g}{\partial n}\,. \tag{1.7.35}$$

If we introduce δ-time derivative by

$$\frac{\delta}{\delta t} = \frac{\partial}{\partial t} + u_n \frac{\partial}{\partial n}\,, \tag{1.7.36}$$

then from (1.7.35) we obtain

$$\frac{\delta g}{\delta t} = \frac{dg}{dt} - a^{\alpha\beta} g_{;\alpha}\varphi_{i;\beta}\nu_i\,, \tag{1.7.37}$$

and

$$\frac{\partial g}{\partial t} = \frac{\delta g}{\delta t} - u_n \frac{\partial g}{\partial n}\,. \tag{1.7.38}$$

If g is a known function on σ, then the derivatives dg/dt and $g_{;\alpha}$ can be calculated. It follows from (1.7.37) that $\delta g/\delta t$ can be also calculated on σ. The relations (1.7.34) and (1.7.38) are called *compatibility conditions for first partial derivatives* of g on σ. If g and $\partial g/\partial n$ are given on σ then (1.7.34) and (1.7.38) determine the first partial derivatives of g on σ.

Taking into account (1.7.33), (1.7.37) and (1.7.27) we can write

$$\frac{\delta \varphi_{i;\alpha}}{\delta t} = \nu_{i;\alpha} - a^{\varrho\beta}\varphi_{j;\beta}\nu_j\varphi_{i;\alpha\varrho}\,.$$

Thus, in view of (1.7.23), we find that

$$n_i\frac{\delta \varphi_{i;\alpha}}{\delta t} = \nu_{i;\alpha}n_i + \nu_j n_{j;\alpha} = u_{n;\alpha}\,. \tag{1.7.39}$$

By (1.7.19) and (1.7.39),

$$n_i\frac{\delta n_i}{\delta t} = 0\,, \qquad \varphi_{i;\alpha}\frac{\delta n_i}{\delta t} = -u_{n;\alpha}\,. \tag{1.7.40}$$

If we multiply each side of the equation $(1.7.40)_2$ by $a^{\alpha\beta}\varphi_{j;\beta}$ and sum on the index α, then, in view of (1.7.26) and $(1.7.40)_1$, we obtain

$$\frac{\delta n_i}{\delta t} = -a^{\alpha\beta}\varphi_{i;\alpha}u_{n;\beta} \tag{1.7.41}$$

We now assume that $g \in C^2(\Omega)$. The relations (1.7.34) and (1.7.38) can be applied to $\delta g/\delta t$ and $\partial g/\partial x_i$. Thus, we get

$$\frac{\partial^2 g}{\partial x_i \partial x_j} = a^{\alpha\beta} \left(\frac{\partial g}{\partial x_j} \right)_{;\alpha} \varphi_{i;\beta} + n_i \frac{\partial}{\partial n} \left(\frac{\partial g}{\partial x_j} \right), \tag{1.7.42}$$

so that

$$\frac{\partial}{\partial n} \left(\frac{\partial g}{\partial x_i} \right) = a^{\alpha\beta} \left(\frac{\partial g}{\partial x_j} \right)_{;\alpha} \varphi_{i;\beta} n_j + n_i \frac{\partial^2 g}{\partial n^2}, \tag{1.7.43}$$

where

$$\frac{\partial^2 g}{\partial n^2} = n_j \frac{\partial}{\partial n} \left(\frac{\partial g}{\partial x_i} \right) = \frac{\partial^2 g}{\partial x_i \partial x_j} n_i n_j . \tag{1.7.44}$$

By using the relations (1.7.19), (1.7.22), (1.7.23), (1.7.34) and (1.7.43), from (1.7.42) we obtain

$$\begin{aligned} \frac{\partial^2 g}{\partial x_i \partial x_j} =& a^{\alpha\beta} \left\{ \left(\frac{\partial g}{\partial n} \right)_{;\alpha} + a^{\nu\varrho} b_{\alpha\nu} g_{;\varrho} \right\} (n_i \varphi_{j;\beta} + n_j \varphi_{i;\beta}) + \\ & + a^{\alpha\beta} a^{\nu\varrho} \left(g_{;\alpha\nu} - b_{\alpha\nu} \frac{\partial g}{\partial n} \right) \varphi_{i;\beta} \varphi_{j;\varrho} + n_i n_j \frac{\partial^2 g}{\partial n^2} . \end{aligned} \tag{1.7.45}$$

With the help of relations (1.7.9) and (1.7.19) we obtain

$$\frac{\partial^2 g}{\partial x_i \partial x_i} = a^{\alpha\beta} g_{;\alpha\beta} - 2H \frac{\partial g}{\partial n} + \frac{\partial^2 g}{\partial n^2} . \tag{1.7.46}$$

If we replace g by $\partial g/\partial t$ in equation (1.7.38) then we get

$$\frac{\partial^2 g}{\partial x_i \partial t} = a^{\alpha\beta} \left(\frac{\partial g}{\partial t} \right)_{;\alpha} \varphi_{i;\beta} + n_i \frac{\partial}{\partial n} \left(\frac{\partial g}{\partial t} \right) . \tag{1.7.47}$$

From (1.7.34), (1.7.38), (1.7.41) and (1.7.44) we find

$$\begin{aligned} \frac{\delta}{\delta t} \left(\frac{\partial g}{\partial n} \right) &= n_i \frac{\delta}{\delta t} \left(\frac{\partial g}{\partial x_i} \right) + \frac{\partial g}{\partial x_i} \frac{\delta n_i}{\delta t} = \\ &= \frac{\partial}{\partial n} \left(\frac{\partial g}{\partial t} \right) + u_n \frac{\partial^2 g}{\partial n^2} - a^{\alpha\beta} g_{;\alpha} u_{n;\beta} . \end{aligned} \tag{1.7.48}$$

From (1.7.47) and (1.7.8) we obtain

$$\begin{aligned}\frac{\partial^2 g}{\partial x_i \partial t} =& a^{\alpha\beta}\left(\frac{\delta g}{\delta t} - u_n \frac{\partial g}{\partial n}\right)_{;\alpha} \varphi_{i;\beta} + \\ &+ \left\{\frac{\delta}{\delta t}\left(\frac{\partial g}{\partial n}\right) + a^{\varrho\lambda} g_{;\varrho} u_{n;\lambda} - u_n \frac{\partial^2 g}{\partial n^2}\right\} n_i \,.\end{aligned} \tag{1.7.49}$$

By (1.7.38) and (1.7.48),

$$\begin{aligned}\frac{\partial^2 g}{\partial t^2} =& \frac{\delta}{\delta t}\left(\frac{\delta g}{\delta t} - u_n \frac{\partial g}{\partial n}\right) - u_n \left\{\frac{\delta}{\delta t}\left(\frac{\partial g}{\partial n}\right) + \right. \\ &\left. + a^{\alpha\beta} g_{;\alpha} u_{n;\beta} - u_n \frac{\partial^2 g}{\partial n^2}\right\} .\end{aligned} \tag{1.7.50}$$

The relations (1.7.45), (1.7.49) and (1.7.50) are called *the compatibility conditions for second partial derivatives of g on σ.*

c) **Singular surfaces. Geometrical compatibility conditions. Kinematical compatibility conditions.** First, we extend the concept of a differentiability class to functions whose domains are not open sets. We say that the function $f : Q \to \mathbb{R}$ is of class C^r on Q if there exists a function $h : W \to \mathbb{R}$, where W is an open subset of $\mathcal{E}$ containing Q, such that $h \in C^r(W)$ and the restriction of h to Q coincides with f. The extension h of the function f is not unique but we shall not have occasion to consider an extension outside the closure of the domain Q.

We denote by B_t^+ and B_t^- the disjoint open subsets of B whose union is $B - \sigma(t)$. We assume that B_t^+ is the subset towards which the normal vector $\mathbf{n}$ on $\sigma(t)$ is directed. We introduce the notations

$$\Omega^+ = \{(x,t) : x \in B_t^+,\ \ t \in I\}\,, \qquad \Omega^- = \{(x,t) : x \in B_t^-,\ \ t \in I\}\,.$$

We assume that g is continuous on $\Omega^+ \cup \sigma$ and on $\Omega^- \cup \sigma$, but not on σ. We denote by g^+ and g^- the extensions of g on open subsets of $\mathcal{E}$ containing $\Omega^+ \cup \sigma$ and $\Omega^- \cup \sigma$, respectively.

The jump of g across σ is defined by

$$[g] = g^+ - g^- .$$

If $[g] \neq 0$ almost everywhere on $\sigma(t)$, then $\sigma(t)$ is called a *singular surface* of g. We assume now that $g \in C^{r-1}(\Omega)$, $(r \geq 1)$, $g \in C^r(\Omega^+ \cup \sigma)$ and $g \in C^r(\Omega^- \cup \sigma)$. There exist extensions of g in terms of which the jumps across σ of the partial derivatives of g can be defined. If one of these jumps is different from zero almost everywhere on $\sigma(t)$, then $\sigma(t)$ is called a singular surface of g of order r.

We assume that g is of class C^1 on $\Omega^+ \cup \sigma$ and on $\Omega^- \cup \sigma$. Then, the compatibility conditions (1.7.34) and (1.7.38) are applicable on σ to both g^+ and g^-. Taking into account that

$$[g]_{;\alpha} = (g^+ - g^-)_{;\alpha} = g^+_{;\alpha} - g^-_{;\alpha}\,, \qquad \frac{\delta[g]}{\delta t} = \frac{\delta g^+}{\delta t} - \frac{\delta g^-}{\delta t}\,,$$

we find that

$$\left[\frac{\partial g}{\partial x_i}\right] = a^{\alpha\beta} A_{;\alpha}\varphi_{i;\beta} + \hat{B} n_i\,,$$
$$\left[\frac{\partial g}{\partial t}\right] = \frac{\delta A}{\delta t} - u_n \hat{B}\,, \tag{1.7.51}$$

where

$$A = [g]\,, \qquad \hat{B} = \left[\frac{\partial g}{\partial n}\right]. \tag{1.7.52}$$

If $\sigma(t)$ is a singular surface of g of order $r = 1$, then $A = 0$ and (1.7.51) reduces to

$$\left[\frac{\partial g}{\partial x_i}\right] = \hat{B} n_i\,, \qquad \left[\frac{\partial g}{\partial t}\right] = -u_n \hat{B}\,. \tag{1.7.53}$$

We now assume that g is of class C^2 on $\Omega^+ \cup \sigma$ and on $\Omega^- \cup \sigma$. It follows from (1.7.45), (1.7.49) and (1.7.50) that

$$\left[\frac{\partial^2 g}{\partial x_i x_j}\right] = a^{\alpha\beta}(\hat{B}_{;\alpha} + a^{\lambda\varrho} b_{\alpha\lambda} A_{;\varrho})(n_i \varphi_{j;\beta} + n_j \varphi_{i;\beta}) +$$
$$+ a^{\alpha\beta} a^{\nu\varrho}(A_{;\alpha\nu} - b_{\alpha\nu}\hat{B})\varphi_{i;\beta}\varphi_{j;\varrho} + C n_i n_j\,,$$
$$\left[\frac{\partial^2 g}{\partial x_i \partial t}\right] = a^{\alpha\beta}(\frac{\delta A}{\delta t} - u_n \hat{B})_{;\alpha}\,\varphi_{i;\beta} + (\frac{\delta \hat{B}}{\delta t} + a^{\alpha\beta} A_{;\alpha} u_{n;\beta} - u_n C) n_i\,, \tag{1.7.54}$$
$$\left[\frac{\partial^2 g}{\partial t^2}\right] = \frac{\delta}{\delta t}(\frac{\delta A}{\delta t} - u_n \hat{B}) - u_n(\frac{\delta \hat{B}}{\delta t} + a^{\alpha\beta} A_{;\alpha} u_{n;\beta} - u_n C)\,,$$

where

$$C = \left[\frac{\partial^2 g}{\partial n^2}\right]. \tag{1.7.55}$$

We note that from (1.7.54) we obtain

$$\left[\frac{\partial^2 g}{\partial x_i x_i}\right] = a^{\alpha\beta} A_{;\alpha\beta} - 2H\hat{B} + C\,. \tag{1.7.56}$$

If $\sigma(t)$ is a singular surface of g of order two, then $A = \hat{B} = 0$ and (1.7.54) reduces to

$$\left[\frac{\partial^2 g}{\partial x_i x_j}\right] = C n_i n_j\,, \qquad \left[\frac{\partial^2 g}{\partial x_i t}\right] = -u_n C n_i\,, \qquad \left[\frac{\partial^2 g}{\partial t^2}\right] = C u_n^2\,. \tag{1.7.57}$$

The relations (1.7.51) represent the compatibility conditions for discontinuities in the first partial derivatives of g on σ. The relations (1.7.54) represent the compatibility conditions for discontinuities in second partial derivatives of g on σ. The conditions $(1.7.51)_1$ and $(1.7.54)_1$ are called *geometrical compatibility conditions,* and $(1.7.51)_2$, $(1.7.54)_{2,3}$ are called *kinematical compatibility conditions.*

We assume now that $\sigma(t)$ is a singular surface of g of order $r \geq 2$. The equations (1.7.53) apply to the $(r-1)$-th partial derivatives of g. We arrive to

$$\left[\frac{\partial^r g}{\partial x_i \partial x_j \ldots \partial x_m \partial t^{r-s}}\right] = (-u_n)^{r-s}\left[\frac{\partial^r g}{\partial n^r}\right] n_i n_j \ldots n_m\,, \quad (0 \leq s \leq r)\,, \tag{1.7.58}$$

where

$$\left[\frac{\partial^r g}{\partial n^r}\right] = \frac{\partial^r g}{\partial x_p \ldots \partial x_s} n_p \ldots n_s\,. \tag{1.7.59}$$

Let us suppose that g is of class C^{r+1} on $\Omega^+ \cup \sigma$ and on $\Omega^- \cup \sigma$. Then the compatibility conditions (1.7.54), (1.7.56) apply to the $(r-1)$-th partial derivatives of g. If we replace in these conditions g by $\partial^{r-1} g/\partial x_m \ldots \partial x_s$ and take into account (1.7.58), then we obtain

$$\begin{aligned}
\left[\frac{\partial^{r+1} g}{\partial x_p \ldots \partial x_s \partial x_i \partial x_j}\right] &= a^{\alpha\beta}\left(n_p \ldots n_s \left[\frac{\partial^r g}{\partial n^r}\right]\right)_{;\alpha}(n_i \varphi_{j;\beta} + n_j \varphi_{i;\beta}) - \\
&- a^{\alpha\beta} a^{\nu\varrho} b_{\alpha\nu}\left[\frac{\partial^r g}{\partial n^r}\right] n_p \ldots n_s \varphi_{i;\beta}\varphi_{j;\beta} + \left[\frac{\partial^{r+1} g}{\partial x_p \ldots \partial x_s \partial x_k \partial x_m}\right] n_k n_m n_i n_j\,,\\
\left[\frac{\partial^{r+1} g}{\partial x_p \ldots \partial x_s \partial x_i \partial t}\right] &= -a^{\alpha\beta}\left(u_n \left[\frac{\partial^r g}{\partial n^r}\right] n_p \ldots n_s\right)_{;\alpha} \varphi_{i;\beta} + \\
&+ \frac{\delta}{\delta t}\left(\left[\frac{\partial^r g}{\partial n^r}\right] n_p \ldots n_s\right) n_i - u_n \left[\frac{\partial^{r+1} g}{\partial x_p \ldots \partial x_s \partial x_k \partial x_m}\right] n_k n_m n_i\,,\\
\left[\frac{\partial^{r+1} g}{\partial x_p \ldots \partial x_s \partial t^2}\right] &= -\left[\frac{\partial^r g}{\partial n^r}\right] n_p \ldots n_s \frac{\delta u_n}{\delta t} - 2u_n \frac{\delta}{\delta t}\left(\left[\frac{\partial^r g}{\partial n^r}\right] n_p \ldots n_s\right) + \\
&+ u_n^2 n_k n_m \left[\frac{\partial^{r+1} g}{\partial x_p \ldots \partial x_s \partial x_k \partial x_m}\right],\\
\left[\frac{\partial^{r+1} g}{\partial x_p \ldots \partial x_s \partial x_k \partial x_k}\right] &= -2H\left[\frac{\partial^r g}{\partial n^r}\right] n_p \ldots n_s + \left[\frac{\partial^{r+1} g}{\partial x_p \ldots \partial x_s \partial x_k \partial x_m}\right] n_k n_m\,.
\end{aligned} \tag{1.7.60}$$

d) **Waves.** We now consider a moving surface which propagates through the body considered in Section 1.1. Let $\sigma(t)$ be the surface which admits the representation (1.7.29), in the spatial description. The material description of the surface is provided by

$$F(X_1, X_2, X_3, t) = f(x_i(X_K, t), t) = 0 . \tag{1.7.61}$$

The equation (1.7.61) represents, in the material coordinates, the surface $\Sigma(t)$. The alternative forms (1.7.29) and (1.7.61) represent the same moving surface. It follows from (1.7.15) and (1.1.1) that the surface $\Sigma(t)$ admits the representation

$$X_K = X_K\big(\varphi_i(\theta^1, \theta^2, t), t\big) = \Phi_K(\theta^1, \theta^2, t) . \tag{1.7.62}$$

The metric tensor of $\Sigma(t)$ is

$$A_{\alpha\beta} = \varphi_{K;\alpha}\varphi_{K;\beta} . \tag{1.7.63}$$

The unit normal $\boldsymbol{N}$ of the surface $\Sigma(t)$ is given by

$$\boldsymbol{N} = \frac{1}{|\text{grad}_X F|}\text{grad}_X F , \tag{1.7.64}$$

where

$$|\text{grad}_X F| = (F_{,K}F_{,K})^{\frac{1}{2}} .$$

With the help of the relations

$$f_{,i} = F_{,K}X_{K,i} , \qquad F_{,K} = f_{,j}x_{j;K} ,$$

we obtain

$$N_K = \frac{1}{C_n}n_i x_{i,K} , \qquad n_i = C_n X_{K,i} N_K , \tag{1.7.65}$$

where

$$C_n^2 = x_{j,L}x_{r,L}n_j n_r . \tag{1.7.66}$$

The *speed of propagation* of the surface $\Sigma(t)$ is defined by

$$U_N = -\frac{1}{|\text{grad}_X F|}\frac{\partial F}{\partial t} . \tag{1.7.67}$$

We have

$$\dot{F} = \frac{\partial F}{\partial t} = \dot{f} = \frac{\partial f}{\partial t} + f_{,i}v_i ,$$

where $\boldsymbol{v} = \dot{\boldsymbol{x}}$. In view of (1.7.30) and (1.7.32) we get

$$\dot{F} = -(u_n - v_n)|\mathrm{grad}_x f|\,, \tag{1.7.68}$$

where

$$v_n = v_i n_i\,. \tag{1.7.69}$$

We define *the local speed of propagation* by

$$U = u_n - v_n\,. \tag{1.7.70}$$

It follows from (1.7.68) - (1.7.70) that

$$U_N = U C_n^{-1}. \tag{1.7.71}$$

Using the compatibility conditions obtained for the function $g : \Omega \to \mathbb{R}$ we can derive results of the same type for discontinuities in partial derivatives of the function $G(\boldsymbol{X}, t) = g(\boldsymbol{x}(\boldsymbol{X}, t), t)$. Thus, we obtain the following compatibility conditions corresponding to (1.7.51) and (1.7.54),

$$\begin{aligned}
\left[\frac{\partial G}{\partial X_K}\right] &= A^{\alpha\beta}\tilde{A}_{|\beta}\Phi_{K|\beta} + \tilde{B}N_K\,,\\
\left[\frac{\partial G}{\partial t}\right] &= \frac{\delta_D \tilde{A}}{\delta t} - U_N\tilde{B}\,,\\
\left[\frac{\partial^2 G}{\partial X_K \partial X_L}\right] &= A^{\alpha\beta}\left(\tilde{B}_{|\alpha} + A^{\nu\varrho}B_{\alpha\nu}\tilde{A}_{|\varrho}\right)\left(N_K\Phi_{L|\beta} + N_L\Phi_{K|\beta}\right) + \\
&\quad + A^{\alpha\beta}A^{\nu\varrho}\left(\tilde{A}_{|\alpha\nu} - B_{\alpha\nu}\tilde{B}\right)\Phi_{K|\beta}\Phi_{L|\varrho} + \tilde{C}N_K N_L\,, \qquad (1.7.72)\\
\left[\frac{\partial^2 G}{\partial X_K \partial t}\right] &= A^{\alpha\beta}\left(\frac{\delta_D \tilde{A}}{\delta t} - U_N\tilde{B}\right)_{|\alpha}\Phi_{K|\beta} + \\
&\quad + \left(\frac{\delta_D \tilde{B}}{\delta t} + A^{\alpha\beta}\tilde{A}_{|\alpha}U_{N|\beta} - U_N\tilde{C}\right)N_K\,,\\
\left[\frac{\partial^2 G}{\partial t^2}\right] &= \frac{\delta_D}{\delta t}\left(\frac{\delta_D \tilde{A}}{\delta t} - U_N\tilde{B}\right) - U_N\left(\frac{\delta_D \tilde{B}}{\delta t} + A^{\alpha\beta}\tilde{A}_{|\alpha}U_{N|\beta} - U_N\tilde{C}\right),
\end{aligned}$$

where a vertical line denotes covariant differentiation in the θ coordinate system using Christoffel symbols calculated from the metric tensors $A_{\alpha\beta}$, $A^{\alpha\beta}$, and

$$B_{\alpha\beta} = \Phi_{K|\alpha\beta} N_K \,, \qquad \tilde{A} = [G] \,,$$

$$\tilde{B} = \left[\frac{\partial G}{\partial N}\right] = \frac{1}{C_n} x_{i,L} x_{j,L} n_i \left[\frac{\partial g}{\partial x_j}\right], \tag{1.7.73}$$

$$\tilde{C} = \left[\frac{\partial^2 G}{\partial X_K \partial X_L} N_K N_L\right], \quad \frac{\delta_D}{\delta t} = \frac{\partial}{\partial t} + U_N \frac{\partial}{\partial N} \,.$$

A singular surface is said to be a *wave* if U_N is different from zero.

e) **Dynamical compatibility conditions.** The integral forms of the equations of balance are special cases of the equation

$$\frac{d}{dt} \int_{\mathcal{P}} \varrho F dv - \int_{\mathcal{P}} \varrho S dv - \int_{\partial \mathcal{P}} M_i n_i da \geq 0 \,,$$

which, at a singular surface $\sigma(t)$ which intersects $\mathcal{P}$, is equivalent to the condition (see, e.g. Eringen [123])

$$U[\varrho F] + n_i [M_i] \geq 0 \qquad \text{on} \quad \sigma(t). \tag{1.7.74}$$

When (1.7.74) is applied to (1.1.6), (1.1.12) and (1.1.17) we obtain the jump conditions

$$[\varrho U v_i] + [t_{ji}] n_j = 0 \,,$$

$$[\varrho(\varepsilon + \frac{1}{2} v^2) U] + [t_{ji} v_j + q_i] n_i = 0 \,, \tag{1.7.75}$$

$$[\varrho \eta U] + [\frac{1}{T} q_i] n_i \geq 0 \qquad \text{on} \quad \sigma(t).$$

It is convenient to have an alternative form of the conditions (1.7.75) in which the stresses and heat flux are measured with respect to reference configuration. Thus, we can write the following conditions on $\Sigma(t)$,

$$[\varrho_0 \boldsymbol{v} (\nu_K - v_K) + \boldsymbol{T}_K] N_K = 0 \,,$$

$$[\varrho_0 (\varepsilon + \frac{1}{2} \boldsymbol{v}^2)(\nu_K - v_K) + \boldsymbol{T}_K \cdot \boldsymbol{v} + Q_K] N_K = 0 \,, \tag{1.7.76}$$

$$[\varrho_0 \eta (\nu_K - v_K) + \frac{1}{T} Q_K] N_K \geq 0 \,.$$

The relations (1.7.75) and (1.7.76) are called *dynamical compatibility conditions*.

2 The Linear Theory of Thermoelasticity for Bodies with Initial Stress and Initial Heat Flux

2.1 Equations of perturbed motion

In what follows we establish the equations governing infinitesimal thermoelastic deformations superimposed on large deformations. The theory of infinitesimal thermoelastic deformations in a body which has already been subjected to an isothermal elastic deformation was established by England and Green [120] and Green [149]. The equations established in [120], [149] have been employed by Knops and Wilkes [223], p. 187 in a discussion of thermoelastic stability. The theory of infinitesimal thermoelastic deformations superimposed on a large deformation at nonconstant temperature was established by Iesan [182].

We consider three states of the body:

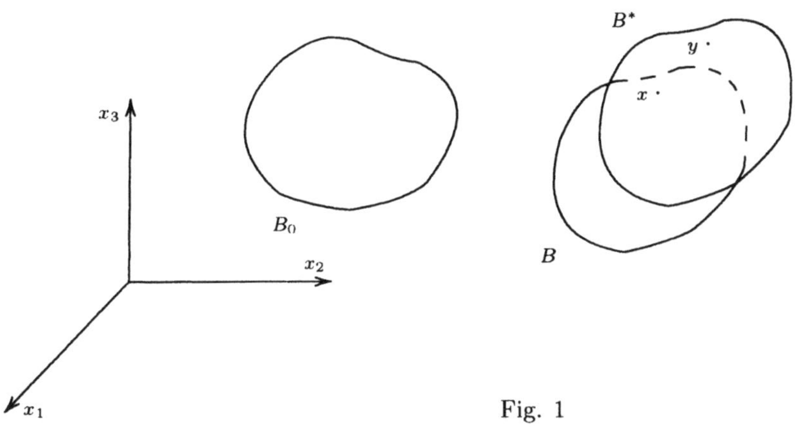

Fig. 1

the state B_0, and two current configurations, B and B^*, corresponding respectively to constant temperature T_0, temperature T, and temperature T^*. We call B the primary state, and B^* the secondary state. We shall designate as incremental those quantities associated with the difference of motions between the secondary and primary states. The quantities associated with the secondary state will be denoted with an asterisk and the position coordinates of the particle X at time t in B^* will be denoted by y_i. We have

$$y_i = y_i(X_A, t) \ , \quad (X_A, t) \in B_0 \times (t_0, t_1) \ , \tag{2.1.1}$$

with

$$J^* = \det (y_{i,A}) > 0 \ . \tag{2.1.2}$$

Now let

$$u_i = y_i - x_i \ , \quad \theta = T^* - T \ . \tag{2.1.3}$$

We assume that u_i and θ are small, i.e.

$$u_i = \varepsilon' u_i' \ , \quad \theta = \varepsilon' \theta' \ , \tag{2.1.4}$$

where ε' is a constant small enough for squares and higher powers to be neglected, and u_i' and θ' are independent of ε'. The problem consists in establishing the equations, boundary conditions, and initial conditions for u_i and θ when the primary deformation and the loadings associated with B and B^* are known.

If we refer all quantities to the configuration B_0, then we have

$$u_i = u_i(X_A, t) \ , \quad \theta = \theta(X_A, t) \ , \quad (X_A, t) \in B_0 \times (t_0, t_1) \ . \tag{2.1.5}$$

It is sometimes convenient to refer all quantities to the configuration of the body in its primary motion at time t instead of the reference configuration at time t_0. This is especially useful when the primary state B is a configuration of equilibrium. Then, we have

$$u_i = u_i(x_j, t) \ , \quad \theta = \theta(x_j, t) \ , \quad (x_j, t) \in B_0 \times (t_0, t_1) \ . \tag{2.1.6}$$

In the secondary state B^* we consider the following stress tensors: t_{ij}^* is the Cauchy stress tensor; T_{Ai}^* and T_{AB}^* are the Piola -Kirchhoff stress tensors measured per unit area in the configuration B_0; $T_{kl}^{*(1)}$ is the first Piola -Kirchhoff stress tensor measured per unit area in the configuration B. We also consider the following heat flux vectors: Q_A^* is the heat flux across surfaces in B^* that in the configuration B_0 were coordinate planes perpendicular to the X_A axes, measured per unit area of these planes and per unit time; $\bar{Q}_i^*$ is the heat flux across surfaces in B^* that in the configuration B were coordinate planes perpendicular to the x_i axes, measured per unit area of these planes and per unit time. Thus, we have

$$\begin{aligned} &J^* t_{ij}^* = y_{i,A} T_{Aj}^* = y_{i,A} y_{j,B} T_{AB}^* \ , \quad J' t_{ij}^* = y_{i,s} T_{sj}^{*(1)} \ , \\ &q_i^* = \frac{1}{J^*} y_{i,A} Q_A^* = \frac{1}{J'} y_{i,j} \bar{Q}_j^* \ , \end{aligned} \tag{2.1.7}$$

where

$$J^* = J'J\ , \quad J' = \det\,(y_{i,s})\ . \tag{2.1.8}$$

Throughout we use the notation $f_{,i} = \partial f/\partial x_i$. To a second order approximation, we get

$$J' = 1 + u_{s,s}\ . \tag{2.1.9}$$

The equations of motion for the secondary state may be written in the forms

$$T^*_{Ai,A} + \varrho_0 f^*_i = \varrho_0 \ddot{y}_i \qquad \text{on } B_0 \times (t_0, t_1)\ , \tag{2.1.10}$$

$$T^{*(1)}_{ji,j} + \varrho f_i = \varrho \ddot{y}_i \qquad \text{on } B \times (t_0, t_1)\ . \tag{2.1.11}$$

The energy equation governing the secondary state can be written in the forms

$$\varrho_0 T^* \dot{\eta}^* = Q^*_{A,A} + \varrho_0 s^* \qquad \text{on } B_0 \times (t_0, t_1)\ , \tag{2.1.12}$$

$$\varrho T^* \dot{\eta}^* = \bar{Q}^*_{i,i} + \varrho s^* \qquad \text{on } B \times (t_0, t_1)\ . \tag{2.1.13}$$

The constitutive equations imply that

$$\begin{aligned} &\psi^* = \hat{\psi}(E^*_{AB},\, T^*,\, X_M)\ , \qquad T^*_{AB} = \frac{\partial \psi^*}{\partial E^*_{AB}}\ ,\\ &\varrho_0 \eta^* = -\frac{\partial \psi^*}{\partial T^*}\ , \qquad Q^*_A = \hat{Q}_A(E^*_{KL},\, T^*,\, T^*_{,N},\, X_M)\ , \end{aligned} \tag{2.1.14}$$

where

$$2E^*_{AB} = y_{i,A} y_{i,B} - \delta_{AB}\ . \tag{2.1.15}$$

The heat conduction inequality leads to

$$q_i T_{,i} \geq 0\ ,\ \ Q_A T_{,A} \geq 0\ ,\ \ Q^*_A T^*_{,A} \geq 0\ . \tag{2.1.16}$$

We consider the following boundary conditions for the secondary deformation

$$\begin{aligned} &y_i = \hat{y}_i \ \text{ on } \bar{\Sigma}_1 \times [t_0, t_1)\ , \qquad T^*_{Ai} N_A = \hat{T}^*_i \ \text{ on } \Sigma_2 \times [t_0, t_1)\ ,\\ &T^* = \hat{T}^* \ \text{ on } \bar{\Sigma}_3 \times [t_0, t_1)\ , \qquad Q^*_A N_A = \hat{Q}^* \ \text{ on } \Sigma_4 \times [t_0, t_1)\ , \end{aligned} \tag{2.1.17}$$

where $\hat{y}_i$, $\hat{T}^*_i$, $\hat{T}^*$ and $\hat{Q}^*$ are prescribed functions. The conditions (2.1.17) may be written in the form

$$\begin{aligned} &y_i = \tilde{y}_i \ \text{ on } \bar{S}_1 \times [t_0, t_1)\,, \qquad T^{*(1)}_{ji} n_j = \tilde{T}^*_i \ \text{ on } S_2 \times [t_0, t_1)\\ &T^* = \tilde{T}^* \ \text{ on } \bar{S}_3 \times [t_0, t_1)\,, \qquad \bar{Q}^*_i n_i = \tilde{q}^* \ \text{ on } S_4 \times [t_0, t_1)\ , \end{aligned} \tag{2.1.18}$$

where $\tilde{y}_i$, $\tilde{T}_i^*$, $\tilde{T}^*$ and $\tilde{q}^*$ are specified functions.

Let us establish the boundary-initial-value problem which characterizes the functions u_i and θ.

Since

$$y_{i,A} = x_{i,A} + u_{i,A} = x_{i,A} + u_{i,j}x_{j,A} ,$$

the relations (1.2.12) and (2.1.15) imply that

$$E^*_{AB} = E_{AB} + \frac{1}{2}(x_{i,B}u_{i,A} + x_{i,A}u_{i,B}) , \tag{2.1.19}$$

or equivalently,

$$E^*_{AB} = E_{AB} + x_{i,A}x_{j,B}e_{ij} . \tag{2.1.20}$$

Here e_{ij} is defined by

$$2e_{ij} = u_{i,j} + u_{j,i} . \tag{2.1.21}$$

If we use (2.1.4) and (2.1.19), then to a second order approximation, we obtain

$$\begin{aligned} \frac{\partial\psi^*}{\partial E^*_{KL}} &= \frac{\partial\psi}{\partial E_{KL}} + A_{KLMN}(E^*_{MN} - E_{MN}) - B_{KL}\theta , \\ \frac{\partial\psi^*}{\partial T^*} &= \frac{\partial\psi}{\partial T} - B_{MN}(E^*_{MN} - E_{MN}) - A\theta , \end{aligned} \tag{2.1.22}$$

where

$$A_{KLMN} = \frac{\partial^2\psi}{\partial E_{KL}\partial E_{MN}} , \quad B_{KL} = -\frac{\partial^2\psi}{\partial E_{KL}\partial T} , \quad A = -\frac{\partial^2\psi}{\partial T^2} . \tag{2.1.23}$$

Clearly,

$$A_{KLMN} = A_{MNKL} = A_{LKMN} , \qquad B_{KL} = B_{LK} . \tag{2.1.24}$$

From (1.3.6), (2.1.14), (2.1.19) or (2.1.20), and (2.1.22) we obtain

$$\begin{aligned} T^*_{KL} &= T_{KL} + A_{KLMN}x_{i,M}u_{i,N} - B_{KL}\theta , \\ \varrho_0\eta^* &= \varrho_0\eta + B_{MN}x_{i,M}u_{i,N} + A\theta , \end{aligned} \tag{2.1.25}$$

or

$$\begin{aligned} T^*_{KL} &= T_{KL} + A_{KLMN}x_{i,M}x_{j,N}e_{ij} - B_{KL}\theta , \\ \varrho_0\eta^* &= \varrho_0\eta + B_{MN}x_{i,M}x_{j,N}e_{ij} + A\theta . \end{aligned} \tag{2.1.26}$$

It follows from (1.2.9) and (2.1.25) that

$$T^*_{Ki} = T_{Ki} + E_{iKjN}u_{j,N} - F_{Ki}\theta + T_{KN}u_{i,N} , \tag{2.1.27}$$

where

$$E_{iKjN} = x_{i,B}x_{j,M}A_{BKMN} , \qquad F_{Ki} = B_{KL}x_{i,L} . \tag{2.1.28}$$

If we now define

$$S_{Ai} = T^*_{Ai} - T_{Ai} , \tag{2.1.29}$$

then (2.1.27) implies

$$S_{Ai} = E_{iAjN}u_{j,N} - F_{Ai}\theta + T_{AN}u_{i,N} . \tag{2.1.30}$$

We now establish a relation between the stress tensors $T^{*(1)}_{ij}$ and t_{ij}. By (2.1.7),

$$JT^{*(1)}_{kj} = x_{k,A}T^*_{Aj} = x_{k,A}x_{r,B}T^*_{AB}y_{j,r} . \tag{2.1.31}$$

It follows from (1.2.15), (2.1.3), (2.1.25) and (2.1.31) that

$$T^{*(1)}_{ij} = t_{ij} + C_{ijrs}e_{rs} - \beta_{ij}\theta + t_{ir}u_{j,r} , \tag{2.1.32}$$

where

$$JC_{ijrs} = x_{i,K}x_{j,L}x_{r,M}x_{s,N}A_{KLMN} , \qquad J\beta_{ij} = x_{i,K}x_{j,L}B_{KL} . \tag{2.1.33}$$

Obviously,

$$C_{ijrs} = C_{jirs} = C_{rsij} , \qquad \beta_{ij} = \beta_{ji} . \tag{2.1.34}$$

If we define

$$s_{ij} = T^{*(1)}_{ij} - t_{ij} , \tag{2.1.35}$$

then (2.1.32) implies

$$s_{ij} = C_{ijrs}u_{r,s} - \beta_{ij}\theta + t_{ir}u_{j,r} . \tag{2.1.36}$$

Next, we define

$$\gamma = \varrho_0(\eta^* - \eta) , \qquad \Phi_A = Q^*_A - Q_A , \qquad \varphi_i = \bar{Q}^*_i - q_i . \tag{2.1.37}$$

By (2.1.25), (2.1.26), (2.1.28) and (2.1.33) we obtain

$$\gamma = F_{Ai}u_{i,A} + A\theta , \tag{2.1.38}$$

$$\varrho(\eta^* - \eta) = \frac{1}{J}\gamma = \beta_{ij}e_{ij} + a\theta = \beta_{ij}u_{i,j} + a\theta , \tag{2.1.39}$$

where

$$a = \frac{1}{J}A . \tag{2.1.40}$$

We now consider the heat flux vectors. In view of (2.1.4) and (2.1.14),

$$Q_A^* = Q_A + R_{AMN}(E_{MN}^* - E_{MN}) + D_A\theta + K_{AM}\theta_{,M} \,, \tag{2.1.41}$$

where

$$R_{AMN} = \frac{\partial Q_A}{\partial E_{MN}} = R_{ANM} \,, \quad D_A = \frac{\partial Q_A}{\partial T} \,, \quad K_{AB} = \frac{\partial Q_A}{\partial T_{,B}} \,. \tag{2.1.42}$$

We note that (1.3.7) implies

$$R_{AMN} = 0 \,, \qquad D_A = 0 \qquad \text{if} \quad T_{,K} = 0 \,. \tag{2.1.43}$$

From (2.1.19), (2.1.37) and (2.1.41) we find

$$\Phi_A = G_{AiN}u_{i,N} + D_A\theta + K_{AM}\theta_{,M} \,, \tag{2.1.44}$$

where

$$G_{AiN} = R_{AMN}x_{i,M} \,. \tag{2.1.45}$$

By (2.1.20) and (2.1.41),

$$Q_A^* = Q_A + R_{AMN}x_{i,M}x_{j,N}e_{ij} + D_A\theta + K_{AM}x_{i,M}\theta_{,i} \,. \tag{2.1.46}$$

The relation (2.1.7) implies that

$$\bar{Q}_i^* = \frac{1}{J}x_{i,A}Q_A^* \,. \tag{2.1.47}$$

Using (1.2.16), (2.1.37), (2.1.46) and (2.1.47) we obtain

$$\varphi_i = h_{irs}u_{r,s} + a_i\theta + k_{ij}\theta_{,j} \,, \tag{2.1.48}$$

where

$$\begin{aligned} h_{irs} &= \frac{1}{J}x_{i,K}x_{r,L}x_{s,M}R_{KLM} \,, \\ a_i &= \frac{1}{J}x_{i,K}D_K \,, \qquad k_{ij} = \frac{1}{J}x_{i,A}x_{j,B}K_{AB} \,. \end{aligned} \tag{2.1.49}$$

It follows from (2.1.43) and (2.1.49) that

$$h_{irs} = 0 \,, \quad a_i = 0 \qquad \text{if} \quad T_{,j} = 0 \,. \tag{2.1.50}$$

In view of (2.1.16) and (2.1.47),

$$\bar{Q}_i^* T_{,i}^* \geq 0 \,,$$

so that we have

$$q_iT_i + h_{irs}e_{rs}T_{,i} + a_i\theta T_{,i} + k_{ij}\theta_{,j}T_{,i} + q_i\theta_{,i} + h_{irs}e_{rs}\theta_{,i} + \\ + a_i\theta\theta_{,i} + k_{ij}\theta_{,i}\theta_{,j} \geq 0 \, , \tag{2.1.51}$$

for any small quantities e_{ij} and θ. From (2.1.50) and (2.1.51) we find that

$$k_{ij}\theta_{,i}\theta_{,j} \geq 0 \qquad \text{if} \quad T_{,s} = 0 \, . \tag{2.1.52}$$

a) **The equations of thermoelastic deformations referred to the configuration B_0.** If we refer all quantities to the configuration B_0, then we have

$$u_i = u_i(X_A, t) \, , \quad \theta = \theta(X_A, t) \, , \quad (X_A, t) \in B_0 \times (t_0, t_1) \, . \tag{2.1.53}$$

Forming differences of corresponding terms in the two sets of equations (1.2.3) and (2.1.10), we obtain

$$S_{Ai,A} + \varrho_0 F_i = \varrho_0 \ddot{u}_i \quad \text{on } B_0 \times (t_0, t_1) \, , \tag{2.1.54}$$

where

$$F_i = f_i^* - f_i \, . \tag{2.1.55}$$

Similarly, forming differences of corresponding terms in the two equations (1.3.8) and (2.1.12), with the aid of (2.1.37), we get

$$T\dot{\gamma} + \varrho_0 \dot{\eta}\theta = \Phi_{A,A} + \varrho_0 S \quad \text{on } B_0 \times (t_0, t_1) \, , \tag{2.1.56}$$

where

$$S = s^* - s \, . \tag{2.1.57}$$

Thus, the basic equations consist of the equations of motion (2.1.54), the energy equation (2.1.56), and the "constitutive" equations (2.1.30), (2.1.38) and (2.1.44). It is important to realize that the coefficients E_{iAjN} , F_{Ai} , A , G_{AiK} , D_A and K_{MN} depend on the state of strain and temperature of the primary state. From (1.5.1) and (2.1.17) we obtain the boundary conditions

$$\begin{aligned} u_i &= \hat{u}_i \quad \text{on } \bar{\Sigma}_1 \times [t_0, t_1) \, , \qquad & S_{Ai}N_A &= \hat{P}_i \quad \text{on } \Sigma_2 \times [t_0, t_1) \, , \\ \theta &= \hat{\theta} \quad \text{on } \bar{\Sigma}_3 \times [t_0, t_1) \, , \qquad & \Phi_A N_A &= \hat{\Phi} \quad \text{on } \Sigma_4 \times [t_0, t_1) \, , \end{aligned} \tag{2.1.58}$$

where

$$\hat{u}_i = \hat{y}_i - \hat{x}_i \, , \quad \hat{P}_i = \hat{T}_i^* - \hat{T}_i \, , \quad \hat{\theta} = \hat{T}^* - \hat{T} \, , \quad \hat{\Phi} = \hat{Q}^* - \hat{Q} \, .$$

To complete the specification of the problem we adjoin the initial conditions

$$u_i(\boldsymbol{X},0) = \hat{a}_i(\boldsymbol{X}) \,, \quad \dot{u}_i(\boldsymbol{X},0) = \hat{b}_i(\boldsymbol{X}) \,, \quad \theta(\boldsymbol{X},0) = \hat{\theta}_0(\boldsymbol{X}) \,, \quad \boldsymbol{X} \in B_0 \,,$$

where $\hat{a}_i$, $\hat{b}_i$ and $\hat{\theta}_0$ are prescribed functions.

b) **The equations of thermoelasticity referred to the primary state.** If we refer the thermodynamic quantities to the primary state, then we have

$$u_i = u_i(x_j,t) \,, \quad \theta = \theta(x_j,t) \,, \quad (x_j,t) \in B_0 \times (t_0,t_1) \,. \tag{2.1.59}$$

It follows from (1.1.9), (2.1.3), (2.1.11) and (2.1.35) that

$$s_{ji,j} + \varrho F_i = \varrho \ddot{u}_i \quad \text{on } B \times (t_0,t_1) \,. \tag{2.1.60}$$

Forming differences of corresponding terms in the two equations (1.5.3) and (2.1.13), we obtain

$$\frac{1}{J} T\dot{\gamma} + \varrho \dot{\eta} \theta = \varphi_{i,i} + \varrho S \qquad \text{on } B \times (t_0,t_1) \,. \tag{2.1.61}$$

In this case, the basic equations consist of the equations of motion (2.1.60), the energy equation (2.1.61), and the "constitutive" equations (2.1.36), (2.1.39) and (2.1.48).

By using (1.5.4) and (2.1.18) we derive the boundary conditions

$$\begin{aligned} u_i = \tilde{u}_i \quad \text{on } \bar{S}_1 \times [t_0,t_1) \,, \qquad & s_{ji} n_j = \tilde{p}_i \quad \text{on } S_2 \times [t_0,t_1) \,, \\ \theta = \tilde{\theta} \quad \text{on } \bar{S}_3 \times [t_0,t_1) \,, \qquad & \varphi_i n_i = \tilde{\varphi} \quad \text{on } S_4 \times [t_0,t_1) \,, \end{aligned} \tag{2.1.62}$$

where

$$\tilde{u}_i = \tilde{y}_i - \tilde{x}_i \,, \quad \tilde{p}_i = \tilde{T}^*_i - \tilde{T}_i \,, \quad \tilde{\theta} = \tilde{T}^* - \tilde{T} \,, \quad \tilde{\varphi} = \tilde{q}^* - \tilde{q} \,.$$

To the basic equations and boundary conditions we adjoin the initial conditions

$$u_i(\boldsymbol{x},0) = \tilde{a}_i(\boldsymbol{x}) \,, \quad \dot{u}_i(\boldsymbol{x},0) = \tilde{b}_i(\boldsymbol{x}) \,, \quad \theta(\boldsymbol{x},0) = \tilde{\theta}_0(\boldsymbol{x}) \,, \quad \boldsymbol{x} \in \bar{B}_0 \,, \tag{2.1.63}$$

where $\tilde{a}_i, \tilde{b}_i$ and $\tilde{\theta}_0$ are specified functions.

Finally, we add an expression for the free energy function ψ^* corresponding to the secondary state B^*. Since ψ^* is a function of $T+\theta$ and the strain tensor E^*_{KL}, retaining only linear and quadratic terms, we find that

$$\begin{aligned} \psi^* = \psi + J[t_{ij}e_{ij} - \varrho\eta\theta + \frac{1}{2}(C_{ijrs}e_{ij}e_{rs} - 2\beta_{ij}e_{ij}\theta - \\ - a\theta^2 + t_{ij}u_{r,i}u_{r,j})] \,. \end{aligned} \tag{2.1.64}$$

The function ψ^* may be written in the form

$$\begin{aligned}\psi^* = \psi + T_{Ai}u_{i,A} - \varrho_0\eta\theta + \frac{1}{2}(E_{iKjN}u_{i,K}u_{j,N} - \\ - 2F_{Ai}u_{i,A}\theta - A\theta^2 + T_{AB}u_{i,A}u_{i,B}) \,. \end{aligned} \tag{2.1.65}$$

c) **Isotropic materials.** The constitutive equations for isotropic materials have been presented in Section 1.4. Our next objective is to determine the form of the coefficients A_{KLMN}, B_{MN}, R_{AMN}, D_A and K_{MN} for isotropic bodies. From (1.3.5), (2.1.23) and (1.4.5), we have

$$A_{ABMN} = \frac{1}{2}\Big(\frac{\partial}{\partial E_{MN}} + \frac{\partial}{\partial E_{NM}}\Big)(\gamma_0\delta_{AB} + \gamma_1 C_{AB} + \gamma_2 C_{AB}^{(-1)}) \,. \tag{2.1.66}$$

By (1.4.3) and

$$\frac{\partial C_{AB}^{(-1)}}{\partial E_{MN}} = -2C_{AM}^{(-1)}C_{BN}^{(-1)}, \tag{2.1.67}$$

it follows from (2.1.66) that

$$\begin{aligned}A_{ABMN} = A_{MN}^{(-1)}C_{AB}^{(-1)} + A_{MN}^{(0)}\delta_{AB} + A_{MN}^{(1)}C_{AB} - \gamma_2(C_{AM}^{(-1)}C_{BN}^{(-1)} + \\ + C_{AN}^{(-1)}C_{BM}^{(-1)}) + \gamma_1(\delta_{AM}\delta_{BN} + \delta_{AN}\delta_{BM}) \,, \end{aligned} \tag{2.1.68}$$

$$B_{KL} = -\Big(\frac{\partial\gamma_0}{\partial T}\delta_{KL} + \frac{\partial\gamma_1}{\partial T}C_{KL} + \frac{\partial\gamma_2}{\partial T}C_{KL}^{(-1)}\Big) \,,$$

where

$$A_{MN}^{(s)} = 2\Big(\frac{\partial\gamma_s}{\partial I_1}\delta_{MN} + \frac{\partial\gamma_s}{\partial I_2}\Lambda_{MN} + \frac{\partial\gamma_s}{\partial I_3}I_3C_{MN}^{(-1)}\Big) \,, \qquad (s = 0, 1, 2) \,.$$

By (2.1.33) and (2.1.68), we get

$$\begin{aligned}C_{ijrs} = \delta_{ij}\alpha_{rs}^{(2)} + b_{ij}\alpha_{rs}^{(0)} + b_{ip}b_{jp}\alpha_{rs}^{(1)} - \\ - p(\delta_{ir}\delta_{js} + \delta_{is}\delta_{jr}) - \Psi(b_{ir}b_{js} + b_{is}b_{jr}) \,, \end{aligned} \tag{2.1.69}$$

$$\beta_{ij} = -\frac{1}{J}\Big(\frac{\partial\gamma_2}{\partial T}\delta_{ij} + \frac{\partial\gamma_0}{\partial T}b_{ij} + \frac{\partial\gamma_1}{\partial T}b_{is}b_{js}\Big) \,,$$

where

$$\alpha_{rs}^{(p)} = \frac{2}{J}\Big(\frac{\partial\gamma_p}{\partial I_1}b_{rs} + \frac{\partial\gamma_p}{\partial I_2}B_{rs} + I_3\frac{\partial\gamma_p}{\partial I_3}\delta_{rs}\Big) \,, \qquad p = (0, 1, 2) \,.$$

If we define

$$A_{ij} = \frac{2}{J}\frac{\partial^2 \hat{\psi}}{\partial I_i \partial I_j}, \quad L_i = \frac{2}{J}\frac{\partial^2 \hat{\psi}}{\partial I_i \partial T}, \tag{2.1.70}$$

then

$$\beta_{ij} = -I_3 L_3 \delta_{ij} - (L_1 + I_1 L_2) b_{ij} + L_2 b_{is} b_{js}, \tag{2.1.71}$$

and $\alpha_{rs}^{(k)}$ become

$$\begin{aligned}
\alpha_{rs}^{(0)} &= 2[(A_{11} + I_1 A_{12} + \Psi) b_{rs} + (A_{12} + I_1 A_{22}) B_{rs} + (A_{13} + I_1 A_{23}) I_3 \delta_{rs}], \\
\alpha_{rs}^{(1)} &= -2(A_{12} b_{rs} + A_{22} B_{rs} + A_{23} I_3 \delta_{rs}), \\
\alpha_{rs}^{(2)} &= 2I_3[A_{13} b_{rs} + A_{23} B_{rs} + \left(\frac{p}{I_3} + A_{33} I_3\right)\delta_{rs}].
\end{aligned} \tag{2.1.72}$$

It follows from (1.4.11), (2.1.42) and (2.1.71) that

$$\begin{aligned}
R_{KMN} &= R_{MN}^{(1)} T_{,K} + R_{MN}^{(2)} E_{KL} T_{,L} + R_{MN}^{(3)} E_{KA} E_{AL} T_{,L} + \\
&+ \frac{1}{2} K_2 (\delta_{KM}\delta_{LN} + \delta_{KN}\delta_{LM}) T_{,L} + \frac{1}{2} K_3 (\delta_{KM} E_{NL} + \delta_{LN} E_{KM} + \\
&+ \delta_{KN} E_{MN} + \delta_{ML} E_{KN}) T_{,L}, \\
D_K &= (D^{(1)}\delta_{KL} + D^{(2)} E_{KL} + D^{(3)} E_{KM} E_{ML}) T_{,L}, \\
K_{MN} &= (J_N^{(1)}\delta_{ML} + J_N^{(2)} E_{ML} + J_N^{(3)} E_{MA} E_{AL}) T_{,L} + K_1 \delta_{MN} + \\
&+ K_2 E_{MN} + K_3 E_{MA} E_{AN},
\end{aligned} \tag{2.1.73}$$

where

$$\begin{aligned}
R_{MN}^{(i)} &= 2\frac{\partial K_i}{\partial I_1}\delta_{MN} + 2\frac{\partial K_i}{\partial I_2}\Lambda_{MN} + 2I_3 C_{MN}^{(-1)}\frac{\partial K_i}{\partial I_3} + \\
&\quad + T_{,M} T_{,N}\frac{\partial K_i}{\partial I_5} + (E_{MA} T_{,N} + E_{NA} T_{,M}) T_{,A}\frac{\partial K_i}{\partial I_6}, \\
D^{(i)} &= \frac{\partial K_i}{\partial T}, \quad J_N^{(i)} = 2\left(T_{,N}\frac{\partial K_i}{\partial I_4} + E_{NA} T_{,A}\frac{\partial K_i}{\partial I_5} + E_{NA} E_{AL} T_{,L}\frac{\partial K_i}{\partial I_6}\right).
\end{aligned} \tag{2.1.74}$$

Using (2.1.49) and (2.1.73) we can find the expressions for h_{irs}, a_i and k_{ij}. When the primary state is at constant temperature we obtain

$$\begin{aligned}
&Q_A = 0, \qquad R_{KMN} = 0, \qquad D_K = 0, \\
&K_{MN} = K_1 \delta_{MN} + K_2 E_{MN} + K_3 E_{MA} E_{AN} = K_{NM},
\end{aligned} \tag{2.1.75}$$

where K_i are evaluated for the constant temperature T.

d) **Thermoelastic deformations with initial stress and initial heat flux.** We now assume that the primary state is identical with that of the initial body B_0 so that $x_1 = X_1$, $x_2 = X_2$, $x_3 = X_3$. We assume that B_0 is subjected to initial stress and is at nonuniform temperature $\bar{T}$. We then consider that the secondary state is obtained from B by an infinitesimal thermoelastic deformation. In this case,

$$\begin{aligned} &J = 1\,, \quad E_{AB} = 0\,, \quad T = \bar{T}\,, \quad T_{AB} = \frac{\partial \hat{\psi}}{\partial E_{AB}}\,, \\ &\varrho_0 \eta = -\frac{\partial \hat{\psi}}{\partial T}, \quad Q_A = \hat{Q}_A(0, \bar{T}, \bar{T}_{,A})\,, \end{aligned} \tag{2.1.76}$$

the derivatives being evaluated at $E_{AB} = 0$ and $T = \bar{T}$. Clearly, T_{AB}, T_{Ai} and t_{ij} coincide. Moreover, $Q_A = q_i \delta_{iA}$. The initial stresses satisfy the conditions that the body B is in equilibrium. The initial heat flux Q_A is an arbitrary function of $\boldsymbol{X}$ apart from satisfying the energy equation.

It follows from (2.1.33), (2.1.40) and (2.1.48) that C_{ijrs}, β_{ij}, a, h_{irs}, a_i, k_{ij} are identical with A_{KLMN}, B_{KL}, A, R_{MNL}, D_K, K_{AB}, respectively. The coefficients defined by (2.1.23) and (2.1.42) are now evaluated at $E_{AB} = 0$ and $T = \bar{T}$.

The equations governing the infinitesimal thermoelastic deformations are the equations of motion

$$s_{ji,j} + \varrho F_i = \varrho \ddot{u}_i \quad \text{on } B \times (t_0, t_1)\,, \tag{2.1.77}$$

the energy equation

$$\bar{T}\dot{\gamma} = \varphi_{i,i} + \varrho S \quad \text{on } B \times (t_0, t_1)\,, \tag{2.1.78}$$

and the constitutive equations

$$\begin{aligned} s_{ij} &= (C_{ijrs} + t_{is}\delta_{jr}) u_{r,s} - \beta_{ij}\theta\,, \\ \gamma &= \beta_{ij} u_{i,j} + a\theta\,, \\ \varphi_i &= h_{irs} u_{r,s} + a_i \theta + k_{ij}\theta_{,j}\,. \end{aligned} \tag{2.1.79}$$

If $\bar{T}$ is constant, then $q_i = 0$, $h_{irs} = 0$, $a_i = 0$, and the theory reduces to that established by Green [149].

To the system of equations (2.1.77)-(2.1.79) we adjoin the boundary conditions (2.1.62) and the initial conditions (2.1.63). When the body B is isotropic, then (2.1.49), (2.1.69) and (2.1.74) yield

$$\begin{aligned} &C_{ijrs} = \lambda \delta_{ij}\delta_{rs} + \mu(\delta_{ir}\delta_{js} + \delta_{is}\delta_{jr})\,, \qquad \beta_{ij} = \beta\delta_{ij}\,, \\ &h_{irs} = h_1 \bar{T}_{,i}\delta_{rs} + h_2 \bar{T}_{,r}\bar{T}_{,s}\bar{T}_{,i} + h_3(\delta_{ir}\delta_{ps} + \delta_{is}\delta_{rp})\,\bar{T}_{,p}\,, \\ &a_i = m\bar{T}_{,i}\,, \qquad k_{ij} = k_0 \bar{T}_{,i}\bar{T}_{,j} + k\delta_{ij}\,, \qquad t_{ij} = p\delta_{ij}\,, \end{aligned} \tag{2.1.80}$$

where λ, μ, β, h_i, m, k_0, k and p are given functions.

The form of the free energy ψ^* may be written down immediately as a special case of (2.1.64) when $E_{AB} = 0$, $T = \bar{T}$ and is

$$\begin{aligned} \psi^* = \psi_0 + t_{ij}e_{ij} - f\theta + \frac{1}{2}(C_{ijrs}e_{ij}e_{rs} - \\ - 2\beta_{ij}e_{ij}\theta - a\theta^2 + t_{ij}u_{r,i}u_{r,j}) \,, \end{aligned} \tag{2.1.81}$$

where the coefficients ψ_0, t_{ij}, f, C_{ijrs}, β_{ij} and a are, in general, functions of the coordinates x_i.

2.2 Dynamic Theory

The basic system of field equations describing the infinitesimal thermoelastic deformations of a strained body, *referred to the configuration* B_0, consists of the equations of motion

$$S_{Ai,A} + \varrho_0 F_i = \varrho_0 \ddot{u}_i \,, \tag{2.2.1}$$

the energy equation

$$T\dot{\gamma} + \varrho_0 \dot{\eta}\theta = \Phi_{A,A} + \varrho_0 S \,, \tag{2.2.2}$$

and the constitutive equations

$$\begin{aligned} S_{Ai} &= D_{iAjK}u_{j,K} - F_{Ai}\theta \,, \\ \gamma &= F_{Ai}u_{i,A} + A\theta \,, \\ \Phi_A &= G_{AiK}u_{i,K} + D_A\theta + K_{AM}\theta_{,M} \,, \end{aligned} \tag{2.2.3}$$

on $B_0 \times (0, t_1)$. In (2.2.3) we have used the notation

$$D_{iAjN} = E_{iAjN} + \delta_{ij}T_{AN} \,, \tag{2.2.4}$$

and we have supposed that $t_0 = 0$. In view of (2.1.23) and (2.1.18) we obtain

$$D_{iAjB} = D_{jBiA} \,. \tag{2.2.5}$$

The coefficients D_{iAjK}, F_{Ai}, A, G_{AiK}, D_A and K_{MN} are evaluated in the primary state and, in general, are functions of both position and time. These quantities are determined explicitly by the deformation of the primary state.

We suppose that D_{iAjB}, F_{Ai}, A, G_{AiK}, D_A, K_{MN}, T, η, ϱ_0, F_i and S are prescribed functions with the properties: (i) D_{iAjB}, F_{Ai}, A, G_{AiK}, D_A, K_{MN}, η, and T are continuously differentiable on $\bar{B}_0 \times [0, t_1)$; (ii) ϱ_0 is continuous and strictly positive on $\bar{B}_0$; (iii) D_{iAjB} satisfy the conditions (2.2.5); (iv) F_i and S are continuous on $\bar{B}_0 \times [0, t_1)$.

We say that the vector field $\boldsymbol{u}$ is an admissible displacement field on $\bar{B}_0 \times [0, t_1)$

provided: (*i*) $\boldsymbol{u}$ is of class C^2 on $B_0 \times (0, t_1)$; (*ii*) $\boldsymbol{u}$, $\dot{\boldsymbol{u}}$, $\ddot{\boldsymbol{u}}$, $\boldsymbol{u}_{,A}$ and $\dot{\boldsymbol{u}}_{,A}$ are continuous on $\bar{B}_0 \times [0, t_1)$. We say that θ is an admissible temperature field on $\bar{B}_0 \times [0, t_1)$ provided: (*i*) θ is of class $C^{2,1}$ on $B_0 \times (0, t_1)$; (*ii*) θ, $\theta_{,A}$ and $\dot{\theta}_{,A}$ are continuous on $B_0 \times [0, t_1)$. An admissible stress field on $\bar{B}_0 \times [0, t_1)$ is a tensor field S_{Ai} with the properties: (*i*) S_{Ai} are of class $C^{1,0}$ on $B_0 \times (0, t_1)$; (*ii*) S_{Ai} and $S_{Ai,A}$ are continuous on $\bar{B}_0 \times [0, t_1)$. We say that Φ_A is an admissible heat flux vector on $\bar{B}_0 \times [0, t_1)$ provided: (*i*) Φ_A is of class $C^{1,0}$ on $B_0 \times (0, t_1)$; (*ii*) Φ_A and $\Phi_{M,M}$ are continuous on $\bar{B}_0 \times [0, t_1)$. By an admissible process on $\bar{B}_0 \times [0, t_1)$ we mean an ordered array of functions $z = (u_i,\ \theta,\ S_{Aj},\ \gamma,\ \Phi_M)$ with the following properties: (*i*) $\boldsymbol{u}$ is an admissible displacement field on $\bar{B}_0 \times [0, t_1)$; (*ii*) θ is an admissible temperature field on $\bar{B}_0 \times [0, t_1)$; (*iii*) S_{Ai} is an admissible stress field on $\bar{B}_0 \times [0, t_1)$; (*iv*) γ is of class $C^{0,1}$ on $B_0 \times (0, t_1)$; (*v*) Φ_A is an admissible heat flux vector on $\bar{B}_0 \times [0, t_1)$. When we omit mention of the domain of definition of an admissible process, it will always be understood to be $\bar{B}_0 \times [0, t_1)$. We say that $z = (u_i,\ \theta,\ S_{Aj},\ \gamma,\ \Phi_K)$ is a thermoelastic process corresponding to the body force $\boldsymbol{F}$ and the heat supply S if z is an admissible process that satisfies the fundamental system of field equations (2.2.1)-(2.1.3) on $B_0 \times (0, t_1)$. The corresponding surface traction $\boldsymbol{P}$ and the surface heat flux Φ are defined at every regular point of $\partial B_0 \times [0, t_1)$ by

$$P_i = S_{Ai} N_A\ , \qquad \Phi = \Phi_K N_K\ . \tag{2.2.6}$$

Let $z = (u_i,\ \theta,\ S_{Ai},\ \gamma,\ \Phi_M)$ be a thermoelastic process corresponding to the body force $\boldsymbol{F}$, and the heat supply S. Then, we define the functions w and Π on $[0, t_1)$ by

$$\begin{aligned} w &= \frac{1}{2} \int\limits_{B_0} (\varrho_0 \dot{u}_i \dot{u}_i + D_{iAjK} u_{i,A} u_{j,K} + A\theta^2)\, dV\ , \\ \Pi &= w - \frac{1}{2} \int_0^t \int\limits_{B_0} \Big[(\dot{D}_{iAjK} u_{i,A} u_{j,K} - \dot{A}\theta^2 - 2\dot{F}_{Ai} u_{i,A}\theta - \\ &\quad - \frac{2}{T} \varrho_0 \dot{\eta} \theta^2 - 2\Big(\frac{1}{T}\theta\Big)_{,A} \Phi_A \Big]\, dV dt\ . \end{aligned} \tag{2.2.7}$$

Theorem 2.2.1. *Let $(u_i\,,\ \theta\,,\ S_{Ai}\,,\ \gamma\,,\ \Phi_K)$ be a thermoelastic process corresponding to the body force $\boldsymbol{F}$, and the heat supply S. Then*

$$\dot{\Pi} = \int\limits_{\partial B_0} (P_i \dot{u}_i + \frac{1}{T}\Phi\theta) dA + \int\limits_{B_0} \varrho_0 (F_i \dot{u}_i + \frac{1}{T} S\theta)\, dV\ . \tag{2.2.8}$$

Proof. In view of (2.2.3) and (2.2.5), we get

$$S_{Ai}\dot{u}_{i,A} + \dot{\gamma}\theta = \frac{1}{2}\frac{\partial}{\partial t}(D_{iAjK}u_{i,A}u_{j,K} + A\theta^2) - \\ - \frac{1}{2}\dot{D}_{iAjK}u_{i,A}u_{j,K} + \frac{1}{2}\dot{A}\theta^2 + \dot{F}_{Ki}u_{i,K}\theta \,. \tag{2.2.9}$$

On the other hand, by (2.2.1) and (2.2.2), we can write

$$S_{Ai}\dot{u}_{i,A} + \dot{\gamma}\theta = (S_{Ai}\dot{u}_i + \frac{1}{T}\theta\Phi_A)_{,A} - S_{Ai,A}\dot{u}_i - \\ - \left(\frac{1}{T}\theta\right)_{,A}\Phi_A + \frac{1}{T}\varrho_0 S\theta - \frac{1}{T}\varrho_0\dot{\eta}\theta^2 = \\ = (S_{Ai}\dot{u}_i + \frac{1}{T}\theta\Phi_A)_{,A} + \varrho_0(F_i\dot{u}_i + \frac{1}{T}S\theta) - \\ - \varrho_0\dot{u}_i\ddot{u}_i - \frac{1}{T}\varrho_0\dot{\eta}\theta^2 - \left(\frac{1}{T}\theta\right)_{,A}\Phi_A \,. \tag{2.2.10}$$

Using the divergence theorem, from (2.2.10) and (2.2.6) we get

$$\int_{B_0}(S_{Ai}\dot{u}_{i,A} + \dot{\gamma}\theta)\, dV = \int_{\partial B_0}(P_i\dot{u}_i + \frac{1}{T}\Phi\theta)dA + \\ + \int_{B_0}\varrho_0(F_i\dot{u}_i + \frac{1}{T}S\theta)\, dV - \int_{B_0}[\varrho_0\dot{u}_i\ddot{u}_i + \frac{1}{T}\varrho_0\dot{\eta}\theta^2 + \\ + (\frac{1}{T}\theta)_{,A}\Phi_A]\, dV \,. \tag{2.2.11}$$

From (2.2.7), (2.2.9) and (2.2.11) we obtain (2.2.8). □

We note that (2.2.8) may be written in the form

$$\Pi(t) - \Pi(0) = \int_0^t\int_{\partial B_0}(P_i\dot{u}_i + \frac{1}{T}\Phi\theta)dA dt + \int_0^t\int_{B_0}\varrho_0(F_i\dot{u}_i + \frac{1}{T}S\theta)\, dV dt \,. \tag{2.2.12}$$

Theorem 2.2.1 has the following consequence

Theorem 2.2.2. *Let $(u_i\,,\ \theta\,,\ S_{Ai}\,,\ \gamma\,,\ \Phi_K)$ be a thermoelastic process corresponding to the body force $\boldsymbol{F}$ and the heat supply S, and assume that*

$$\boldsymbol{F} = 0,\ S = 0 \quad \text{on } \bar{B}_0 \times [0, t_1)\,, \quad P_i\dot{u}_i = 0, \quad \Phi\theta = 0 \quad \text{on } \partial B_0 \times [0, t_1)\,.$$

Then Π *is conserved, i.e.*

$$\Pi(t) = \Pi(0)\ , \quad 0 \le t < t_1\ . \tag{2.2.13}$$

We now assume that *the primary state is a configuration of equilibrium.* In this case we refer the deformation of the body to the configuration B. The basic system of field equations consists of the equations of motion

$$s_{ji,j} + \varrho F_i = \varrho \ddot{u}_i\ , \tag{2.2.14}$$

the energy equation

$$\frac{1}{J} T \dot{\gamma} = \varphi_{i,i} + \varrho S\ , \tag{2.2.15}$$

and the constitutive equations

$$\begin{aligned} s_{ji} &= d_{ijrs} u_{r,s} - \beta_{ij}\theta\ , \\ \gamma &= J(\beta_{ij} u_{i,j} + a\theta)\ , \\ \varphi_i &= h_{irs} u_{r,s} + a_i\theta + k_{ij}\theta_{,j}\ , \end{aligned} \tag{2.2.16}$$

on $B \times (0, t_1)$. Here we have used the notation

$$d_{ijrs} = C_{ijrs} + \delta_{ir} t_{js}\ . \tag{2.2.17}$$

We note that

$$d_{ijrs} = d_{rsij}\ , \quad \beta_{ij} = \beta_{ji}\ . \tag{2.2.18}$$

The coefficients d_{ijrs}, β_{ij}, a, h_{ijr} , a_i , k_{ij} and J depend not only upon the material but also upon the initial stress and the initial heat flux and, in general, are functions of position. When the initial strain and the initial heat flux are absent then (2.2.16) reduces to the constitutive equations of the classical linear thermoelasticity. We assume that: (*i*) d_{ijrs}, β_{ij}, a, h_{ijr}, a_i, k_{ij}, J and T are continuously differentiable on $\bar{B}$; (*ii*) ϱ is continuous and strictly positive on $\bar{B}$; (*iii*) $\boldsymbol{F}$ and S are continuous on $\bar{B} \times [0, t_1)$; (*iv*) d_{ijrs} and β_{ij} satisfy the conditions (2.2.18).

By an admissible process on $B \times [0, t_1)$ we mean an ordered array of functions $(u_i,\ \theta,\ s_{ij},\ \gamma,\ \varphi_i)$ with the following properties: (*i*) $\boldsymbol{u}$ is an admissible displacement field on $\bar{B} \times [0, t_1)$; (*ii*) θ is an admissible temperature field on $\bar{B} \times [0, t_1)$; (*iii*) s_{ji} is an admissible stress field on $\bar{B} \times [0, t_1)$; (*iv*) γ is of class $C^{0,1}$ on $B \times (0, t_1)$; (*v*) φ_i is an admissible heat flux vector on $\bar{B} \times [0, t_1)$. We say that $\pi = (u_i,\ \theta,\ s_{ij},\ \gamma,\ \varphi_i)$ is a thermoelastic process on $\bar{B} \times [0, t_1)$ corresponding to the body force $\boldsymbol{F}$ and the heat supply S if π is an admissible process on $\bar{B} \times [0, t_1)$ that satisfies the equations

(2.2.14)-(2.2.16) on $B \times [0, t_1)$. The surface traction $\boldsymbol{s}$ and the surface heat flux q are defined at every regular point of $\partial B \times [0, t_1)$ by

$$s_i = s_{ji} n_j \ , \qquad q = \varphi_i n_i \ . \tag{2.2.19}$$

From (2.1.28), (2.1.33), (2.1.40), (2.1.49), (2.2.4) and (2.2.7) we find that

$$\begin{aligned} w &= \frac{1}{2} \int_B (\varrho \dot{u}_i \dot{u}_i + d_{ijrs} u_{i,j} u_{r,s} + a\theta^2) dv \ , \\ \Pi &= w + \int_0^t \int_B \left(\frac{1}{T} \theta \right)_{,i} \varphi_i dv dt \ . \end{aligned} \tag{2.2.20}$$

We note that if T is constant, then (2.1.50) leads to $\varphi_i = k_{ij}\theta_{,j}$. The following proposition is an immediate consequence of Theorem 2.2.2.

Theorem 2.2.3. *Let B be a configuration of equilibrium at constant temperature. Let $\pi = (u_i,\ \theta,\ s_{ij},\ \gamma,\ \varphi_i)$ be a thermoelastic process on $\bar{B} \times [0, t_1)$ and assume that*

$$F_i = 0 \ , \quad S = 0 \quad \text{on } \bar{B} \times [0, t_1) \ , \quad s_i \dot{u}_i = 0, \quad q\theta = 0 \quad \text{on } \partial B \times [0, t_1) \ . \tag{2.2.21}$$

Then

$$w(t) - w(0) = -\frac{1}{T} \int_0^t \int_B k_{ij} \theta_{,i} \theta_{,j} dv dt \ , \quad 0 \le t < t_1 \ . \tag{2.2.22}$$

The relation (2.2.22) was used in the framework of the classical theory of thermoelasticity to derive uniqueness results (see Weiner [362] and Ionescu- Cazimir [201]).

2.3 Characterizations of the solutions to the boundary-initial-value problems

We first assume that the primary state is a configuration of equilibrium. Thus, the fundamental system of field equations is given by (2.2.14)-(2.2.16) on $B \times [0, t_1)$. In addition to the specification of d_{ijrs}, β_{ij}, a, h_{ijk}, a_i, k_{ij}, J, ϱ and T we assume that the following data are given: (i) the initial displacement, velocity and temperature $\tilde{\boldsymbol{a}}$, $\tilde{\boldsymbol{b}}$ and $\tilde{\theta}_0$, all continuous on $\bar{B}$; (ii) the surface displacement $\tilde{\boldsymbol{u}}$ continuous on $\bar{S}_1 \times [0, t_1)$; (iii) the surface traction $\tilde{\boldsymbol{p}}$, piecewise regular on $S_2 \times [0, t_1)$, and continuous in time; (iv) the surface temperature $\tilde{\theta}$, continuous on $\bar{S}_3 \times [0, t_1)$; (v) the surface heat flux $\tilde{\varphi}$, piecewise regular on $S_4 \times [0, t_1)$ and continuous in time; (vi) the body force $\boldsymbol{F}$ and heat supply S both continuous on $\bar{B} \times [0, t_1)$.

By an *external data system* we mean an ordered array $L = (\boldsymbol{F}\ ,\ S\ ,\ \tilde{\boldsymbol{u}}\ ,\ \tilde{\boldsymbol{p}}\ ,\ \tilde{\theta}\ ,$

$\tilde{\varphi}$, $\tilde{\boldsymbol{a}}$, $\tilde{\boldsymbol{b}}$, $\tilde{\theta}_0$) with the properties (i) – (vi).

The mixed problem of the dynamic theory of thermoelasticity consists in finding a thermoelastic process (u_i , θ , s_{ij} , γ , φ_i) on $\bar{B} \times [0, t_1)$ corresponding to the body force $\boldsymbol{F}$ and heat supply S, that satisfies the initial conditions

$$\boldsymbol{u}(\cdot, 0) = \tilde{\boldsymbol{a}} \,, \quad \dot{\boldsymbol{u}}(\cdot, 0) = \tilde{\boldsymbol{b}} \,, \quad \theta(\cdot, 0) = \tilde{\theta}_0 \quad \text{on } \bar{B} \,, \tag{2.3.1}$$

and the boundary conditions

$$\begin{aligned} \boldsymbol{u} = \tilde{\boldsymbol{u}} \quad \text{on } \bar{S}_1 \times [0, t_1) \,, \quad \boldsymbol{s} = \tilde{\boldsymbol{p}} \quad \text{on } S_2 \times [0, t_1) \,, \\ \theta = \tilde{\theta} \quad \text{on } \bar{S}_3 \times [0, t_1) \,, \quad q = \tilde{\varphi} \quad \text{on } S_4 \times [0, t_1) \,. \end{aligned} \tag{2.3.2}$$

Substitution of (2.2.16) into the equations (2.2.15), (2.2.14) and (2.3.2) yields the equations

$$\begin{aligned} &(d_{ijrs} u_{r,s})_{,j} - (\beta_{ij}\theta)_{,j} + \varrho F_i = \varrho \ddot{u}_i \,, \\ &(k_{ij}\theta_{,j} + a_i\theta)_{,i} + (h_{irs} u_{r,s})_{,i} - T\beta_{ij}\dot{u}_{i,j} - aT\dot{\theta} = -\varrho S \,, \end{aligned} \tag{2.3.3}$$

on $B \times [0, t_1)$ and the boundary conditions

$$\begin{aligned} u_i = \tilde{u}_i \quad \text{on } \bar{S}_1 \times [0, t_1) \,, \quad (d_{ijrs} u_{r,s} - \beta_{ij}\theta) n_j = \tilde{p}_i \quad \text{on } S_2 \times [0, t_1) \,, \\ \theta = \tilde{\theta} \quad \text{on } \bar{S}_3 \times [0, t_1) \,, \quad (k_{ij}\theta_{,j} + a_i\theta + h_{irs} u_{r,s}) n_i = \tilde{\varphi} \quad \text{on } S_4 \times [0, t_1) \,. \end{aligned} \tag{2.3.4}$$

By a solution of the mixed problem we mean the ordered pair $(\boldsymbol{u}, \theta)$ with the following properties: (i) $\boldsymbol{u}$ is an admissible displacement field on $\bar{B} \times [0, t_1)$; (ii) θ is an admissible temperature field on $\bar{B} \times [0, t_1)$; (iii) $\boldsymbol{u}$ and θ satisfy the equations (2.3.3) on $B \times (0, t_1)$, the initial conditions (2.3.1) and the boundary conditions (2.3.4).

We now present an alternative characterization of the solution to the mixed problem in which the initial conditions are incorporated into the field equations.

Let f_1 and f_2 be scalar fields on $B \times [0, t_1)$ that are continuous in time. By the convolution of f_1 and f_2 we mean the function $f_1 * f_2$ defined on $B \times [0, t_1)$ by

$$f_1 * f_2(\boldsymbol{x} \,, t) = \int_0^t f_1(\boldsymbol{x} \,, t - s) f_2(\boldsymbol{x} \,, s) \, ds \,.$$

Henceforth (unless otherwise specified) i will denote the function on $[0, t_1)$ defined by

$$i(t) = t \,. \tag{2.3.5}$$

Let h_j and g be functions on $\bar{B} \times [0, t_1)$ defined by

$$h_j = \varrho(i * F_j + \tilde{a}_j + t\tilde{b}_j) ,$$
$$g = 1 * \varrho S + aT\tilde{\theta}_0 + T\beta_{rs}\tilde{a}_{r,s} . \tag{2.3.6}$$

Theorem 2.3.1. *Let $\pi = (u_i, \theta, s_{ij}, \gamma, \varphi_i)$ be an admissible process on $\bar{B} \times [0, t_1)$. Then π satisfies the equations of motion (2.2.14), the equation of energy (2.2.15) and the initial conditions (2.3.1) if and only if*

$$i * s_{rj,r} + h_j = \varrho u_j ,$$
$$T\gamma = J(1 * \varphi_{i,i} + g) \quad \text{on } B \times [0, t_1) . \tag{2.3.7}$$

Proof. Assume that (2.2.14), (2.2.15) and the initial conditions (2.3.1) hold. Taking the convolution of the equation (2.2.14) with i, and the convolution of the equation (2.2.15) with 1, we conclude with the aid of (2.3.6) that the equations (2.3.7) hold. The converse is verified by reversing the foregoing argument. □

In the isothermal theory this result was established by Ignaczak [191] and Gurtin [157]. In the framework of thermoelasticity theory the result was derived in [173], [176] (see also Carlson [46], p.338 and Lebon [233], p.370).

The next proposition is an immediate consequence of Theorem 2.3.1.

Theorem 2.3.2. *Let $\boldsymbol{u}$ be an admissible displacement field on $\bar{B} \times [0, t_1)$, and let θ be an admissible temperature field on $\bar{B} \times [0, t_1)$. Then $(\boldsymbol{u}, \theta)$ is a solution of the mixed problem if and only if $\boldsymbol{u}$ and θ satisfy the field equations*

$$i * (d_{kjrs}u_{r,s} - \beta_{kj}\theta)_{,k} + h_j = \varrho u_j ,$$
$$1 * (h_{irs}u_{r,s} + a_i\theta + k_{ij}\theta_{,j})_{,i} - T(\beta_{ij}u_{i,j} + a\theta) = -g , \tag{2.3.8}$$

on $B \times [0, t_1)$, and the boundary conditions (2.3.4).

Theorem 2.3.2 furnishes an alternative characterization of the solution to the mixed problem in which the initial conditions are incorporated into the field equations.

We now assume that the primary state is a time-dependent configuration. Then we take B_0 as reference configuration. The fundamental system of field equations consists of the equations (2.2.1)-(2.2.3) on $B_0 \times (0, t_1)$. In this system we adjoin the initial conditions

$$\boldsymbol{u}(\cdot, 0) = \hat{\boldsymbol{a}} , \quad \dot{\boldsymbol{u}}(\cdot, 0) = \hat{\boldsymbol{b}} , \quad \theta(\cdot, 0) = \hat{\theta}_0 \quad \text{on } \bar{B}_0 , \tag{2.3.9}$$

and the boundary conditions

$$\boldsymbol{u} = \hat{\boldsymbol{u}} \quad \text{on } \bar{\Sigma}_1 \times [0,t_1)\,, \qquad \boldsymbol{P} = \hat{\boldsymbol{P}} \quad \text{on } \Sigma_2 \times [0,t_1)\,,$$
$$\theta = \hat{\theta} \quad \text{on } \bar{\Sigma}_3 \times [0,t_1)\,, \qquad \Phi = \hat{\Phi} \quad \text{on } \Sigma_4 \times [0,t_1)\,. \tag{2.3.10}$$

We assume that: (i) $\hat{\boldsymbol{a}}$, $\hat{\boldsymbol{b}}$ and $\hat{\theta}_0$ are continuous on $\bar{B}_0$; (ii) $\hat{\boldsymbol{u}}$ is continuous on $\bar{\Sigma}_1 \times [0,t_1)$; (iii) $\hat{\boldsymbol{P}}$ is continuous in time and piecewise regular on $\Sigma_2 \times [0,t_1)$; (iv) $\hat{\theta}$ is continuous on $\Sigma_3 \times [0,t_1)$; (v) $\hat{\Phi}$ is continuous in time and piecewise regular on $\Sigma_4 \times [0,t_1)$; (vi) $\boldsymbol{F}$ and S are continuous on $\bar{B}_0 \times [0,t_1)$. By an external data system on $\bar{B}_0 \times [0,t_1)$ we mean an ordered array $L^0 = (\boldsymbol{F}\,,\, S\,,\, \hat{\boldsymbol{u}}\,,\, \hat{\boldsymbol{P}}\,,\, \hat{\theta}\,,\, \hat{\Phi}\,,\, \hat{\boldsymbol{a}}\,,\, \hat{\boldsymbol{b}}\,, \hat{\theta}_0)$ with the properties $(i) - (vi)$.

Substitution of (2.2.3) into the equations (2.2.1), (2.2.2) and (2.3.10) yields the equations

$$(D_{iAjK}u_{j,K})_{,A} - (F_{Ai}\theta)_{,A} + \varrho_0 F_i = \varrho_0 \ddot{u}_i\,,$$
$$(K_{AM}\theta_{,M})_{,A} + (D_A\theta + G_{AiK}u_{i,K})_{,A} - \varrho_0\dot{\eta}\theta - T(F_{Ai}\dot{u}_{i,A} + A\dot{\theta} + \tag{2.3.11}$$
$$+ \dot{F}_{Ai}u_{i,A} + \dot{A}\theta) = -\varrho_0 S\,,$$

and the boundary conditions

$$u_i = \hat{u}_i \qquad \text{on } \bar{\Sigma}_1 \times [0,t_1)\,,$$
$$(D_{iAjK}u_{j,K} - F_{Ai}\theta)N_A = \hat{P}_i \qquad \text{on } \Sigma_2 \times [0,t_1)\,,$$
$$\theta = \hat{\theta} \qquad \text{on } \bar{\Sigma}_3 \times [0,t_1)\,, \tag{2.3.12}$$
$$(K_{AM}\theta_{,M} + G_{AiK}u_{i,K} + D_A\theta)N_A = \hat{\Phi} \qquad \text{on } \Sigma_4 \times [0,t_1)\,.$$

By a solution of the problem we mean the ordered pair $(\boldsymbol{u},\theta)$ with the properties: (i) $\boldsymbol{u}$ is an admissible displacement field on $\bar{B}_0 \times [0,t_1)$; (ii) θ is an admissible temperature field on $\bar{B}_0 \times [0,t_1)$; (iii) $\boldsymbol{u}$ and θ satisfy the equations (2.3.11) on $B_0 \times (0,t_1)$, the initial conditions (2.3.9) and the boundary conditions (2.3.12).

2.4 Continuous dependence results

a) **Time-dependent non-isothermal primary states.** We assume that the primary state B is obtained from B_0 by a non-isothermal motion. The fundamental system of field equations is given by (2.2.1)-(2.2.3) on $B_0 \times [0,t_1)$. To this system we adjoin the initial conditions (2.3.9) and the boundary conditions (2.3.10). We examine the continuous dependence of the solutions upon initial data and supply terms. We present the results established by Chiriţa in [68] by the method of Dafermos [92], [93]. Let $(\boldsymbol{u}^{(1)},\ \theta^{(1)})$ and $(\boldsymbol{u}^{(2)},\ \theta^{(2)})$ be solutions corresponding to the external data systems $E^{(1)} = (\boldsymbol{F}^{(1)},\ S^{(1)},\ \hat{\boldsymbol{u}},\ \hat{\boldsymbol{P}},\ \hat{\theta},\ \hat{\Phi},\ \hat{\boldsymbol{a}}^{(1)},\ \hat{\boldsymbol{b}}^{(1)},\ \hat{\theta}_0^{(1)})$ and $E^{(2)} = (\boldsymbol{F}^{(2)},\ S^{(2)},\ \hat{\boldsymbol{u}},\ \hat{\boldsymbol{P}},\ \hat{\theta},\ \hat{\Phi},\ \hat{\boldsymbol{a}}^{(2)},\ \hat{\boldsymbol{b}}^{(2)},\ \hat{\theta}_0^{(2)})$, respectively. If we define $\boldsymbol{u} = \boldsymbol{u}^{(1)} - \boldsymbol{u}^{(2)}$ and $\theta = \theta^{(1)} - \theta^{(2)}$, then $(\boldsymbol{u},\ \theta)$ is a solution of the mixed problem corresponding to the external data system $L = (\boldsymbol{F},\ S,\ \boldsymbol{0},\ \boldsymbol{0},\ 0,\ 0,\ \hat{\boldsymbol{a}},\ \hat{\boldsymbol{b}},\ \theta_0)$ where $\boldsymbol{F} = \boldsymbol{F}^{(1)} - \boldsymbol{F}^{(2)}$, $S = S^{(1)} - S^{(2)}$, $\hat{\boldsymbol{a}} = \hat{\boldsymbol{a}}^{(1)} - \hat{\boldsymbol{a}}^{(2)}$, $\hat{\boldsymbol{b}} = \hat{\boldsymbol{b}}^{(1)} - \hat{\boldsymbol{b}}^{(2)}$, $\theta_0 = \hat{\theta}^{(1)} - \hat{\theta}^{(2)}$. We denote this problem by (P). We define the "distance" between the solutions $(\boldsymbol{u}^{(1)}\ ,\ \theta^{(1)})$ and $(\boldsymbol{u}^{(2)}\ ,\ \theta^{(2)})$ by

$$y = \frac{1}{2}\int_{B_0} (\varrho_0 \dot{u}_i \dot{u}_i + D_{iAjK} u_{i,A} u_{j,B} + A\theta^2)\, dV\ . \tag{2.4.1}$$

In what follows we study the behavior of the body on the finite time interval $[0, t_1]$. We assume that

(i) $\varrho_0 \geq \varrho_1 > 0$, $A \geq A_0 > 0$, where ϱ_1 and A_0 are constants;

(ii) D_{iAjB} is positive definite in the sense that there exists a positive constant D_0 such that

$$\int_{B_0} D_{iAjB} v_{iA} v_{jB}\, dV \geq D_0 \int_{B_0} v_{iA} v_{iA}\, dV\ ,$$

for all tensors v_{iA} and any $t \in [0, t_1]$;

(iii) K_{MN} is positive definite in the sense that there exists a positive constant K_0 such that

$$\int_{B_0} \frac{1}{T} K_{MN} \xi_M \xi_N dV \geq K_0 \int_{B_0} \xi_M \xi_M dV\ ,$$

for all vectors ξ_M and any $t \in [0, t_1]$;

(iv) Σ_2 and Σ_4 are empty.

The condition (iii) is the defining property of a definite elastic conductor introduced by Coleman and Gurtin [85]. The following proposition is an immediate consequence of Theorem 2.2.1.

Lemma 2.4.1. *Let $(\boldsymbol{u}\ ,\ \theta)$ be a solution of the problem (P). Then*

$$\begin{aligned} \dot{y} = \int_{B_0} \varrho_0 (F_i \dot{u}_i + \frac{1}{T} S\theta)\, dV + \frac{1}{2}\int_{B_0} (\dot{D}_{iAjK} u_{i,A} u_{j,K} - \\ - \dot{A}\theta^2 - 2\dot{F}_{Ki} u_{i,K}\theta - \frac{2}{T}\varrho_0 \dot{\eta}\theta^2)\, dV - \\ - \int_{B_0} \Phi_A \Big(\frac{1}{T}\theta\Big)_{,A}\, dV\ . \end{aligned} \tag{2.4.2}$$

This lemma gives the evolution in time of the "distance" defined by (2.4.1).

Lemma 2.4.2. *Let* ($\boldsymbol{u}$, θ) *be a solution of the problem* (P). *Then there exists a positive constant* c_0 *such that*

$$-\int_{B_0}\left(\frac{1}{T}\theta\right)_{,A}\Phi_A dV \leq c_0\int_{B_0}(u_{i,K}u_{i,K}+\theta^2)\,dV\,. \tag{2.4.3}$$

Proof. By (2.2.3),

$$-\int_{B_0}\left(\frac{1}{T}\theta\right)_{,A}\Phi_A dV = -\int_{B_0}\frac{1}{T^2}\Phi_A(\theta_{,A}T-\theta T_{,A})\,dV =$$

$$= -\int_{B_0}\frac{1}{T}K_{AM}\theta_{,A}\theta_{,M}dV + \int_{B_0}(C\theta^2+L_A\theta\theta_{,A}+V_{iK}u_{i,K}\theta +$$

$$+P_{AiK}u_{i,K}\theta_{,A})\,dV\,,$$

where

$$C=T^{-2}D_KT_{,K}\,,\qquad L_M=T^{-2}(K_{AM}T_{,A}-TD_M)\,,$$

$$V_{iK}=T^{-2}G_{AiK}T_{,A}\,,\qquad P_{AiK}=-T^{-1}G_{AiK}\,.$$

By using the arithmetic-geometric mean inequality and Schwarz's inequality we obtain

$$-\int_{B_0}\left(\frac{1}{T}\theta\right)_{,A}\Phi_A dV \leq -\int_{B_0}\frac{1}{T}K_{AM}\theta_{,A}\theta_{,M}dV +$$

$$+\frac{1}{2}(\alpha_1+\alpha_2)\int_{B_0}\theta_{,A}\theta_{,A}dV+\frac{1}{2}(m_1^2\alpha_1^{-1}+2m_4^2+1)\int_{B_0}\theta^2dV + \tag{2.4.4}$$

$$+\frac{1}{2}(m_2^2\alpha_2^{-1}+m_3^2)\int_{B_0}u_{i,A}u_{i,A}dV\,,$$

where α_ϱ are arbitrary positive constants and

$$m_1^2=\max\,(L_AL_A)\,,\quad m_2^2=\max\,(P_{AiK}P_{AiK})\,,$$

$$m_3^2=\max\,(V_{iK}V_{iK})\,,\quad m_4^2=\max|C|\,,\quad \text{on } \bar{B}_0\times[0,t_1]\,.$$

It follows from (2.4.4) and the condition (*iii*) that

$$-\int_{B_0}\left(\frac{1}{T}\theta\right)_{,A}\Phi_A dV \leq \frac{1}{2}(\alpha_1+\alpha_2-2K_0)\int_{B_0}\theta_{,A}\theta_{,A}dV +$$

$$+c_0\int_{B_0}(\theta^2+u_{i,A}u_{i,A})\,dV\,, \tag{2.4.5}$$

where

$$c_0 = \frac{1}{2} \max\left(m_1^2 \alpha_1^{-1} + 2m_4^2 + 1\,,\ m_2^2 \alpha_2^{-1} + m_3^2\right).$$

If α_1 and α_2 are chosen such that

$$\alpha_1 + \alpha_2 - 2K_0 \le 0,$$

then the inequality (2.4.5) implies the desired result. □

With a view toward deriving the continuous dependence result, we record the following Gronwall-type inequality (Dafermos [92]).

Lemma 2.4.3. *Assume that the nonnegative functions $f \in L^\infty([0,t_1])$ and $g \in L^1([0,t_1])$ satisfy the inequality*

$$f^2(t) \le M^2 f^2(0) + \int_0^t [(2\gamma + 4\beta t) f^2(\tau) + 2N g(\tau) f(\tau)]d\tau\ ,\ t \in [0,t_1]\,, \qquad (2.4.6)$$

where β, γ, M and N are nonnegative constants. Then

$$f(t) \le [M f(0) + N \int_0^t g(s)\, ds] \exp\left(\alpha t + \beta t^2\right),\ t \in [0,t_1]\,,$$

where $\alpha = \gamma + \beta/\gamma$.

Let Y and Z be the functions on $[0, t_1]$ defined by

$$Y = \left\{ \int_{B_0} (\dot{u}_i \dot{u}_i + u_{i,A} u_{i,A} + \theta^2)\, dV \right\}^{1/2},$$

$$Z = \left\{ \int_{B_0} (F_i F_i + S^2)\, dV \right\}^{1/2}.$$

Theorem 2.4.1. *Let $(\boldsymbol{u}, \theta)$ be a solution of the problem (P). Then there exist positive constants m, M, and N such that*

$$Y(t) \le [M Y(0) + N \int_0^t Z(s)\, ds] \exp(mt)\,, \qquad t \in [0,t_1]\,.$$

Proof. By (2.4.2) and (2.4.3), we obtain

$$\dot{y} \le \int_{B_0} \varrho_0 (F_i \dot{u}_i + \frac{1}{T} S\theta)\, dV + \frac{1}{2} \int_{B_0} (\dot{D}_{iAjK} u_{i,A} u_{j,K} -$$

$$- \dot{A}\theta^2 - 2\dot{F}_{Ki} u_{i,K} \theta - \frac{2}{T} \varrho_0 \dot{\eta} \theta^2)\, dV + c_0 \int_{B_0} (u_{i,K} u_{i,K} + \theta^2)\, dV\,.$$

By using the arithmetic-geometric mean inequality and the Schwarz inequality, we arrive at

$$\dot{y} \leq \nu Z \left[\int_{B_0} (\dot{u}_i \dot{u}_i + \theta^2) \, dV \right]^{1/2} + c_1 \int_{B_0} (u_{i,A} u_{i,A} + \theta^2) \, dV \ , \tag{2.4.7}$$

where ν and c_1 are positive constants. It follows from (2.4.7), and the hypotheses (i), (ii) that

$$Y^2(t) \leq M^2 Y^2(0) + 2 \int_0^t [m Y^2(s) + N Z(s) Y(s)] \, ds \ , \tag{2.4.8}$$

where

$$M^2 = \frac{m_0}{c_2} \ , \qquad m^2 = \frac{c_1}{2c_2} \ , \qquad N = \frac{\nu}{2c_2} \ ,$$

$$m_0 = \frac{1}{2} \max \left(\varrho_0, D_{iAjK}, A \right) \ ,$$

$$c_2 = \frac{1}{2} \min \left(\varrho_1, D_0, A_0 \right) \ , \quad \text{on } \bar{B}_0 \times [0, t_1] \ .$$

The inequality (2.4.8) and Lemma 2.4.3 imply the desired result. □

Other types of boundary conditions may also be considered. The following proposition is a direct consequence of the above result.

Theorem 2.4.2. *Assume that the hypotheses $(i) - (iv)$ hold. Then the initial-boundary-value problem of thermoelasticity has at most one solution.*

b) **Isothermal primary states.** We now assume that the primary state is a configuration of equilibrium at constant temperature T. In this case the fundamental system of field equations (2.3.3) reduces to

$$\begin{aligned} &(d_{ijrs} u_{r,s})_{,j} - (\beta_{ij} \theta)_{,j} + \varrho F_i = \varrho \ddot{u}_i \ , \\ &(k_{ij} \theta_{,j})_{,i} + \varrho S = T(\beta_{ij} \dot{u}_{i,j} + a \dot{\theta}) \ , \end{aligned} \tag{2.4.9}$$

on $B \times [0, t_1)$. To this system we add the initial conditions (2.3.1) and boundary conditions (2.3.2).

Continuous dependence results for this theory have been established in various papers (see, e.g. Payne [293], Knops and Payne [220], Wilkes [363], Levine [239], Dafermos [91], Rionero and Chiriţa [311], Ames and Straughan [4]).

We first present the continuous dependence results presented by Ames and Straughan [4]. To facilitate the solution of the problem often is made a simplifying approximation which consists in the neglecting of the term $T\beta_{ij}\dot{u}_{i,j}$ in the energy equation $(2.4.9)_2$; the resulting theory is referred to as being *uncoupled.* In [4], by using the logarithmic convexity method, Ames and Straughan have established that the solution depends continuously on changes in the body force, in the heat supply, and in the coupling coefficients β_{ij}. We assume that

(*i*) $\varrho \geq \varrho_m > 0$, $a \geq a_0 > 0$, where ϱ_m and a_0 are constants;

(*ii*) there exists the constant M such that

$$\sup_B |\beta_{ij}\beta_{ij}|, \ \sup_B |\beta_{ij,j}\beta_{im,m}| \leq M ;$$

(*iii*) k_{ij} is symmetric on $\bar{B}$;

(*iv*) k_{ij} is positive definite in the sense that there exists a positive constant κ_0 such that

$$\frac{1}{T}k_{ij}\xi_i\xi_j \geq \kappa_0\xi_i\xi_i \qquad \text{for any } \xi_i .$$

(*v*) S_2 and S_4 are empty;

(*vi*) the domain B is bounded.

We denote by $(u_i^{(1)}, \theta^{(1)})$ the solution of the problem corresponding to the external data system $L^{(1)} = (\boldsymbol{F}^{(1)}, S^{(1)}, \tilde{\boldsymbol{u}}, \tilde{\theta}, \tilde{\boldsymbol{a}}, \tilde{\boldsymbol{b}}, \tilde{\theta}_0)$ and coupling coefficients β_{ij}. Let $(u_i^{(2)}, \theta^{(2)})$ be solution corresponding to the external data system $L^{(2)} = (\boldsymbol{F}^{(2)}, S^{(2)}, \underset{\sim}{\boldsymbol{u}}, \tilde{\theta}, \tilde{\boldsymbol{a}}, \tilde{\boldsymbol{b}}, \tilde{\theta}_0)$ and coupling coefficients β^*_{ij}. We introduce the notations

$$w_i = u_i^{(2)} - u_i^{(1)}, \quad \theta = \theta^{(2)} - \theta^{(1)}, \quad \varepsilon_{ij} = \beta^*_{ij} - \beta_{ij},$$
$$F_i = \varrho(F_i^{(1)} - F_i^{(2)}), \qquad Q = \frac{1}{T}\varrho(S^{(1)} - S^{(2)}). \tag{2.4.10}$$

It follows from (2.4.9) that $(\boldsymbol{w}, \theta)$ is a solution of the following system

$$\varrho\ddot{w}_i = (d_{ijrs}w_{r,s})_{,j} - (\beta_{ij}\theta)_{,j} - (\varepsilon_{ij}\theta^{(2)})_{,j} + F_i,$$
$$a\dot{\theta} + \beta_{ij}\dot{w}_{i,j} + \varepsilon_{ij}\dot{u}_{i,j}^{(2)} = (\kappa_{ij}\theta_{,j})_{,i} + Q, \tag{2.4.11}$$

on $B \times [0, t_1]$, where

$$\kappa_{ij} = \frac{1}{T}k_{ij}. \tag{2.4.12}$$

Moreover, $(\boldsymbol{w}, \theta)$ satisfies the initial conditions

$$w_i(\boldsymbol{x}, 0) = 0, \quad \dot{w}_i(\boldsymbol{x}, 0) = 0, \quad \theta(\boldsymbol{x}, 0) = 0, \quad \boldsymbol{x} \in \bar{B}, \tag{2.4.13}$$

and the boundary conditions

$$w_i = 0, \qquad \theta = 0 \quad \text{on } \partial B \times [0, t_1]. \tag{2.4.14}$$

We denote by (P_0) the boundary-initial-value problem (2.4.11), (2.4.13), (2.4.14).

Let V and Λ be the functions on $[0, t_1]$ defined by

$$\begin{aligned} V &= \int_0^t \int_B (\varrho \dot{w}_i \dot{w}_i + a\theta^2 + d_{ijrs} w_{r,s} w_{i,j}) dv \,, \\ \Lambda &= \int_B \varrho w_i w_i dv \,. \end{aligned} \tag{2.4.15}$$

Lemma 2.4.4. *Let $(\boldsymbol{w}, \theta)$ be a solution of the problem (P_0). Then*

$$\begin{aligned} V &= 2\int_0^t \int_B (t-s)\big[F_i \dot{w}_i + Q\theta - \kappa_{ij}\theta_{,i}\theta_{,j} + \varepsilon_{ij}(\theta^{(2)}\dot{w}_{i,j} - \dot{u}_{i,j}^{(2)}\theta)\big]\, ds dv \,, \\ \ddot{\Lambda} &= 2\int_B [\varrho \dot{w}_i \dot{w}_i + F_i w_i - d_{ijrs} w_{r,s} w_{i,j} + \beta_{ij}\theta w_{i,j} + \varepsilon_{ij}\theta^{(2)} w_{i,j}] dv \,, \\ \int_B \kappa_{ij}\tau_{,j}\theta_{,i} dv &= \int_B [W\theta - a\theta^2 - \beta_{ij} w_{i,j}\theta - \varepsilon_{ij}(u_{i,j}^{(2)} - \tilde{a}_{i,j})\theta] dv \,, \end{aligned} \tag{2.4.16}$$

where

$$\tau(\boldsymbol{x}, t) = \int_0^t \theta(\boldsymbol{x}, s)\, ds\,, \quad W(\boldsymbol{x}, t) = \int_0^t Q(\boldsymbol{x}, s)\, ds\,, \quad (\boldsymbol{x}, t) \in B \times [0, t_1]. \tag{2.4.17}$$

Proof. In view of (2.4.11),

$$\begin{aligned} \varrho \ddot{w}_i \dot{w}_i &= F_i \dot{w}_i - d_{ijrs} w_{r,s} \dot{w}_{i,j} + \beta_{ij}\theta \dot{w}_{i,j} + \varepsilon_{ij}\theta^{(2)}\dot{w}_{i,j} + (d_{ijrs} w_{r,s}\dot{w}_i - \\ &\quad - \beta_{ij}\dot{w}_i\theta - \varepsilon_{ij}\theta^{(2)}\dot{w}_i)_{,j} \,, \\ a\theta\dot{\theta} + \beta_{ij}\dot{w}_{i,j}\theta + \varepsilon_{ij}\dot{u}_{i,j}^{(2)}\theta &= Q\theta - \kappa_{ij}\theta_{,j}\theta_{,i} + (\kappa_{ij}\theta_{,j}\theta)_{,i} \,. \end{aligned} \tag{2.4.18}$$

By the divergence theorem, (2.4.18) and (2.4.14), we obtain

$$\frac{1}{2}\ddot{V} = \int_B \big[F_i \dot{w}_i + Q\theta - \kappa_{ij}\theta_{,j}\theta_{,i} + \varepsilon_{ij}(\theta^{(2)}\dot{w}_{i,j} - \dot{u}_{i,j}^{(2)}\theta)\big]\, dv \,. \tag{2.4.19}$$

We note that

$$\int_0^t \Big(\int_0^s f(r) dr\Big)\, ds = \int_0^t (t-s) f(s)\, ds = i * f(t)\,, \tag{2.4.20}$$

where i is defined by (2.3.5). It follows from (2.4.15), (2.4.19) and (2.4.20) that the first relation of (2.4.16) holds.

If we integrate the equation $(2.4.11)_2$ with respect to the time we find that

$$a\theta + \beta_{ij}w_{i,j} + \varepsilon_{ij}u^{(2)}_{i,j} = (\kappa_{ij}\tau_{,j})_{,i} + W + \varepsilon_{ij}\tilde{a}_{i,j} \ . \tag{2.4.21}$$

Next we multiply the equation $(2.4.11)_1$ by w_i and the equation (2.4.21) by θ. We obtain

$$\begin{aligned} \varrho\ddot{w}_i w_i &= F_i w_i - d_{ijrs}w_{r,s}w_{i,j} + \beta_{ij}\theta w_{i,j} + \varepsilon_{ij}\theta^{(2)}w_{i,j} + \\ &\quad + (d_{ijrs}w_{r,s}w_i - \beta_{ij}w_i\theta - \varepsilon_{ij}\theta^{(2)}w_i)_{,j} \ , \\ a\theta^2 &+ \beta_{ij}w_{i,j}\theta + \varepsilon_{ij}u^{(2)}_{i,j}\theta = (\kappa_{ij}\tau_{,j}\theta)_{,i} - \kappa_{ij}\tau_{,j}\theta_{,i} + W\theta + \varepsilon_{ij}\tilde{a}_{i,j} \ . \end{aligned} \tag{2.4.22}$$

Clearly,

$$w_i w_i = 2\int_0^t w_i\dot{w}_i ds \ , \qquad \frac{\partial^2}{\partial t^2}(w_i w_i) = 2\dot{w}_i\dot{w}_i + 2w_i\ddot{w}_i \ . \tag{2.4.23}$$

It follows from (2.4.22), (2.4.23), the divergence theorem and the boundary conditions (2.4.14) that (2.4.16) holds. □

Theorem 2.4.3. *Assume that the hypotheses (i) – (iv) hold. Let $(\boldsymbol{w}, \theta)$ be a solution of the problem (P_0). Then there exists a positive constant M^* such that*

$$\begin{aligned} \int_0^t \int_B [\varrho w_i w_i + (t-s)\kappa_{ij}\tau_{,i}\tau_{,j}]\, dsdv \le M^* \Bigg[\int_0^{t_1} (F_i F_i + 2\varepsilon_{ij}\varepsilon_{ij} + \\ + \varepsilon_{is,s}\varepsilon_{ij,j} + W^2 + Q^2 \Bigg]^{1-t/t_1} dt \ . \end{aligned} \tag{2.4.24}$$

Proof. We denote

$$G(t) = \int_0^t \int_B [\varrho w_i w_i + (t-s)\kappa_{ij}\tau_{,i}\tau_{,j}]\, dsdv + C \ , \tag{2.4.25}$$

where C is a constant term to be selected. If we take into account $(2.4.23)_1$, (2.4.20) and (2.4.17), then from (2.4.25) we obtain

$$\begin{aligned} \dot{G} &= \int_0^t \int_B [2\varrho w_i \frac{\partial w_i}{\partial s} + \kappa_{ij}\tau_{,i}\tau_{,j}]\, dsdv = \\ &= 2\int_0^t \int_B [\varrho w_i \frac{\partial w_i}{\partial s} + (t-s)\kappa_{ij}\tau_{,i}\theta_{,j}]\, dsdv \ , \end{aligned} \tag{2.4.26}$$

and

$$\ddot{G} = \int_0^t \int_B [\varrho \frac{\partial^2}{\partial s^2}(w_i w_i) + 2\kappa_{ij}\tau_{,i}\theta_{,j}]\, dsdv\,. \tag{2.4.27}$$

By (2.4.15), (2.4.16) and (2.4.27),

$$\begin{aligned}\ddot{G} = 2\int_0^t \int_B [&2\varrho\dot{w}_i\dot{w}_i + 2(t-s)\kappa_{ij}\theta_{,i}\theta_{,j} - \varepsilon_{ij}(u^{(2)}_{i,j} - \tilde{a}_{i,j})\theta + \\ &+ W\theta + F_i w_i + \varepsilon_{ij}\theta^{(2)} w_{i,j}]\, dsdv \\ &- 4\int_0^t \int_B (t-s)[F_i\dot{w}_i + Q\theta + \varepsilon_{ij}\theta^{(2)}\dot{w}_{i,j} - \varepsilon_{ij}\dot{u}^{(2)}_{i,j}\theta]\, dsdv\,.\end{aligned}$$

If we use the divergence theorem and the conditions (2.4.14), then we find that

$$\begin{aligned}\ddot{G} = 2\int_0^t \int_B [&2\varrho\dot{w}_i\dot{w}_i + 2(t-s)\kappa_{ij}\theta_{,i}\theta_{,j} - \varepsilon_{ij}(u^{(2)}_{i,j} - \tilde{a}_{i,j})\theta + \\ &+ W\theta + F_i w_i + \varepsilon_{ij}\theta^{(2)} w_{i,j}]\, dsdv - 4\int_0^t \int_B (t-s)[F_i\dot{w}_i + \\ &+ Q\theta - (\varepsilon_{ij,j}\theta^{(2)} + \varepsilon_{ij}\theta^{(2)}_{,j})\dot{w}_i + (\varepsilon_{ij,j}\theta + \varepsilon_{ij}\theta_{,j})\dot{u}^{(2)}_i]\, dsdv\,.\end{aligned} \tag{2.4.28}$$

It follows from (2.4.25), (2.4.26) and (2.4.28) that

$$\begin{aligned}G\ddot{G} - \dot{G}^2 = 4N^2 &+ 4C\int_0^t \int_B [\varrho\dot{w}_i\dot{w}_i + (t-s)\kappa_{ij}\theta_{,i}\theta_{,j}]\, dsdv - \\ &- 2G\int_0^t \{2(t-s)[F_i\dot{w}_i + Q\theta - (\varepsilon_{ij,j}\theta^{(2)} + \varepsilon_{ij}\theta^{(2)}_{,j})\dot{w}_i + \\ &+ (\varepsilon_{ij,j}\theta + \varepsilon_{ij}\theta_{,j})\dot{u}^{(2)}_i] + \varepsilon_{ij}(u^{(2)}_{i,j} - \tilde{a}_{i,j})\theta - F_i w_i - W\theta - \varepsilon_{ij}\theta^{(2)} w_{i,j}\}\, dsdv\,,\end{aligned} \tag{2.4.29}$$

where

$$\begin{aligned}N^2 = 4\int_0^t \int_B [\varrho w_i w_i + (t-s)\kappa_{ij}\tau_{,i}\tau_{,j}]\, dsdv \int_0^t \int_B [\varrho\dot{w}_i\dot{w}_i + (t-s)\kappa_{ij}\theta_{,i}\theta_{,j}]\, dsdv - \\ - \left\{\int_0^t \int_B [\varrho w_i\dot{w}_i + (t-s)\kappa_{ij}\tau_{,i}\theta_{,j}]\, dsdv\right\}^2.\end{aligned}$$

In view of the Schwarz inequality we have $N^2 \geq 0$. If we use the arithmetic-geometric mean inequality, the Schwarz inequality, the Poincaré inequality, (i) and (iv), then we find

$$4G \int_0^t (t-s)[F_i\dot{w}_i + Q\theta - (\varepsilon_{ij,j}\theta^{(2)} + \varepsilon_{ij}\theta^{(2)}_{,j})\dot{w}_i + \\ + (\varepsilon_{ij,j}\theta + \varepsilon_{ij}\theta_{,j})\dot{u}_i^{(2)}]\, ds \leq AF^2 + 4\left[\int_0^t \int_B [\varrho\dot{w}_i\dot{w}_i + \right. \\ \left. + (t-s)\kappa_{ij}\theta_{,i}\theta_{,j}]dsdt\right]\left[\int_0^{t_1}\int_B (F_iF_i + Q^2 + \varepsilon_{ij}\varepsilon_{ij} + \varepsilon_{ir,r}\varepsilon_{is,s})\, dsdv\right], \tag{2.4.30}$$

where

$$A = t_1^2(1+2M_1)\varrho_0^{-1} + t_1\kappa_0^{-1}\lambda_1^{-1}(1 + M_1 + M_1\lambda_1)\,,$$

M_1 being a bound for

$$\sup_D |\theta^{(2)}|^2, \quad \sup_D |\operatorname{grad}\theta^{(2)}|^2, \quad \sup_D |\dot{\boldsymbol{u}}^{(2)}|^2, \quad \sup_D |\operatorname{grad}(\boldsymbol{u}^{(2)} - \tilde{\boldsymbol{a}})|, \quad D = B\times[0,t_1],$$

and λ_1 is the constant in the Poincaré inequality for the domain B. Similarly,

$$2G \int_0^t \int_B (F_iw_i + \varepsilon_{ij}\theta^{(2)}w_{i,j})\, dsdv \leq \varrho_0^{-1}(1+2M_1)\, G^2 + \\ + G\int_0^{t_1}\int_B (F_iF_i + \varepsilon_{ij}\varepsilon_{ij} + \varepsilon_{ir,r}\varepsilon_{ik,k})\, dtdv\,. \tag{2.4.31}$$

We also note that

$$2G \int_0^t \int_B W\theta dsdv \leq \frac{1}{c_1}G^2 + c_1\left(\int_0^t\int_B \theta^2 dsdv\right)\left(\int_0^{t_1}\int_B W^2 dsdv\right), \\ 2G\int_0^t\int_B \varepsilon_{ij}(u^{(2)}_{i,j} - \tilde{a}_{i,j})\theta dsdv \leq \frac{1}{c_2}G^2 + \\ + c_2M_1\left(\int_0^{t_1}\int_B \varepsilon_{ij}\varepsilon_{ij}dsdv\right)\left(\int_0^t\int_B \theta^2 dsdv\right), \tag{2.4.32}$$

for arbitrary positive constants c_1 and c_2. In view of (i), (2.4.11), (2.4.13) and (2.4.14) we have

$$\int_0^t \int_B \theta^2 dsdv \le \frac{1}{a_0} \int_0^t \int_B a\theta^2 dsdv = -\frac{1}{a_0} \int_0^t \int_B \frac{\partial}{\partial s}[(t-s)a\theta^2]\, dsdv \; +$$
$$+ \frac{2}{a_0} \int_0^t \int_B (t-s)a\theta \frac{\partial \theta}{\partial s} dsdv = \frac{2}{a_0} \int_0^t \int_B (t-s)\theta[Q + (\kappa_{ij}\theta_{,j})_{,i} -$$
$$- \beta_{ij}\dot{w}_{i,j} - \varepsilon_{ij}\dot{u}^{(2)}_{i,j}]\, dsdv \le \frac{2}{a_0} \int_0^t \int_B (t-s)[Q\theta + (\beta_{ij}\theta)_{,j}\dot{w}_i \; +$$
$$+ (\varepsilon_{ij}\theta)_{,j}\dot{u}^{(2)}_i - \kappa_{ij}\theta_{,i}\theta_{,j}]\, dsdv \, .$$

If we assume that $M \le M_1$ and apply the arithmetic-geometric mean inequality and the Poincaré inequality, then we get

$$\int_0^t \int_B \theta^2 dsdv \le m \int_0^{t_1} \int_B [\lambda_1 \varrho_0 \varepsilon_{ij}\varepsilon_{ij} + \varrho_0 \varepsilon_{ir,r}\varepsilon_{ij,j} \; + \\ + M_1 \varrho_0 Q^2 + (1+\lambda_1)\varrho \dot{w}_i \dot{w}_i]\, dsdv \, , \tag{2.4.33}$$

where $m = 5t_1 M_1 / 2\varrho_0 \lambda_1 \kappa_0 a_0$.

We use the bound (2.4.33) in (2.4.32) and set

$$c_2 = 4/mM_1(1+\lambda_1)\, , \quad c_1 = M_1 c_2 \, . \tag{2.4.34}$$

If we take

$$C = \int_0^{t_1} \int_B (F_i F_i + 2\varepsilon_{ij}\varepsilon_{ij} + \varepsilon_{ij,j}\varepsilon_{ir,r} + W^2 + Q^2)\, dsdv \, , \tag{2.4.35}$$

then, from (2.4.29)-(2.4.34), we conclude that we can determine explicitly the constant b such that

$$G\ddot{G} - \dot{G}^2 \ge -bG^2 \, . \tag{2.4.36}$$

If we define the function

$$R(t) = \ln\left[G(t)\exp\left(\frac{1}{2}bt^2\right)\right], \qquad t \in [0, t_1]\, , \tag{2.4.37}$$

then

$$\ddot{R} = G^{-2}(G\ddot{G} - \dot{G}^2 + bG^2)\, .$$

Thus, by (2.4.36),

$$\ddot{R} \geq 0 . \tag{2.4.38}$$

Jensen's inequality gives

$$G(t) \leq \{G(0)\}^{1-t/t_1}\{G(t_1)\}^{t/t_1} \exp\left[\frac{1}{2}bt(t_1-t)\right], \tag{2.4.39}$$

for $t \in [0, t_1]$. It follows from (2.4.25), (2.4.35) and (2.4.39) that (2.4.24) holds. □

The inequality (2.4.24) implies the continuous dependence on the body force, the heat supply and the coupling coefficients.

In [311], Rionero and Chiriţa have used the Lagrange identity method to establish the continuous dependence of the solutions to the boundary-initial-value problem (2.4.9), (2.3.1), (2.3.4) with respect to body force and heat supply in bounded and exterior domains. The results imply the continuous dependence on the initial data and the constitutive coefficients. We next present some of the results given in [311]. Let $(\boldsymbol{u}^{(1)}, \theta^{(1)})$ and $(\boldsymbol{u}^{(2)}, \theta^{(2)})$ be solutions to (2.4.9), (2.3.1), (2.3.4) corresponding to the external data systems $L^{(1)} = (\boldsymbol{F}^{(1)}, S^{(1)}, \tilde{\boldsymbol{u}}, \tilde{\boldsymbol{p}}, \tilde{\theta}, \tilde{\varphi}, \tilde{\boldsymbol{a}}, \tilde{\boldsymbol{b}}, \tilde{\theta}_0)$ and $L^{(2)} = (\boldsymbol{F}^{(2)}, S^{(2)}, \tilde{\boldsymbol{u}}, \tilde{\boldsymbol{p}}, \tilde{\theta}, \tilde{\varphi}, \tilde{\boldsymbol{a}}, \tilde{\boldsymbol{b}}, \tilde{\theta}_0)$, respectively. We define $\boldsymbol{u} = \boldsymbol{u}^{(2)} - \boldsymbol{u}^{(1)}$, $\theta = \theta^{(2)} - \theta^{(1)}$. It follows from (2.4.9), (2.3.1) and (2.3.4) that $(\boldsymbol{u}, \theta)$ is a solution of the system

$$\begin{aligned} \varrho \ddot{u}_i &= (d_{ijrs}u_{r,s})_{,j} - (\beta_{ij}\theta)_{,j} + F_i , \\ a\dot{\theta} + \beta_{ij}\dot{u}_{i,j} &= (\kappa_{ij}\theta_{,j})_{,i} + Q , \end{aligned} \tag{2.4.40}$$

on $B \times [0, t_1)$, with the initial conditions

$$\boldsymbol{u}(\boldsymbol{x}, 0) = \boldsymbol{0}, \qquad \dot{\boldsymbol{u}}(\boldsymbol{x}, 0) = 0, \qquad \theta(\boldsymbol{x}, 0) = 0 \quad \boldsymbol{x} \in \bar{B} , \tag{2.4.41}$$

and the boundary conditions

$$\begin{aligned} &u_i = 0 \quad \text{on } \bar{S}_1 \times [0, t_1), \qquad (d_{ijrs}u_{r,s} - \beta_{ij}\theta)n_j = 0 \quad \text{on } S_2 \times [0, t_1), \\ &\theta = 0 \quad \text{on } \bar{S}_3 \times [0, t_1), \qquad \kappa_{ij}\theta_{,j}n_i = 0 \quad \text{on } S_4 \times [0, t_1), \end{aligned} \tag{2.4.42}$$

where F_i , Q and κ_{ij} are given by (2.4.10) and (2.4.12).

We now assume that

(α) $\varrho \geq \varrho_m > 0$ where ϱ_m is a constant, and $a \neq 0$;
(β) k_{ij} is symmetric on $\bar{B}$;
(γ) k_{ij} is a positive definite tensor;
(δ) the domain B is bounded.

Let $\boldsymbol{U}$ and $\boldsymbol{V}$ be vector fields of class $C^{0,2}$ on $B \times [0, t_1)$. Then the following Lagrange identity

$$\int_B \varrho(U_i \dot{V}_i - \dot{U}_i V_i)\, dv = \int_0^t \int_B \varrho(U_i \ddot{V}_i - \ddot{U}_i V_i)\, dsdv + \int_B \varrho[U_i(\boldsymbol{x},0)\dot{V}_i(\boldsymbol{x},0) - \dot{U}_i(\boldsymbol{x},0)V_i(\boldsymbol{x},0)]\, dv\,, \tag{2.4.43}$$

holds. If we take

$$U_i(\boldsymbol{x},s) = u_i(\boldsymbol{x},s)\,, \quad V_i(\boldsymbol{x},s) = u_i(\boldsymbol{x},2t-s)\,, \quad s \in [0,2t]\,,\ t \in [0,\frac{1}{2}t_1)\,, \tag{2.4.44}$$

then (2.4.13), with the help of (2.4.41), yields

$$2\int_B \varrho u_i \dot{u}_i dv = \int_0^t \int_B \varrho[\ddot{u}_i(s)u_i(2t-s) - u_i(s)\ddot{u}_i(2t-s)]\, dsdv\,, \qquad t \in [0,\frac{1}{2}t_1)\,, \tag{2.4.45}$$

where, for convenience, we have suppresed the argument $\boldsymbol{x}$. In view of (2.4.40) we obtain

$$\varrho\ddot{u}_i(2t-s)u_i(s) = [d_{ijmn}u_{m,n}(2t-s)u_i(s) - \beta_{ij}\theta(2t-s)u_i(s)]_{,j} + F_i(2t-s)u_i(s) - d_{ijmn}u_{m,n}(2t-s)u_{i,j}(s) + \beta_{ij}\theta(2t-s)u_{i,j}(s)\,. \tag{2.4.46}$$

In view of (2.2.18), (2.4.42), (2.4.46) and the divergence theorem, from (2.4.45) we get

$$2\int_B \varrho \dot{u}_i \dot{u}_i dv = \int_0^t \int_B [F_i(s)u_i(2t-s) + F_i(2t-s)u_i(s) + \beta_{ij}\theta(s)u_{i,j}(2t-s) - \beta_{ij}\theta(2t-s)u_{i,j}(s)]\, dsdv\,, \qquad t \in [0,\frac{1}{2}t_1)\,. \tag{2.4.47}$$

If we integrate the equation $(2.4.40)_2$ with respect to the time and use (2.4.41), then we get

$$a\theta + \beta_{ij}u_{i,j} = (\kappa_{ij}\tau_{,j})_{,i} + W\,, \tag{2.4.48}$$

where τ and W are given by (2.4.17).

By (2.4.48) and (2.4.42),

$$\int_B \beta_{ij}[\theta(s)u_{i,j}(2t-s) - \theta(2t-s)u_{i,j}(s)]\, dsdv =$$

$$= \int_B \{W(2t-s)\theta(s) - W(s)\theta(2t-s) - \kappa_{ij}[\theta_{,i}(s)\tau_{,j}(2t-s) - \tag{2.4.49}$$

$$- \theta_{,i}(2t-s)\tau_{,j}(s)]\}\, dv\,, \qquad s \in [0,2t]\,,\ t \in [0, \frac{1}{2}t_1)\,.$$

From (2.4.47) and (2.4.49) we obtain

$$2\int_B \varrho u_i \dot{u}_i dv = \int_0^t \int_B [F_i(s)u_i(2t-s) - F_i(2t-s)u_i(s) +$$

$$+ W(2t-s)\theta(s) - W(s)\theta(2t-s)]\, dsdv - \int_0^t \int_B \kappa_{ij}[\theta_{,i}(s)\tau_{,j}(2t-s) - \tag{2.4.50}$$

$$- \theta_{,i}(2t-s)\tau_{,j}(s)]\, ds\,, \qquad t \in [0, \frac{1}{2}t_1)\,.$$

By (2.4.17) and (β),

$$\kappa_{ij}[\theta_{,i}(s)\tau_{,j}(2t-s) - \theta_{,i}(2t-s)\tau_{,j}(s)] = \kappa_{ij}\frac{d}{ds}[\tau_{,i}(s)\tau_{,j}(2t-s)]\,.$$

Thus, (2.4.5) becomes

$$\int_B (2\varrho u_i \dot{u}_i + \kappa_{ij}\tau_{,i}\tau_{,j})\, dv = \int_0^t \int_B [F_i(s)u_i(2t-s) - F_i(2t-s)u_i(s) + \tag{2.4.51}$$

$$+ W(2t-s)\theta(s) - W(s)\theta(2t-s)]\, dsdv\,, \qquad t \in [0, \frac{1}{2}t_1)\,.$$

Theorem 2.4.4. *Assume that the hypotheses (α)–(δ) hold. Suppose, in addition, that there exists $t_2 \in (0, t_1)$ so that*

$$\int_0^{t_2} \int_B \boldsymbol{F}^2 dsdv \le m_1^2\,, \quad \int_0^{t_2} \int_B W^2 dsdv \le m_2^2\,,$$
$$\int_0^{t_2} \int_B \varrho \boldsymbol{u}^2 dsdv \le m_3^2\,, \quad \int_0^{t_2} \int_B \theta^2 dsdv \le m_4^2\,. \tag{2.4.52}$$

Then

$$\int_B \varrho u_i u_i dv + \int_0^t \int_B \kappa_{ij} \tau_{,i} \tau_{,j}\, dt dv \leq m_3 t_2 \left(\int_0^{t_2} \int_B F_i F_i dt dv \right)^{1/2} + \\ + m_4 t_2 \left(\int_0^{t_2} \int_B W^2 dt dv \right)^{1/2} , \quad t \in [0, \frac{1}{2} t_2] . \tag{2.4.53}$$

Proof. To prove (2.4.53) we use the identity (2.4.51). By means of Schwarz's inequality and (2.4.52), we obtain

$$\begin{aligned} &\int_0^t \int_B F_i(s) u_i(2t-s)\, ds dv \leq \left\{ \left(\int_0^t \int_B \boldsymbol{F}^2 ds dv \right) \left(\int_0^t \int_B \boldsymbol{u}^2(2t-s)\, ds dv \right) \right\}^{1/2} = \\ &= \left\{ \left(\int_0^t \int_B \boldsymbol{F}^2 ds dv \right) \left(\int_t^{2t} \int_B \boldsymbol{u}^2(\tau)\, d\tau dv \right) \right\}^{1/2} \leq m_3 \left(\int_0^{t_2} \int_B \boldsymbol{F}^2 dt dv \right)^{1/2} , \\ &- \int_0^t \int_B F_i(2t-s) u_i(s)\, ds dv \leq m_3 \left(\int_0^{t_2} \int_B \boldsymbol{F}^2 dt dv \right)^{1/2} , \\ &\int_0^t \int_B W(2t-s) \theta(s)\, ds dv \leq m_4 \left(\int_0^{t_2} \int_B W^2 dt dv \right)^{1/2} , \\ &- \int_0^t \int_B W(s) \theta(2t-s)\, ds dv \leq m_4 \left(\int_0^{t_2} \int_B W^2 dt dv \right)^{1/2} , \qquad t \in [0, \frac{1}{2} t_2] . \end{aligned} \tag{2.4.54}$$

It follows from (2.4.51) and (2.4.54) that (2.4.53) holds. □

We now study the dependence of solutions on the initial data. Let $(\boldsymbol{v}^{(1)}, \psi^{(1)})$ and $(\boldsymbol{v}^{(2)}, \psi^{(2)})$ be solutions corresponding to the external data systems $L^{(1)} = (\boldsymbol{F}, S, \tilde{\boldsymbol{u}}, \tilde{\boldsymbol{p}}, \tilde{\theta}, \tilde{\varphi}, \tilde{\boldsymbol{a}}^{(1)}, \tilde{\boldsymbol{b}}^{(1)}, \tilde{\theta}_0^{(1)})$ and $L^{(2)} = (\boldsymbol{F}, S, \tilde{\boldsymbol{u}}, \tilde{\boldsymbol{p}}, \tilde{\theta}, \tilde{\varphi}, \tilde{\boldsymbol{a}}^{(2)}, \tilde{\boldsymbol{b}}^{(2)}, \tilde{\theta}_0^{(2)})$, respectively. We denote $\boldsymbol{w} = \boldsymbol{v}^{(2)} - \boldsymbol{v}^{(1)}$, $\psi = \psi^{(2)} - \psi^{(1)}$. Then $(\boldsymbol{w}, \psi)$ is a solution of the equations

$$\begin{aligned} &\varrho \ddot{w}_i = (d_{ijrs} w_{r,s} - \beta_{ij} \psi)_{,j} , \\ &a \dot{\psi} + \beta_{ij} \dot{w}_{i,j} = (\kappa_{ij} \psi_{,j})_{,i} , \qquad \text{on } B \times [0, t_1) , \end{aligned} \tag{2.4.55}$$

which satisfies the initial conditions

$$\boldsymbol{w}(\boldsymbol{x}, 0) = \boldsymbol{p}(\boldsymbol{x}) , \quad \dot{\boldsymbol{w}}(\boldsymbol{x}, 0) = \boldsymbol{q}(\boldsymbol{x}) , \quad \psi(\boldsymbol{x}, 0) = r(\boldsymbol{x}) , \quad \boldsymbol{x} \in \bar{B} , \tag{2.4.56}$$

and the boundary conditions

$$\begin{aligned} &u_i = 0 \quad \text{on } \bar{S}_1 \times [0, t_1)\,, \quad (d_{ijrs} w_{r,s} - \beta_{ij}\psi) n_j = 0 \quad \text{on } S_2 \times [0, t_1)\,, \\ &\psi = 0 \quad \text{on } \bar{S}_3 \times [0, t_1)\,, \quad \kappa_{ij}\psi_{,j} n_i = 0 \quad \text{on } S_4 \times [0, t_1)\,, \end{aligned} \tag{2.4.57}$$

where

$$\boldsymbol{p} = \tilde{\boldsymbol{a}}^{(2)} - \tilde{\boldsymbol{a}}^{(1)}\,, \qquad \boldsymbol{q} = \tilde{\boldsymbol{b}}^{(2)} - \tilde{\boldsymbol{b}}^{(1)}\,, \qquad r = \theta_0^{(2)} - \theta_0^{(1)}\,. \tag{2.4.58}$$

We introduce the notations

$$\begin{aligned} &u_j = i * w_j\,, \quad \theta = i * \psi\,, \quad \tau = 1 * \theta\,, \\ &\gamma_0 = ar + \beta_{ij} p_{i,j}\,. \end{aligned} \tag{2.4.59}$$

We suppose that

$$\int_B (\boldsymbol{p}^2 + \boldsymbol{q}^2)\, dv \le m_5^2\,, \qquad \int_B \gamma_0^2 dv \le m_6^2\,. \tag{2.4.60}$$

Theorem 2.4.4 forms the basis of the following continuous dependence result upon the initial data.

Theorem 2.4.5. *Assume that the hypotheses* $(\alpha) - (\delta)$ *and (2.4.60) hold. Let* $(\boldsymbol{w}, \psi)$ *be a solution to the boundary-initial-value problem (2.4.55)-(2.4.58). Let* $\boldsymbol{u}$ *and* θ *be defined by (2.4.59). If* $\boldsymbol{u}$ *and* θ *satisfy the conditions* $(2.4.52)_2$, *then*

$$\begin{aligned} &\int_B \varrho u_i u_i dv + \int_0^t \int_B \kappa_{ij} \tau_{,i} \tau_{,j}\, dt dv \le m_3 t_2^2 \Big[\Big(\frac{1}{3} t_2^2 + \\ &+ \frac{1}{2} t_2 \Big) \int_B q_i q_i dv + (1 + \frac{1}{2} t_2) \int_B p_i p_i dv \Big]^{1/2} + \\ &+ \frac{1}{2\sqrt{5}} m_4 t_2^{7/2} \Big(\int_B \gamma_0^2 dv \Big)^{1/2}\,, \qquad t \in [0, \frac{1}{2} t_2]\,. \end{aligned} \tag{2.4.61}$$

Proof. The functions u_i and θ defined by (2.4.59) satisfy the equations (2.4.40), the initial conditions (2.4.41) and the boundary conditions (2.4.42) with the supplies

$$F_i = \varrho(t q_i + p_i)\,, \quad Q = \gamma_0 t\,.$$

From (2.4.53) we obtain the desired result. □

Other continuous dependence results are presented in the papers of Knops and Payne [220], Rionero and Chiriţa [311], Levine [239], Bloom [27], Wilkes [363], Navarro and Quintanilla [268].

2.5 Uniqueness. Reciprocity

In this section we present a uniqueness result and a reciprocity relation. We assume that the primary state is a configuration of equilibrium at constant temperature. Weiner [362] was the first to establish a uniqueness theorem in the thermoelastodynamics of homogeneous and isotropic bodies. Ionescu-Cazimir [201] extended this result to anisotropic bodies. These results rest on positive definiteness assumption on the tensor d_{ijrs}. In [37], [38], Brun established uniqueness without the definiteness assumption on the elastic coefficients. Knops and Payne [220], Wilkes [363], Rionero and Chiriţa [311] and Ames and Straughan [4] reached a similar conclusion. We note that the above uniqueness results are based on the assumption that the conductivity tensor k_{ij} is positive definite. Clearly, the second law of thermodynamics implies that the conductivity tensor is only positive semi-definite. In [188], Ieşan has established a uniqueness theorem with no definiteness assumption on the elasticity tensor and in the absence of the restriction that the conductivity tensor is positive definite. First, we present this result. If the primary state is at constant temperature T, then the constitutive equations are

$$\begin{aligned} &s_{ji} = d_{ijrs}u_{r,s} - \beta_{ij}\theta\ , \\ &\chi = \beta_{ij}u_{i,j} + a\theta\ , \qquad \varphi_i = k_{ij}\theta_{,j}\ , \end{aligned} \tag{2.5.1}$$

where $\chi = \gamma/J$. The equations of motion and the energy equations are

$$s_{ji,j} + f_i = \varrho\ddot{u}_i\ , \qquad T\dot{\chi} = \varphi_{i,i} + R\ , \tag{2.5.2}$$

on $B\times I$, where $I = [0,\infty)\ ,\ f_i = \varrho F_i$ and $R = \varrho S$. We consider the initial conditions (2.3.1) and the boundary conditions

$$\begin{aligned} &u_i = \tilde{u}_i \quad \text{on } \bar{S}_1 \times I\ , \qquad s_i = \tilde{p}_i \quad \text{on } S_2 \times I\ , \\ &\theta = \tilde{\theta} \quad \text{on } \bar{S}_3 \times I\ , \qquad q = \tilde{\varphi} \quad \text{on } S_4 \times I\ . \end{aligned} \tag{2.5.3}$$

Let K and U be functions on I defined by

$$K = \frac{1}{2}\int_B \varrho\boldsymbol{u}^2\, dv\ , \qquad U = \frac{1}{2}\int_B (d_{ijrs}u_{i,j}u_{r,s} + a\theta^2)\, dv\ . \tag{2.5.4}$$

Theorem 2.5.1. *Assume that d_{ijrs} and k_{ij} satisfy the symmetry relations*

$$d_{ijrs} = d_{rsij}\ , \qquad k_{ij} = k_{ji}\ . \tag{2.5.5}$$

Let

$$G(\alpha,\beta)=\int_B[f_i(\boldsymbol{x},\alpha)u_i(\boldsymbol{x},\beta)-\frac{1}{T}R(\boldsymbol{x},\alpha)\theta(\boldsymbol{x},\beta)]\,dv+ \\ +\int_{\partial B}[s_i(\boldsymbol{x},\alpha)\dot{u}_i(\boldsymbol{x},\beta)-\frac{1}{T}q(\boldsymbol{x},\alpha)\theta(\boldsymbol{x},\beta)]\,da\,, \tag{2.5.6}$$

for all $\alpha,\beta\in I$. *Then*

$$U(t)-K(t)=\frac{1}{2}\int_0^t[G(t+s,t-s)-G(t-s,t+s)]\,ds+ \\ +\frac{1}{2}\int_B[d_{ijrs}(\boldsymbol{x})u_{i,j}(\boldsymbol{x},2t)u_{r,s}(\boldsymbol{x},0)+a(\boldsymbol{x})\theta(\boldsymbol{x},2t)\theta(\boldsymbol{x},0)- \\ -\varrho(\boldsymbol{x})\dot{u}_i(\boldsymbol{x},0)\dot{u}_i(\boldsymbol{x},2t)]\,dv\,,\qquad 0\le t<\infty\,. \tag{2.5.7}$$

Proof. By (2.5.1) and (2.5.5),

$$s_{ji}(t-s)\dot{u}_{i,j}(t+s)+\dot{\chi}(t+s)\theta(t-s)-[s_{ji}(t+s)\dot{u}_{i,j}(t-s)+ \\ +\dot{\chi}(t-s)\theta(t+s)]=\frac{\partial}{\partial s}[d_{ijpq}u_{i,j}(t+s)u_{p,q}(t-s)+ \\ +a\theta(t+s)\theta(t-s)]\,, \tag{2.5.8}$$

where, for convenience, we have suppressed the argument $\boldsymbol{x}$. In view of (2.5.2), we obtain

$$s_{ji}(t-s)\dot{u}_{i,j}(t+s)+\dot{\chi}(t+s)\theta(t-s)=[s_{ji}(t-s)\dot{u}_i(t+s)+\frac{1}{T}\varphi_j(t+ \\ +s)\theta(t-s)]_{,j}+f_i(t-s)\dot{u}_i(t+s)+\frac{1}{T}R(t+s)\theta(t-s)- \\ -\frac{1}{T}k_{ij}\theta_{,j}(t+s)\theta_{,i}(t-s)+\varrho\frac{\partial}{\partial s}[\dot{u}_i(t-s)\dot{u}_i(t+s)]-\varrho\dot{u}_i(t-s)\ddot{u}_i(t+s)\,. \tag{2.5.9}$$

Similarly,

$$s_{ji}(t+s)\dot{u}_{i,j}(t-s)+\dot{\chi}(t-s)\theta(t+s)=[s_{ji}(t+s)\dot{u}_i(t-s)+ \\ +\frac{1}{T}\varphi_j(t-s)\theta(t+s)]_{,j}+f_i(t+s)\dot{u}_i(t-s)+\frac{1}{T}R(t-s)\theta(t+s)- \\ -\frac{1}{T}k_{ij}\theta_{,j}(t-s)\theta_{,i}(t+s)-\varrho\ddot{u}_i(t+s)\dot{u}_i(t-s)\,. \tag{2.5.10}$$

Next, if we integrate the relation (2.5.8) over B, we conclude, with the aid of (2.5.5), (2.5.9) and (2.5.10) and the divergence theorem, that

$$\begin{aligned}&\int_B \frac{\partial}{\partial s}[d_{ijrp}u_{i,j}(t+s)u_{r,p}(t-s)+a\theta(t+s)\theta(t-s)]\,dv=\\&=G(t-s,t+s)-G(t+s,t-s)+\int_B \varrho\frac{\partial}{\partial s}[\dot{u}_i(t-s)\dot{u}_i(t+s)]\,dv\,.\end{aligned}\tag{2.5.11}$$

If we integrate this relation with respect to the time on $[0,t]$, we arrive at the desired result. □

Theorem 2.5.1 is a counterpart of Brun's theorem in the isothermal theory.

Theorem 2.5.2. *Assume that d_{ijrs} and k_{ij} satisfy (2.5.5). Let*

$$P=\int_B (f_i\dot{u}_i+\frac{1}{T}R\theta)\,dv+\int_{\partial B}(s_i\dot{u}_i+\frac{1}{T}q\theta)\,da\,.\tag{2.5.12}$$

Then

$$\begin{aligned}2U(t)&=U(0)+K(0)+\frac{1}{2}\int_B[d_{ijrs}u_{i,j}(2t)u_{r,s}(0)+a\theta(2t)\theta(0)-\\&-\varrho\dot{u}_i(2t)\dot{u}_i(0)]\,dv+\frac{1}{2}\int_0^t[G(t+s,t-s)-G(t-s,t+s)+\\&+2P(s)]\,ds-\frac{1}{T}\int_0^t\int_B k_{ij}\theta_{,i}\theta_{,j}\,dvdt\,,\end{aligned}\tag{2.5.13}$$

and

$$\begin{aligned}2K(t)&=U(0)+K(0)-\frac{1}{2}\int_B[d_{ijrs}u_{i,j}(2t)u_{r,s}(0)+a\theta(2t)\theta(0)-\\&-\varrho\dot{u}_i(2t)\dot{u}_i(0)]\,dv-\frac{1}{2}\int_0^t[G(t+s,t-s)-G(t-s,t+s)-\\&-2P(s)]\,ds-\frac{1}{T}\int_0^t\int_B k_{ij}\theta_{,i}\theta_{,j}\,dvdt\,,\end{aligned}\tag{2.5.14}$$

for all $t\in I$.

Proof. In view of (2.5.1) and (2.5.2), we obtain

$$d_{ijrs}u_{i,j}\dot{u}_{r,s} + a\theta\dot{\theta} = s_{ji}\dot{u}_{i,j} + \dot{\chi}\theta = (s_{ji}\dot{u}_i + \frac{1}{T}\varphi_j\theta)_{,j} +$$

$$+ f_i\dot{u}_i + \frac{1}{T}R\theta - \varrho\dot{u}_i\ddot{u}_i - \frac{1}{T}k_{ij}\theta_{,j}\theta_{,i} .$$

This relation implies that

$$\dot{U} + \dot{K} = P - \frac{1}{T}\int_B k_{ij}\theta_{,j}\theta_{,i}\, dv . \tag{2.5.15}$$

From (2.5.7) and (2.5.15) we obtain (2.5.13) and (2.5.14). □

Theorems 2.5.1 and 2.5.2 form the basis of the following uniqueness result.

Theorem 2.5.3. *Assume that*

(*i*) d_{ijrs} *and* k_{ij} *satisfy the symmetry relations (2.5.5);*

(*ii*) ϱ *is strictly positive;*

(*iii*) a *is strictly positive (or negative);*

(*iv*) k_{ij} *is positive semi-definite.*

Then the mixed problem of thermoelastodynamics has at most one solution.

Proof. Suppose that there are two solutions. Then their difference $(\boldsymbol{u}, \theta)$ corresponds to null data. Thus, we conclude from (2.5.14) that

$$\int_B \varrho\dot{\boldsymbol{u}}^2(t)\, dv + \frac{1}{T}\int_0^t\int_B k_{ij}\theta_{,i}\theta_{,j}\, dvdt = 0\,, \quad t \in I\,.$$

Since $\varrho > 0$ and k_{ij} is positive semi-definite, this relation implies that

$$\dot{\boldsymbol{u}} = \boldsymbol{0} \quad \text{on} \quad B \times I\,, \tag{2.5.16}$$

and

$$\int_0^t\int_B k_{ij}\theta_{,i}\theta_{,j}\, dvdt = 0\,, \qquad t \in I\,. \tag{2.5.17}$$

But $\boldsymbol{u}$ vanishes initially, so that the relation (2.5.16) yields

$$\boldsymbol{u} = \boldsymbol{0} \quad \text{on} \quad \bar{B} \times I\,. \tag{2.5.18}$$

In view of (2.5.17) and (2.5.18), from (2.5.13) we obtain

$$\int_B a\theta^2\, dv = 0\,.$$

By (iii), this relation yields $\theta = 0$ on $B \times I$ and the proof is complete. □

The first reciprocal theorem in the framework of the linear theory of thermoelastic bodies is due to Ionescu-Cazimir [200]. The proof of this theorem is based on the assumption of null initial data and systematic use of the Laplace transform. Later, Ieşan [176], [178] established a reciprocity relation without using the Laplace transform. The method is based on a characterization of the boundary-initial-value problem in which the initial conditions are incorporated into the field equations (Theorem 2.3.2). In [188], [189] a new reciprocal theorem has been derived. The proof avoids both the use of the Laplace transform and the incorporation of the initial conditions into the equations of motion. In what follows we present this theorem and show that it leads to the classical reciprocal theorem and to variational principles.

If h is a continuous function on $B\times I$, then we denote by $\bar{h}$ the function defined by

$$\bar{h}(\boldsymbol{x},t) = \int_0^t h(\boldsymbol{x},s)\,ds\,, \qquad (\boldsymbol{x},t) \in B \times I\,. \tag{2.5.19}$$

Clearly, the energy equation $(2.5.2)_2$ and the initial conditions imply that

$$T\chi = \bar{\varphi}_{i,i} + W\,, \tag{2.5.20}$$

on $B \times I$, where

$$W = \bar{R} + T(a\tilde{\theta}_0 + \beta_{ij}\tilde{a}_{i,j})\,. \tag{2.5.21}$$

Let $L^{(\alpha)} = (\boldsymbol{f}^{(\alpha)},\ R^{(\alpha)},\ \tilde{\boldsymbol{u}}^{(\alpha)},\ \tilde{\boldsymbol{p}}^{(\alpha)},\ \tilde{\theta}^{(\alpha)},\ \tilde{\varphi}^{(\alpha)},\ \tilde{\boldsymbol{a}}^{(\alpha)},\ \tilde{\boldsymbol{b}}^{(\alpha)},\ \tilde{\theta}_0^{(\alpha)})$, $(\alpha = 1,2)$ be two external data systems. Further, let $P^{(\alpha)} = (u_i^{(\alpha)},\ \theta^{(\alpha)},\ s_{ij}^{(\alpha)},\ \chi^{(\alpha)},\ \varphi_i^{(\alpha)})$ be the corresponding thermoelastic processes. We define

$$s_i^{(\alpha)} = s_{ji}^{(\alpha)}u_j\,, \quad q^{(\alpha)} = \varphi_i^{(\alpha)}n_i\,, \quad W^{(\alpha)} = \bar{R}^{(\alpha)} + T(a\tilde{\theta}_0^{(\alpha)} + \beta_{ij}\tilde{a}_{i,j}^{(\alpha)})\,. \tag{2.5.22}$$

Theorem 2.5.4. *Assume that d_{ijrs} and k_{ij} satisfy the symmetry relations (2.5.5). Let*

$$\begin{aligned} E_{\alpha\beta}(r,s) = & \int_{\partial B} [(s_i^{(\alpha)}(r)u_i^{(\beta)}(s) - \frac{1}{T}\bar{q}^{(\alpha)}(r)\theta^{(\beta)}(s)]\,da + \\ & + \int_B [(f_i^{(\alpha)}(r)u_i^{(\beta)}(s) - \frac{1}{T}W^{(\alpha)}(r)\theta^{(\beta)}(s) - \\ & - \varrho\ddot{u}_i^{(\alpha)}(r)u_i^{(\beta)}(s) + \frac{1}{T}\bar{\varphi}_i^{(\alpha)}(r)\theta_{,i}^{(\beta)}(s)]\,dv\,, \end{aligned} \tag{2.5.23}$$

for all $r, s \in I$. Then

$$E_{\alpha\beta}(r,s) = E_{\beta\alpha}(s,r)\,, \tag{2.5.24}$$

for all $r, s \in I$.

Proof. Let

$$I_{\alpha\beta}(r,s) = s_{ji}^{(\alpha)}(r)u_{i,j}^{(\beta)}(s) - \chi^{(\alpha)}(r)\theta^{(\beta)}(s)\,. \tag{2.5.25}$$

From (2.5.1) and (2.5.25)

$$\begin{aligned} I_{\alpha\beta}(r,s) = d_{ijpq}u_{p,q}^{(\alpha)}(r)u_{i,j}^{(\beta)}(s) - \beta_{ij}[\theta^{(\alpha)}(r)u_{i,j}^{(\beta)}(s) + \\ + u_{i,j}^{(\alpha)}(r)\theta^{(\beta)}(s)] - a\theta^{(\alpha)}(r)\theta^{(\beta)}(s)\,. \end{aligned} \tag{2.5.26}$$

From (2.5.5) and (2.5.26) we get

$$I_{\alpha\beta}(r,s) = I_{\beta\alpha}(s,r)\,. \tag{2.5.27}$$

On the other hand, in view of (2.5.2), (2.5.20) and (2.5.25), we have

$$\begin{aligned} I_{\alpha\beta}(r,s) = [s_{ji}^{(\alpha)}(r)u_i^{(\beta)}(s) - \frac{1}{T}\bar{\varphi}_j^{(\alpha)}(r)\theta^{(\beta)}(s)]_{,j} + f_i^{(\alpha)}(r)u_i^{(\beta)}(s) - \\ - \frac{1}{T}W^{(\alpha)}(r)\theta^{(\beta)}(s) - \varrho\ddot{u}_i^{(\alpha)}(r)u_i^{(\beta)}(s) + \frac{1}{T}\bar{q}_i^{(\alpha)}(r)\theta_{,i}^{(\beta)}(s)\,. \end{aligned}$$

If we integrate the above relation over B and use (2.5.7) and the divergence theorem, then we obtain the desired result. □

Theorem 2.5.4 forms the basis of the following reciprocal theorem.

Theorem 2.5.5. *Assume that d_{ijrs} and k_{ij} satisfy the symmetry relations (2.5.5). Let $P^{(\alpha)}$ be a solution corresponding to the external data system $L^{(\alpha)}$, $(\alpha = 1, 2)$. Then*

$$\begin{aligned} &\int_{\partial B} (i * s_k^{(1)} * u_k^{(2)} - \frac{1}{T} j * q^{(1)} * \theta^{(2)})\, da + \int_B (h_k^{(1)} * u_k^{(2)} - \\ &- \frac{1}{T} i * W^{(1)} * \theta^{(2)})\, dv = \int_{\partial B} (i * s_k^{(2)} * u_k^{(1)} - \frac{1}{T} j * q^{(2)} * \theta^{(1)})\, da + \\ &+ \int_B (h_k^{(2)} * u_k^{(1)} - \frac{1}{T} i * W^{(2)} * \theta^{(1)})\, dv\,, \end{aligned} \tag{2.5.28}$$

where

$$i = t\,, \qquad j = \frac{1}{2}t^2\,, \qquad h_s^{(\alpha)} = i * f_s^{(\alpha)} + \varrho(t\tilde{b}_s^{(\alpha)} + \tilde{a}_s^{(\alpha)})\,. \tag{2.5.29}$$

Proof. If we take in (2.5.24) $r=\tau$ and $s=t-\tau$ and integrate from 0 to t, we arrive with the aid of (2.5.5) at

$$\begin{aligned}&\int_{\partial B}(s_i^{(1)}*u_i^{(2)}-\frac{1}{T}\bar{q}^{(1)}*\theta^{(2)})\,da+\int_B(f_i^{(1)}*u_i^{(2)}-\frac{1}{T}W^{(1)}*\theta^{(2)}-\\&-\varrho\ddot{u}_i^{(1)}*u_i^{(2)})\,dv=\int_{\partial B}(s_i^{(2)}*u_i^{(1)}-\frac{1}{T}\bar{q}^{(2)}*\theta^{(1)})\,da+\\&+\int_B(f_i^{(2)}*u_i^{(1)}-\frac{1}{T}W^{(2)}*\theta^{(1)}-\varrho\ddot{u}_i^{(2)}*u_i^{(1)})\,dv\,.\end{aligned}\tag{2.5.30}$$

Clearly,

$$i*\bar{f}=j*f\,,\qquad i*\ddot{u}_i^{(\alpha)}=u_i^{(\alpha)}-tb_i^{(\alpha)}-a_i^{(\alpha)}\,.\tag{2.5.31}$$

If we take the convolution of relation (2.5.30) with i, then, with the aid of (2.5.31), we conclude that (2.5.28) holds. □

Various applications of the reciprocity relation (2.5.28) have been presented in [200],[279].

Theorem 2.5.5 has been established in [176], [178] by using Theorem 2.3.2. In the case of null boundary data, the relation (2.5.28) shows that the operator of the linear thermoelastodynamics is symmetric in convolution. This fact leads to variational theorems of Gurtin type (cf. Ieşan [178]). Theorem 2.5.5 can be also used to derive minimum principles.

2.6 Variational principles. Minimum principles

Throughout this section we assume that the primary state is a configuration of equilibrium at constant temperature. In the context of the classical theory of thermoelasticity, variational theorems have been established by Ieşan [173] and Nickell and Sackmann [272] (see also Carlson [46], p.338, and Lebon [233], p. 370). We present a variational theorem of Gurtin type for the theory of prestressed bodies.

We now state, without proof, two lemmas due to Gurtin [157] (see also Gurtin 158], Sect. 64).

Lemma 2.6.1. *Let Ω be a finite-dimensional inner product space. Let $\boldsymbol{f}$: $B\times[0,t_1)\to\Omega$ be continuous and satisfy*

$$\int_B \boldsymbol{f}*\boldsymbol{g}dv=0\,,\qquad(0\le t<t_1)\,,$$

for every class C^∞ function $\boldsymbol{g}:B\times[0,t_1)\to\Omega$ that vanishes near ∂B. Then $\boldsymbol{f}=\boldsymbol{0}$ on $\bar{B}\times[0,t_1)$.

Lemma 2.6.2. *Let Ω be a finite-dimensional inner product space. Let $\boldsymbol{g} : S_2 \times [0, t_1) \to \Omega$ be piecewise regular and continuous in time, and assume that*

$$\int_{S_2} \boldsymbol{f} * \boldsymbol{g} da = 0\,, \qquad (0 \leq t < t_1)\,,$$

for every class C^∞ function $\boldsymbol{g} : \bar{B} \times [0, t_1) \to \Omega$ that vanishes near S_1. Then $\boldsymbol{f} = \boldsymbol{0}$ on $S_2 \times [0, t_1)$.

The role of these lemmas is analogous to the role of the fundamental lemma in the calculus of variations.

Let L be a linear space, K a subset of L, and $G(\cdot)$ a functional defined on K. Let

$$y\,,\ y^0 \in L\,, \qquad y + \lambda y^0 \in K \quad \text{for every scalar } \lambda\,, \tag{2.6.1}$$

and define

$$\delta_{y^0} G(y) = \frac{d}{d\lambda} G(y + \lambda y^0)|_{\lambda=0}\,.$$

We say that the variation of $G(\cdot)$ is zero at $y \in K$ and write $\delta G(y) = 0$, if $\delta_{y^0} G(y)$ exists and equals zero for every choice of y^0 consistent with (2.6.1).

We say that $(\boldsymbol{u}, \theta)$ is an admissible four-dimensional vector field if $\boldsymbol{u}$ is an admissible displacement field on $\bar{B} \times [0, t_1)$ and θ is an admissible temperature field on $\bar{B} \times [0, t_1)$. Let Q be the set of all admissible four-dimensional vector fields $(\boldsymbol{u}, \theta)$ that satisfy the boundary conditions

$$\boldsymbol{u} = \tilde{\boldsymbol{u}} \quad \text{on } \bar{S}_1 \times [0, t_1)\,, \qquad \theta = \tilde{\theta} \quad \text{on } \bar{S}_3 \times [0, t_1)\,. \tag{2.6.2}$$

Theorem 2.6.1. *Assume that d_{ijrs} and k_{ij} satisfy the symmetry relations (2.5.5). Let $v = (\boldsymbol{u}, \theta) \in Q$ and for each $t \in [0, t_1)$ define the functional $F_t\{\cdot\}$ on Q by*

$$F_t\{v\} = \int_B \{i * [d_{ijrs} u_{r,s} * u_{i,j} - 2\beta_{ij} u_{i,j} * \theta - a\theta * \theta -$$
$$- \frac{1}{T}(1 * k_{ij}\theta_{,j} * \theta_{,i} - 2g * \theta)] + \varrho u_i * u_i - 2h_i * u_i\}\, dv -$$
$$- 2\int_{S_2} i * \tilde{p}_j * u_j da + \frac{2}{T}\int_{S_4} 1 * i * \tilde{q} * \theta\, da\,.$$

Then

$$\delta F_t\{v\} = 0\,, \quad (0 \leq t < t_1)\,, \tag{2.6.3}$$

at $v \in Q$ if and only if v is a solution of the boundary-initial-value problem.
Proof. Let $v^0 = (\boldsymbol{u}^0, \theta^0)$ be an admissible four-dimensional vector fields, and suppose that

$$v + \lambda v^0 \in Q \qquad \text{for every scalar } \lambda\,. \tag{2.6.4}$$

The condition (2.6.4) is equivalent to the requirement that v^0 meets the boundary conditions

$$\boldsymbol{u}^0 = \mathbf{0} \quad \text{on } \bar{S}_1 \times [0, t_1)\,, \qquad \theta^0 = 0 \quad \text{on } \bar{S}_3 \times [0, t_1)\,. \tag{2.6.5}$$

By (2.5.5), (2.6.5) and the divergence theorem we arrive at

$$\begin{aligned}
\delta_{v^0} F_t\{v\} &= 2\int_B \{[\varrho u_i - h_i - i * (d_{ijrs}u_{r,s})_{,j}] * u_i^0 + \\
&+ i * \left[\frac{1}{T}g - \beta_{ij}u_{i,j} - a\theta + \frac{1}{T} * (k_{ij}\theta_{,j})_{,i}\right] * \theta^0\}\, dv + \\
&+ 2\int_{S_2} i * [(d_{ijrs}u_{r,s} - \beta_{ij}\theta)n_j - \tilde{p}_i] * u_i^0 da + \\
&+ \frac{2}{T}\int_{S_4} i * 1 * (\tilde{\varphi} - k_{ij}\theta_{,j}n_i) * \theta^0\, da\,, \qquad (0 \le t < t_1)\,,
\end{aligned} \tag{2.6.6}$$

for every admissible four-dimensional vector field $v^0 = (\boldsymbol{u}^0, \theta^0)$ that satisfies (2.6.5). If v is a solution of the mixed problem, then (2.6.6), in view of Theorem 2.3.2, yields (2.6.3). On the other hand (2.6.3), (2.6.6), Lemmas 2.6.1 and 2.6.2, and Theorem 2.3.2 imply that $(\boldsymbol{u}, \theta)$ is a solution of the boundary-initial-value problem. □

The form of the functional $F_t\{\cdot\}$ has been obtained by the method given in [178].

The variational principles of Gurtin type, in general, are not extremum principles. In [22], Benthien and Gurtin have presented minimum principles for the Laplace transforms of the functionals associated with some of the variational principles presented by Gurtin in [157]. In [310], Reiss and Haug, making use of an appropriate "weight function" have established minimum principles for linear elastodynamics. In what follows we show that the reciprocal theorem implies a minimum principle in thermoelastodynamics (cf. Ieşan [189]).

Let $I = [0, \infty)$. We say that ψ is bounded at infinity if $\lim_{t\to 0}\psi(\boldsymbol{x}, t)$ exists for each $\boldsymbol{x}$ in the domain of definition of ψ. Let X be the set of all functions ψ on $\bar{B} \times I$ with the following properties

(*i*) ψ is of class C^2 on $B \times (0, \infty)$ and of class $C^{1,0}$ on $\bar{B} \times I$,

(*ii*) ψ, $\psi_{,i}$, $\dot{\psi}$ and $\psi_{,ij}$ are bounded at infinity.

Following [310], we introduce the set Γ of admissible weight functions. We say that $g \in \Gamma$ if g is a function on I with the following properties

(α) $\displaystyle\int_0^\infty \int_0^\infty g^{[k]}(t+s)\, dtds$ exists for $k \geq 0$,

(β) g is the Laplace transform of a continuous and positive function G which has a finite limit at infinity, i.e.

$$g(t) = \int_0^\infty G(p) \exp(-pt) dp\,, \qquad t \in I\,,$$

where $g^{[k]}(t) = \partial^k g / \partial t^k$. An example of weight function is $g(t) = (t+a)^{-n}$, $n > 2$, $a > 0$ with $G(t) = [t^{n-1} \exp(-at)]/(n-1)\,!$.

Let Y be the set of all continuous functions on $\bar{B} \times I$ that are bounded at infinity. Let $u, v \in Y$. We introduce the notation

$$[u,v]_g = \int_0^\infty \int_0^\infty \int_B g(t+s) u(\boldsymbol{x}, t) v(\boldsymbol{x}, s)\, dtdsdv\,, \tag{2.6.7}$$

where g is any fixed function of Γ. We note that

$$[u,v]_g = [v,u]_g\,. \tag{2.6.8}$$

Let A be a linear operator from X into Y. We denote by D_A the domain of definition of the operator A. We say that the operator A is g-symmetric if

$$[Au,v]_g = [v,Au]_g\,, \tag{2.6.9}$$

for any $u, v \in \Gamma$. A g-symmetric operator A is called positive if

$$[Au,u]_g \geq 0 \tag{2.6.10}$$

for any $u \in D_A$. We consider the equation

$$Au = F\,, \tag{2.6.11}$$

where $F \in Y$. Let Λ_g be the functional on D_A defined by

$$\Lambda_g\{u\} = [Au,u]_g - 2[u,F]_g\,, \tag{2.6.12}$$

for every $u \in D_A$. We now record the following

Lemma 2.6.3. *Let A be a g-positive operator and let Λ_g be the functional on D_A defined by (2.6.12). Further, let u be a solution of the equation (2.6.11). Then*

$$\Lambda_g\{u\} \leq \Lambda_g\{u^0\}\,,$$

for every $u^0 \in D_A$.

In what follows we denote by $\hat{\psi}$ the Laplace transform with respect to time of the function ψ. By (β) and (2.6.7),

$$[u, v]_g = \int_0^\infty \int_B G(s)\hat{u}(\boldsymbol{x}, s)\hat{v}(\boldsymbol{x}, s)\, dsdv \,, \tag{2.6.13}$$

for any $u, v \in X$.

In view of (2.5.1), (2.5.2) and (2.5.20) we obtain the following form of the basic equations expressed in terms of the displacement and temperature

$$\begin{aligned} &(d_{ijrs}u_{r,s} - \beta_{ji}\theta)_{,j} + f_i = \varrho\ddot{u}_i \,, spag \\ &T(a\theta + \beta_{ij}u_{i,j}) = (k_{ij}\bar{\theta}_{,j})_{,i} + W \,, \end{aligned} \tag{2.6.14}$$

on $B \times I$. To the system (2.6.14) we adjoin the boundary conditions

$$\begin{aligned} &u_i = 0 \quad \text{on } \bar{S}_1 \times I \,, \quad (d_{ijrs}u_{r,s} - \beta_{ij}\theta)n_j = 0 \quad \text{on } \; S_2 \times I \,, \\ &\theta = 0 \quad \text{on } \bar{S}_3 \times I \,, \quad k_{ij}\theta_{,j}n_i = 0 \quad \text{on } S_4 \times I \,, \end{aligned} \tag{2.6.15}$$

and the initial conditions

$$u_i(\boldsymbol{x}, 0) = 0 \,, \qquad \dot{u}_i(\boldsymbol{x}, 0) = 0 \,, \qquad \theta(\boldsymbol{x}, 0) = 0 \,, \qquad \boldsymbol{x} \in \bar{B} \,. \tag{2.6.16}$$

We say that $U = (\boldsymbol{u}, \theta)$ is an admissible state if: (*i*) u_i, $\theta \in X$; (*ii*) $\boldsymbol{u}$ is an admissible displacement field on $\bar{B} \times I$; (*iii*) θ is an admissible temperature field on $\bar{B} \times I$. Let D be the set of all admissible states $U = (\boldsymbol{u}, \theta)$ that satisfy the conditions (2.6.15) and (2.6.16). We introduce the operators L_i $(i = 1, 2, 3, 4)$ on D defined by

$$\begin{aligned} &L_i U = -\varrho\ddot{u}_i + (d_{ijrs}u_{r,s} - \beta_{ji}\theta)_{,j} \,, \\ &L_4 U = -\frac{1}{T}(k_{ij}\int_0^t \theta_{,j}(\tau)d\tau)_{,i} + a\theta + \beta_{ij}u_{i,j} \,, \end{aligned} \tag{2.6.17}$$

for every $U = (\boldsymbol{u}, \theta) \in D$. If we introduce the notations

$$LU = (L_1U,\; L_2U,\; L_3U,\; L_4U) \,, \qquad F = (-f_1,\; -f_2,\; -f_3,\; \frac{1}{T}W) \,,$$

then the equations (2.6.14) can be written in the form

$$LU = F \,. \tag{2.6.18}$$

Let $U = (\boldsymbol{u},\, \theta)$ and $V = (\boldsymbol{v},\, \tau)$ be vector fields on $\bar{B}\times I$ such that $u_i,\ v_i,\ \theta,\ \tau \in Y$. We introduce the notation

$$[U, V]_g = [\theta, \tau]_g + \sum_{i=1}^{3} [u_i, v_i]_g = [\theta, \tau]_g + [\boldsymbol{u}, \boldsymbol{v}]_g \,. \tag{2.6.19}$$

Theorem 2.6.2. *Assume that d_{ijrs} and k_{ij} satisfy the symmetry relations (2.5.5). Let $P^{(\alpha)} = (u_i^{(\alpha)},\ \theta^{(\alpha)},\ s_{ij}^{(\alpha)},\ \chi^{(\alpha)},\ \varphi_i^{(\alpha)})$ be solution corresponding to the external data system $L^{(\alpha)} = (\boldsymbol{f}^{(\alpha)},\ W^{(\alpha)},\ \tilde{\boldsymbol{u}}^{(\alpha)} = \boldsymbol{0},\ \tilde{\boldsymbol{p}}^{(\alpha)} = \boldsymbol{0},\ \tilde{\theta}^{(\alpha)} = 0,\ \tilde{\varphi}^{(\alpha)} = 0,\ \tilde{\boldsymbol{a}}^{(\alpha)} = \boldsymbol{0},\ \tilde{\boldsymbol{b}}^{(\alpha)} = \boldsymbol{0},\ \tilde{\theta}_0^{(\alpha)} = 0)$. Then*

$$[\boldsymbol{f}^{(1)}, \boldsymbol{u}^{(2)}]_g - \frac{1}{T}[W^{(1)}, \theta^{(2)}]_g = [\boldsymbol{f}^{(2)}, \boldsymbol{u}^{(1)}]_g - \frac{1}{T}[W^{(2)}, \theta^{(1)}]_g \,, \tag{2.6.20}$$

$$\begin{aligned}
[\boldsymbol{f}^{(1)}, \boldsymbol{u}^{(1)}]_g - \frac{1}{T}[W^{(1)}, \theta^{(2)}]_g = \int_0^\infty \int_0^\infty \int_B g(t+s)[d_{ijmn} u_{m,n}^{(1)}(t) u_{i,j}^{(1)}(s) - \\
- 2\beta_{ij}\theta^{(1)}(t) u_{i,j}^{(1)}(s) - a\theta^{(1)}(t)\theta^{(1)}(s)]\, dtdsdv - \\
- \frac{1}{T} k_{ij} [\bar{\theta}_{,j}^{(1)}\,,\ \theta_{,i}^{(1)}]_g + [\varrho \dot{\boldsymbol{u}}^{(1)}, \dot{\boldsymbol{u}}^{(1)}]_g \,.
\end{aligned} \tag{2.6.21}$$

Proof. We introduce the notation

$$K_{\alpha\beta} = \int_0^\infty \int_0^\infty g(t+s) E_{\alpha\beta}(t,s)\, dtds \,, \tag{2.6.22}$$

where $E_{\alpha\beta}(t,s)$ are defined by (2.5.23). It follows from (2.5.24) and (2.6.1) that

$$K_{\alpha\beta} = K_{\beta\alpha} \,. \tag{2.6.23}$$

Since $\boldsymbol{u}^{(\alpha)}$ and $\theta^{(\alpha)}$ correspond to null data, from (2.5.23) and (2.6.22), we obtain

$$\begin{aligned}
K_{\alpha\beta} = [\boldsymbol{f}^{(\alpha)}, \boldsymbol{u}^{(\beta)}]_g - \frac{1}{T}[W^{(\alpha)}, \theta^{(\beta)}]_g - [\varrho \dot{\boldsymbol{u}}^{(\alpha)}, \dot{\boldsymbol{u}}^{(\beta)}]_g + \\
+ [k_{ij} \bar{\theta}_{,j}^{(\alpha)}\,,\ \theta_{,i}^{(\beta)}]_g \,.
\end{aligned} \tag{2.6.24}$$

It follows from (2.6.23) and (2.6.24) that (2.6.20) holds.

By (2.5.26), (2.6.22) and

$$E_{\alpha\beta}(r,s) = \int_B I_{\alpha\beta}(r,s)\, dv\ ,$$

we obtain

$$\begin{aligned} K_{11} = \int_0^\infty \int_0^\infty \int_B g(t+s)\big\{d_{ijmn}u^{(1)}_{m,n}(r)u^{(1)}_{i,j}(s) - \\ - \beta_{ij}[u^{(1)}_{i,j}(t)\theta^{(1)}(s) + u^{(1)}_{i,j}(s)\theta^{(1)}(t)] - a\theta^{(1)}(t)\theta^{(1)}(s)\big\}\, dtdsdv\,. \end{aligned} \tag{2.6.25}$$

From (2.6.24) and (2.6.25) we obtain (2.6.21). □

Let $U = (\boldsymbol{u},\, \theta) \in D$. Then we define $E(U)$ by

$$\begin{aligned} E(U) = \int_0^\infty \int_0^\infty \int_B g(t+s)\Big[\frac{1}{T}k_{ij}\bar{\theta}_{,j}(t)\theta_{,i}(s) + a\theta(t)\theta(s) + \\ + 2\beta_{ij}\theta(t)u_{i,j}(s) - d_{ijmn}u_{m,n}(t)u_{i,j}(s) - \\ - \varrho\dot{u}_i(t)\dot{u}_i(s)\Big]\, dtdsdv = \int_0^\infty \int_B G(p)\Big[\frac{1}{pT}k_{ij}\hat{\theta}_{,j}(p)\hat{\theta}_{,i}(p) + \\ + a\hat{\theta}^2(p) + 2\beta_{ij}\hat{\theta}(p)\hat{u}_{i,j}(p) - d_{ijmn}\hat{u}_{m,n}(p)\hat{u}_{i,j}(p) - \\ - \varrho p^2\hat{u}_i(p)\hat{u}_i(p)\Big]\, dpdv\ . \end{aligned} \tag{2.6.26}$$

Let Q be the set of all vector fields $U = (\boldsymbol{u},\, \theta) \in D$ that satisfy $E(U) \geq 0$. We contract the operator L by replacing its domain of definition D with the more restricted domain Q. We denote this contracted operator by L^*.

Lemma 2.6.4. *The operator L^* is g-positive.*

Proof. Let $L^{(\alpha)}$ $(\alpha = 1, 2)$ be as in Theorem 2.6.2. Let $U = (\boldsymbol{u},\, \theta)$ and $V = (\boldsymbol{v},\, \tau)$ be the solutions corresponding to $L^{(1)}$ and $L^{(2)}$, respectively. Then

$$LU = F^{(1)}\,, \qquad LV = F^{(2)}\,,$$

where $F^{(\alpha)} = (-\boldsymbol{f}^{(\alpha)},\, \frac{1}{T}W^{(\alpha)})$. It follows from (2.6.20) and (2.6.21) that

$$[L^*U, V]_g = [U, L^*V]_g\,, \qquad [L^*U, U]_g = E(U)\,, \tag{2.6.27}$$

for every U, $V \in Q$. The relations (2.6.27) imply the desired result. □

Let Ψ_g be the functional on Q defined by

$$\Psi_g\{U\} = [L^*U, U]_g - 2[U, F]_g .$$

Clearly,

$$\begin{aligned}\Psi_g\{U\} = \int_0^\infty \int_0^\infty \int_B g(t+s)\Big[\frac{1}{T}k_{ij}\bar{\theta}_{,j}(t)\theta_{,i}(s) + a\theta(t)\theta(s) + \\ + 2\beta_{ij}\theta(t)u_{i,j}(s) - d_{ijmn}u_{m,n}(t)u_{i,j}(s) - \\ - \varrho\dot{u}_i(t)\dot{u}_i(s) + 2f_i(t)u_i(s) - \frac{2}{T}W(t)\theta(s)\Big]\, dtdsdv ,\end{aligned} \tag{2.6.28}$$

for any $U = (u_i, \theta) \in Q$.
In view of Lemmas 2.6.3 and 2.6.4, we obtain

Theorem 2.6.3. *Let Ψ_g be the functional on Q defined by (2.6.28), and let $U \in Q$ be a solution of the equations (2.6.14). Then*

$$\Psi_g\{U\} \leq \Psi_g\{U^*\} ,$$

for every $U^* \in Q$.

Similarly, we are led to the following result.

Theorem 2.6.4. *Let Q^0 be the set of all four-dimensional vector fields $U = (u, \theta) \in D$ that satisfy $E(U) \leq 0$, and let $\mathcal{M}_g$ be the functional on Q^0 defined by*

$$\begin{aligned}\mathcal{M}_g\{U\} = \int_0^\infty \int_0^\infty \int_B g(t+s)\big[d_{ijmn}u_{m,n}(t)u_{i,j}(s) - 2\beta_{ij}\theta(t)u_{i,j}(s) - \\ - a\theta(t)\theta(s) + \varrho\dot{u}_i(t)\dot{u}_i(s) - \frac{1}{T}k_{ij}\bar{\theta}_{,j}(t)\theta_{,i}(s) - \\ - 2f_i(t)u_i(s) + \frac{2}{T}W(t)\theta(s)\big]\, dtdsdv ,\end{aligned}$$

for any $U = (\boldsymbol{u}, \theta) \in Q^0$. Further, let $U \in Q^0$ be a solution of the equations (2.6.14). Then

$$\mathcal{M}_g\{U\} \leq \mathcal{M}_g\{U^*\} ,$$

for every $U^* \in Q^0$.

As in [178], we can derive minimum principles for nonhomogeneous boundary conditions.

Numerical investigations, in the case when the primary state is a configuration of equilibrium at constant temperature and without stresses, have been presented in various papers (see, e.g. Nickell and Sackmann [273], Chou and Wang [72], Li and Chen [240], Lin and Chang [242], Oden and Poe [285]).

2.7 Existence theorem

Dafermos [91] was the first to establish an existence theorem in thermoelastodynamics. For a detailed analysis of various existence results we refer to the work of Racke [309].

In this section we apply the theory of semigroups of linear operators to establish an existence and uniqueness result for the boundary-initial-value problems of thermoelasticity. The results we give here are due to Navarro and Quintanilla [268].

We consider that the primary state is a time-dependent configuration. The behavior of the body is studied on a finite time interval $[0, t_1]$. We consider the equations (2.2.1)-(2.2.3) on $B_0 \times (0, t_1)$. To these equations we adjoin the initial conditions (2.3.9) and the boundary conditions

$$\boldsymbol{u} = \boldsymbol{0}, \qquad \theta = 0 \quad \text{on } \partial B_0 \times [0, t_1]. \tag{2.7.1}$$

The fields equations can be written in the form (2.3.11). We assume that $A \neq 0$ on $B_0 \times [0, t_1]$. Then, the equations (2.3.11) may be written in the form

$$\begin{aligned} \ddot{u}_i &= \frac{1}{\varrho_0}(D_{iAjK}u_{j,K} - F_{Ai}\theta)_{,A} + F_i, \\ \dot{\theta} &= \frac{1}{AT}\left[(K_{AM}\theta_{,M} + D_A\theta + G_{AiK}u_{i,K})_{,A} - \varrho_0\dot{\eta}\theta\right] - \\ &\quad - \frac{1}{A}(F_{Ai}\dot{u}_{i,A} + \dot{F}_{Ai}u_{i,A} + \dot{A}\theta) + \frac{\varrho_0}{AT}S, \end{aligned} \tag{2.7.2}$$

on $B_0 \times (0, t_1)$. We first transform the boundary-initial-value problem into a temporally inhomogeneous abstract equation in a Hilbert space. Then we use results of the semigroups theory of linear operators to derive an existence and uniqueness theorem. We assume that: (*i*) D_{iAjK}, G_{AiK}, F_{Ai}, K_{AM}, D_A, A, $\dot{A}$ and $\dot{F}_{Ai}$ are continuously differentiable in time on $[0, t_1]$, and Lebesgue measurable and essentially bounded on B_0, for each $t \in [0, t_1]$; (*ii*) $0 < \varrho_1 \leq \operatorname*{ess\,inf}_{X \in B_0} \varrho_0 \leq \operatorname*{ess\,sup}_{X \in B_0} \varrho_0 \leq \varrho_2$, where ϱ_1 and ϱ_2 are constants; (*iii*) there exists a constant A_0

such that $\operatorname*{ess\ inf}_{X\in B_0} A \geq A_0 > 0$ on $[0,t_1]$; (iv) there exists a positive constant d such that

$$\int_{B_0} D_{iAjB}u_{i,A}u_{j,B}dV \geq d \int_{B_0} u_{i,K}u_{i,K}dV ,$$

for all $u_i \in C_0^\infty(B_0)$ and any $t \in [0,t_1]$; (v) K_{MN} is positive definite in the sense that there exists a positive constant K_0 such that

$$\int_{B_0} \frac{1}{T}K_{AB}\theta_{,A}\theta_{,B}dV \geq K_0 \int_{B_0} \theta_{,A}\theta_{,A}dV ,$$

for all $\theta \in C_0^\infty(B_0)$ and any $t \in [0,t_1]$.

Let $v_i = \dot{u}_i$ and define

$$X = \{w = (u_1,\, u_2,\, u_3,\, v_1,\, v_2,\, v_3,\, \theta) \equiv (u_i,\, v_i,\, \theta)\,; \qquad u_i \in W_0^{1,2}(B_0)\,,$$
$$v_i \in L_2(B_0)\,, \qquad \theta \in L_2(B_0)\}\,,$$

where $W_0^{1,2}(B_0)$ is the well-known Sobolev space [2], p.44.

Let

$$A_i w = v_i\,, \quad B_i w = \frac{1}{\varrho_0}(D_{iAjK}u_{j,K} - F_{Ai}\theta)_{,A}\,,$$

$$Cw = \frac{1}{AT}\left[(K_{LM}\theta_{,M} + D_L\theta + G_{LiK}u_{i,K})_{,L} - \varrho_0\dot{\eta}\theta\right] - \tag{2.7.3}$$
$$- \frac{1}{A}(F_{Ki}v_{i,K} + \dot{F}_{Ki}u_{i,K} + \dot{A}\theta)\,.$$

We introduce the operator Q defined by

$$Qw = (A_i w,\, B_i w,\, Cw)\,, \tag{2.7.4}$$

with the domain

$$D = \{w = (u_i,\, v_i,\, \theta) \in X\,; \quad Qw \in X\,, \quad \theta = 0 \quad \text{on } \partial B_0 \times [0,t_1]\}\,. \tag{2.7.5}$$

The boundary-initial-value problem characterized by the relations (2.7.1) , (2.7.2) and (2.3.9) can be reduced to the following abstract evolutionary equation

$$\frac{dw(t)}{dt} = Q(t)w(t) + G(t)\,, \qquad t \in [0,t_1]\,,$$
$$w(0) = w_0\,, \tag{2.7.6}$$

where G and w_0 are seven-dimensional vector fields defined by

$$G = (0,\ 0,\ 0,\ F_1,\ F_2,\ F_3,\ \varrho_0 S/(AT))\,, \qquad w_0 = (\hat{a}_i,\ \hat{b}_i,\ \hat{\theta}_0)\,.$$

Let X_t be the Hilbert space X equipped with the norm $\|\cdot\|_t$, induced by the inner product

$$< w, \bar{w} >_t = \frac{1}{2}\int\limits_{B_0} \left[D_{iAjB}u_{i,A}\bar{u}_{j,B} + \varrho_0 v_i \bar{v}_i + A\theta\bar{\theta}\right] dv\,, \tag{2.7.7}$$

for each $t \in [0, t_1]$. By the hypotheses $(ii) - (iv)$ and the Poincarè inequality we conclude that the norm $\|\cdot\|_t$ is equivalent to the original norm $\|\cdot\|$ in X.

Lemma 2.7.1. *For each $t \in [0, t_1]$ the operator $Q(t)$ is the generator of a quasi-contractive semigroup on X_t .*

Proof. By (2.2.3), (2.2.4) and (2.2.7),

$$\begin{aligned} < Qw, w >_t = \frac{1}{2}\int\limits_{B_0} \{&D_{iAjB}v_{i,A}u_{j,B} + v_i(D_{iAjK}u_{j,K} - F_{Ai}\theta)_{,A} + \\ &+ \frac{1}{T}\theta[(K_{AM}\theta_{,M} + D_A\theta + G_{AiK}u_{i,K})_{,A} - \\ &- \varrho_0\dot{\eta}\theta - T(F_{Ki}v_{i,K} + \dot{F}_{Ki}u_{i,K} + \dot{A}\theta)]\}\, dV\,. \end{aligned}$$

The divergence theorem and the boundary conditions imply

$$\begin{aligned} < Qw, w >_t = &-\frac{1}{2}\int\limits_{B_0} \Phi_A\left(\frac{\theta}{T}\right)_{,A} dv - \frac{1}{2}\int\limits_{B_0} \dot{F}_{Ai}u_{i,A}\theta dv - \\ &-\frac{1}{2}\int\limits_{B_0} (\varrho_0\frac{1}{T}\dot{\eta} + \dot{A})\theta^2\, dV\,, \end{aligned}$$

where Φ_A is given by (2.2.3). Using the arithmetic-geometric mean inequality, Schwarz's inequality and (2.4.3) we conclude that there exists a positive constant β such that

$$< Qw, w >_t \ \le \beta \ < w, w >_t \ . \tag{2.7.8}$$

Our next objective is to show that the operator Q satisfies the range condition $R(\lambda I - Q) = X$. Let $\bar{w} = (\bar{u}_i,\ \bar{v}_i,\ \bar{\theta}) \in X$. We must prove that the equation

$$\lambda w - Qw = \bar{w}\,, \qquad (\lambda > 0)\,, \tag{2.7.9}$$

has a solution $w = (u_i,\ v_i,\ \theta) \in D$. By eliminating v_i, (2.7.9) yields the following system for u_i and θ

$$J_i z \equiv \lambda^2 u_i - \frac{1}{\varrho_0}(D_{iAjK} u_{j,K} - F_{Ai}\theta)_{,A} = m_i\,,$$

$$J z \equiv \lambda\theta - \frac{1}{AT}[(K_{LM}\theta_{,M} + D_L\theta + G_{LiK}u_{i,K})_{,L} - \qquad (2.7.10)$$

$$- \varrho_0\dot{\eta}\theta] - \frac{1}{A}(\lambda F_{Ki}u_{i,K} + \dot{F}_{Ki}u_{i,K} + \dot{A}\theta) = m\,,$$

where

$$z = (u_1,\ u_2,\ u_3,\ \theta)\,, \qquad m_i = \bar{v}_i + \lambda\bar{u}_i\,,$$
$$m = \bar{\theta} - \frac{1}{A}F_{Ki}\bar{u}_{i,K}\,. \qquad (2.7.11)$$

Let $[\cdot,\cdot]$ denote a conveniently weighted L_2 inner product, and consider the bilinear form

$$f(z,\bar{z}) = \left[(J_i z, Jz)\,,\ (\bar{u}_i,\bar{\theta})\right] = \int_{B_0} (k\varrho_0\bar{u}_i J_i z + A\bar{\theta}Jz)\,dV\,, \qquad (2.7.12)$$

where k is an arbitrary positive constant. The divergence theorem and the boundary conditions imply

$$f(z,z) = \int_{B_0} \Big[k(\varrho_0 u_i u_i + D_{iAjK}u_{i,A}u_{j,K} - F_{Ai}u_{i,A}\theta) +$$
$$+ (K_{AM}\theta_{,M} + D_A\theta + G_{AiK}u_{i,A})\Big(\frac{\theta}{T}\Big)_{,A} - \lambda F_{Ai}u_{i,A}\theta -$$
$$- \dot{F}_{Ai}u_{i,A}\theta + (A\lambda + \frac{1}{T}\varrho_0\dot{\eta} - \dot{A})\theta^2\Big]\,dV\,.$$

By use of the Schwarz and Young inequalities, and choosing conveniently the weighting constant k, it is not difficult to show that the bilinear form $f(z,\bar{z})$ determines a norm equivalent to

$$\|z\| = \|\boldsymbol{u}\|_{[W_0^{1,2}]^3} + \|\theta\|_{W_0^{1,2}}\,.$$

Next, we can deduce that $(\boldsymbol{m}, m) \in [W^{-1,2}]^3 \times W^{-1,2}$ (see Navarro and Quintanilla [268]). The Riesz theorem shows that there exists a unique pair $(\boldsymbol{u},\theta) \in \left[W_0^{1,2}\right]^3 \times W_0^{1,2}$. From (2.7.9) we obtain $v_i = \lambda u_i - \bar{u}_i$ so that $v_i \in W_0^{1,2}$. We conclude that $w = (u_i, v_i, \theta) \in D$. The desired result is an immediate consequence of the Lummer-Phillips corollary to the Hille-Yosida theorem [365], p.343. □

Lemma 2.7.2. *There exists a positive constant α such that*

$$\|w\|_t \le e^{\alpha|t-s|}\|w\|_s\,, \qquad (2.7.13)$$

holds for any $w = (u_i, v_i, \theta) \in X$ *and each* $s, t \in [0, t_1]$.

Proof. By (2.4.1) and (2.2.7), $y(t) = \|w\|_t^2$. It follows from (2.4.7) that there exists a positive constant α such that

$$|\dot{y}| \leq 2\alpha y \, .$$

Thus

$$\dot{y} \leq 2\alpha y \, . \tag{2.7.14}$$

The desired result follows from (2.7.14). □

The preceding lemmas imply the following result.

Lemma 2.7.3. *The family of operators* $\{Q(t)\}$, $t \in [0, t_1]$, *is stable in the sense of Kato, with stability constants* $\exp(2\alpha t_1)$ *and* β.

We now consider the problem (2.2.6) where w_0 and G are given elements of a Banach space Y, continuously and densely embedded in X.

The above lemmas and the Kato theorem [216] lead to the following proposition.

Theorem 2.7.1. *Assume that the hypotheses (i) – (v) hold. Let* $F_i, S \in C^1([0, t_1];\ L_2(B_0))$. *Then, for any* $w_0 \in D$, *there exists a unique solution to the boundary-initial-value problem (2.2.1), (2.2.2) and (2.3.9) such that* $(u_i,\ \dot{u}_i,\ \theta) \in C^1([0, t_1];\ X) \cap C([0, t_1];\ D)$.

Remark. Existence theorems and asymptotic stability results, in the case when the primary state is a configuration of equilibrium at constant temperature and without stresses, were first established by Dafermos [91]. Other studies on the linear well-posedness and asymptotic stability have been made by Leis [234] and Racke [304], [307].

2.8 Thermoelastic waves

We assume that the body B_0 is homogeneous and isotropic. Further, we suppose that the primary state is obtained from B_0 by the isothermal uniform finite extensions

$$x_1 = \lambda_1 X_1 \, , \qquad x_2 = \lambda_1 X_2 \, , \qquad x_3 = \lambda_3 X_3, \qquad T = T_0 \, , \tag{2.8.1}$$

where λ_i and T_0 are given constants. The special expression assumed by the formulae of Section 2.1 can be obtained without difficulty. Consequently only the final formulae will be recorded here. It follows from (2.1.36), (2.1.39), (2.1.48)-(2.1.50), (2.1.69) and (2.1.75) that

$$
\begin{aligned}
&s_{11} = a_{11}u_{1,1} + a_{12}u_{2,2} + a_{13}u_{3,3} - \beta_1\theta\,,\\
&s_{22} = a_{12}u_{1,1} + a_{11}u_{2,2} + a_{13}u_{3,3} - \beta_1\theta\,,\\
&s_{33} = a_{13}(u_{1,1} + u_{2,2}) + a_{33}u_{3,3} - \beta_3\theta\,,\\
&s_{23} = a_{44}u_{2,3} + a_{45}u_{3,2}\,, \qquad s_{32} = a_{54}u_{2,3} + a_{44}u_{3,2}\,,\\
&s_{31} = a_{54}u_{1,3} + a_{44}u_{3,1}\,, \qquad s_{13} = a_{44}u_{1,3} + a_{45}u_{3,1}\,,\\
&s_{12} = a_{66}u_{1,2} + a_{65}u_{2,1}\,, \qquad s_{21} = a_{65}u_{1,2} + a_{66}u_{2,1}\,,\\
&\gamma = J(\beta_1 u_{\alpha,\alpha} + \beta_3 u_{3,3} + a\theta)\,,\\
&\varphi_1 = k_1\theta_{,1}\,, \qquad \varphi_2 = k_1\theta_{,2}\,, \qquad \varphi_3 = k_3\theta_{,3}\,,
\end{aligned} \tag{2.8.2}
$$

where

$$
\begin{aligned}
&a_{11} = c_{11} + t_{11}\,, \quad a_{12} = c_{12}\,, \quad a_{13} = c_{13}\,, \quad a_{33} = c_{33} + t_{33}\,, \quad a_{44} = c_{44}\,,\\
&a_{45} = c_{44} + t_{11}\,, \quad a_{54} = c_{44} + t_{33}\,, \quad 2a_{66} = a_{11} - a_{12} - t_{11}\,, \quad a_{65} = a_{66} + t_{11}\,,\\
&c_{11} = 2\lambda_1^4 A_{11} + 2\lambda_1^4(\lambda_2^2 + \lambda_3^2)^2 A_{22} + 2\lambda_1^4\lambda_2^4\lambda_3^4 A_{33} +\\
&\qquad + 4\lambda_1^4(\lambda_2^2 + \lambda_3^2)A_{12} + 4\lambda_1^4\lambda_2^2\lambda_3^2 A_{13} + 4\lambda_1^4\lambda_2^2\lambda_3^2(\lambda_2^2 + \lambda_3^2)A_{23}\,,\\
&c_{12} = 2p + 2\Psi\lambda_1^2\lambda_2^2 + 2\lambda_1^2\lambda_2^2 A_{11} + 2\lambda_1^2\lambda_2^2(\lambda_1^2 + \lambda_3^2)(\lambda_2^2 + \lambda_3^2)A_{22} +\\
&\qquad + 2\lambda_1^4\lambda_2^4\lambda_3^4 A_{33} + 2\lambda_1^2\lambda_2^2\lambda_3^2(2\lambda_1^2\lambda_2^2 + \lambda_2^2\lambda_3^2 + \lambda_1^2\lambda_3^2)A_{23} +\\
&\qquad + 2\lambda_1^2\lambda_2^2\lambda_3^2(\lambda_1^2 + \lambda_2^2)A_{13} + 2\lambda_1^2\lambda_2^2(\lambda_1^2 + \lambda_2^2 + 2\lambda_3^2)A_{12}\,,\\
&c_{44} = -p - \lambda_2^2\lambda_3^2\Psi\,, \quad c_{55} = -p - \lambda_1^2\lambda_3^2\Psi\,, \quad c_{66} = -p - \lambda_1^2\lambda_2^2\Psi\,,\\
&\beta_1 = -\lambda_1^2 L_1 - \lambda_1^2(\lambda_1^2 + \lambda_3^2)L_2 - \lambda_1^4\lambda_3^2 L_3\,,\\
&\beta_3 = -\lambda_3^2 L_1 - 2\lambda_1^2\lambda_3^2 L_2 - \lambda_1^4\lambda_3^2 L_3\,,\\
&k_1 = \frac{1}{\lambda_3}\left[K_1 + \frac{1}{2}K_2(\lambda_1^2 - 1) + \frac{1}{4}K_3(\lambda_1^2 - 1)\right],\\
&k_3 = \frac{\lambda_3}{\lambda_1^2}\left[K_1 + \frac{1}{2}K_2(\lambda_3^2 - 1) + \frac{1}{4}K_3(\lambda_3^2 - 1)\right],\\
&t_{11} = t_{22} = \lambda_1^2\Phi + \lambda_1^2(\lambda_1^2 + \lambda_3^2)\Psi + p\,,\\
&t_{33} = \lambda_3^2\Phi + 2\lambda_1^2\lambda_3^2\Psi + p\,, \qquad t_{ij} = 0 \;\; (i \neq j)\,.
\end{aligned} \tag{2.8.3}
$$

Clearly, the coefficients $a_{rs}(r,s=1,2,\dots,6)$, β_i, a and k_i are constants. Substitution of (2.8.2) into the equations of motion and the energy equation yields the equation

$$
\begin{aligned}
&a_{11}u_{1,11}+\frac{1}{2}(a_{11}-d_{12})u_{1,22}+a_{54}u_{1,33}+\frac{1}{2}(a_{11}+d_{12})u_{2,12}+\\
&+(d_{13}+a_{45})u_{3,13}-\beta_1\theta_{,1}+\varrho F_1=\varrho\ddot{u}_1\,,\\
&\frac{1}{2}(a_{11}+d_{12})u_{1,12}+\frac{1}{2}(a_{11}-d_{12})u_{2,11}+a_{11}u_{2,22}+a_{54}u_{2,33}+\\
&+(d_{13}+a_{45})u_{3,23}-\beta_1\theta_{,2}+\varrho F_2=\varrho\ddot{u}_2\,,\\
&(d_{13}+a_{45})u_{\alpha,\alpha 3}+a_{45}\Delta_0 u_3+a_{33}u_{3,33}-\\
&-\beta_3\theta_{,3}+\varrho F_3=\varrho\ddot{u}_3\,,\\
&k_1\Delta_0\theta+k_3\theta_{,33}-aT_0\dot{\theta}-T_0(\beta_1\dot{u}_{\alpha,\alpha}+\beta_3\dot{u}_{3,3})=-\varrho S\,,
\end{aligned}
\tag{2.8.4}
$$

where Δ_0 is the two-dimensional Laplace operator and

$$
d_{12}=a_{12}-t_{11}\,,\qquad d_{13}=a_{13}-t_{11}\,,\qquad \varrho\lambda_1^2\lambda_3=\varrho_0\,.
$$

It follows from (2.1.52) that

$$
k_1\geq 0\,,\qquad k_3\geq 0\,.
\tag{2.8.5}
$$

a) **Plane waves.** We now apply the equations (2.8.4) to the study of plane harmonic waves. The results we give here are due to Flavin and Green [130]. These results generalize those given by Chadwick [50], p.280 in the classical theory of thermoelasticity.

Following [130], we define

$$
e=u_{\alpha,\alpha}\,,\qquad W=u_{1,2}-u_{2,1}\,,\qquad f=u_{3,3}\,.
\tag{2.8.6}
$$

We assume that $F_i=0$ and $S=0$. The equations (2.8.4) imply

$$
\frac{1}{2}(a_{11}-d_{12})\Delta_0 W+a_{54}W_{,33}=\varrho\ddot{W}\,,
\tag{2.8.7}
$$

$$
\begin{aligned}
&a_{11}\Delta_0 e+a_{54}e_{,33}-\varrho\ddot{e}+(d_{13}+a_{45})\Delta_0 f-\beta_1\Delta_0\theta=0\,,\\
&a_{45}\Delta_0 f+a_{33}f_{,33}-\varrho\ddot{f}+(d_{13}+a_{45})e_{,33}-\beta_3\theta_{,33}=0\,,\\
&k_1\Delta_0\theta+k_3\theta_{,33}-m\dot{\theta}-T_0(\beta_1\dot{e}+\beta_3\dot{f})=0\,,
\end{aligned}
\tag{2.8.8}
$$

where $m = aT_0$. From (2.8.3) we obtain

$$a_{11} - d_{12} = 2\lambda_1^2(\Phi + \lambda_3^2\Psi), \qquad a_{54} = \lambda_3^2(\Phi + \lambda_1^2\Psi). \tag{2.8.9}$$

We seek W of the form

$$W = \mathrm{Re}\,\{w_0 \exp\,[i(\eta l_j x_j - \omega t)]\}, \tag{2.8.10}$$

which represents a plane wave. Here, Re $\{f\}$ is the real part of f, $\boldsymbol{l}$ is a unit vector, η is the wave number, and w_0 a constant amplitude. We define wave velocity by $v_2 = \omega/\eta$. It follows from (2.8.7) and (2.8.10) that

$$v_2^2 = \lambda_1^2\lambda_3\{[\lambda_1^2(1 - l_3^2) + \lambda_3^2 l_3^2]\Phi + \lambda_1^2\lambda_3^2\Psi\}/\varrho_0. \tag{2.8.11}$$

Clearly, (2.8.10) represents a transverse wave, with vibrations in planes parallel to the plane x_1Ox_2. This wave is independent of thermal effects. If the initial deformation is zero, then v_2 reduces to $(\mu/\varrho_0)^{\frac{1}{2}}$, where μ is the second Lamé constant. We now consider the equations (2.8.8). Let

$$u_\alpha = (f_1 + f_2)_{,\alpha}, \qquad u_3 = (h_1 f_1 + h_2 f_2)_{,3},$$

wheref_α are class C^2 fields on $B \times (0, t_1)$, and h_α are given by

$$h_\alpha = (a_{11}m_\alpha - a_{54})/(d_{13} + a_{45}) = (d_{13} + a_{45})m_\alpha/(a_{33} - a_{45}m_\alpha),$$

m_1 and m_2 being the roots of the equation

$$a_{11}a_{45}m^2 + [(d_{13} + a_{45})^2 - a_{11}a_{33} - a_{45}a_{54}]m + a_{33}a_{54} = 0. \tag{2.8.12}$$

It follows from (2.8.6) that

$$W = 0, \qquad e = \Delta_0(f_1 + f_2), \qquad f = (h_1 f_1 + h_2 f_2)_{,33}.$$

The equations (2.8.8) are satisfied if

$$\begin{gathered}
a_{11}(\Delta_0 f_1 + m_1 f_{1,33}) - \varrho\ddot{f}_1 + a_{11}(\Delta_0 f_2 + m_2 f_{2,33}) - \varrho\ddot{f}_2 - \beta_1\theta = 0, \\
r_2(\Delta_0 f_1 + m_1 f_{1,33}) - \varrho h_1\ddot{f}_1 + r_1(\Delta_0 f_2 + m_2 f_{2,33}) - \varrho h_2\ddot{f}_2 - \beta_3\theta = 0, \\
k_1(\Delta_0\theta + b\theta_{,33}) - m\dot{\theta} - T_0(\beta_1\Delta_0\dot{f}_1 + \beta_3 h_1\dot{f}_{1,33} + \beta_1\Delta_0\dot{f}_2 + \beta_3 h_2\dot{f}_{2,33}) = 0,
\end{gathered} \tag{2.8.13}$$

where

$$b = k_3/k_1, \qquad r_\alpha = m_\alpha a_{11}/h_\alpha. \tag{2.8.14}$$

We seek solutions to (2.8.13) of the type

$$(f_1,\ f_2,\ \theta) = \mathrm{Re}\left\{(a_1,\ a_2,\ a_3)\exp\left[i(\eta l_s x_s - \omega t)\right]\right\}, \qquad l^2 = 1\,. \tag{2.8.15}$$

If we put

$$\begin{gathered}
A_\alpha = 1 - l_3^2 + m_\alpha l_3^2\,, \qquad A_3 = 1 - l_3^2 + b l_3^2\,, \\
\varrho v_1^2 = a_{11}\,, \qquad k_1\omega^* = m v_1^2\,, \qquad \chi = \omega/\omega^*\,, \\
\xi = v_1\eta/\omega^*\,, \qquad g_i = -\beta_i/m \quad (i = 1,3)\,, \qquad h = mT_0/a_{11}\,,
\end{gathered} \tag{2.8.16}$$

the condition for the plane waves is

$$\det\,(\tilde{A}_{ij}) = 0\,, \tag{2.8.17}$$

where

$$\begin{gathered}
\tilde{A}_{11} = A_1\xi^2 - \chi^2\,, \quad \tilde{A}_{12} = A_2\xi^2 - \chi^2\,, \quad \tilde{A}_{13} = -g_1\,, \quad \tilde{A}_{23} = -g_2\,, \\
\tilde{A}_{21} = \frac{m_2}{h_2}A_1\xi^2 - h_1\chi^2\,, \quad \tilde{A}_{22} = \frac{m_1}{h_1}A_1\xi^2 - h_2\chi^2\,, \\
\tilde{A}_{3\alpha} = -ih\xi^2\chi[g_1(1 - l_3^2) + g_3 h_\alpha l_3^2]\,, \quad \tilde{A}_{33} = A_3\xi^2 - i\chi\,.
\end{gathered} \tag{2.8.18}$$

We examine the roots of equation (2.8.17) when $\chi \ll 1$. It follows that one root of (2.8.17) is given by

$$\xi^2 = iA_3^{-1}(1 + s)\chi + O(\chi^2)\,, \tag{2.8.19}$$

where

$$\begin{aligned}
s = h(A_1A_2)^{-1}\{&l_3^2(1 - l_3^2)(g_3 - m_1g_1/h_1)(g_3 - m_2g_2/h_2)a_{11}/a_{55} - \\
&+ [g_1(1 - l_3^2) + l_3^2 h_1 g_3][g_1(1 - l_3^2) + l_3^2 h_2 g_3]\}\,.
\end{aligned}$$

The remaining roots are given by

$$\xi^2 = z\chi^2 + O(\chi^3)\,, \tag{2.8.20}$$

where z is defined by the equation

$$\begin{aligned}
&(1 + s)A_1A_2z^2 - \left\{1 - l_3^2(1 - h_1h_2) + (a_{11}/a_{45})\left[(1 + hg_1^2)(1 - l_3^2) + \right.\right. \\
&\left.\left. + (hg_3^2 + m_1m_2/h_1h_2)l_3^2\right]\right\}z + a_{11}/a_{45} = 0\,.
\end{aligned} \tag{2.8.21}$$

In the limiting case of no initial deformation the value of ξ^2 given by (2.8.19) reduces to the quasi-thermal mode of classical theory of thermoelasticity (see Section 3.2).

Moreover, the roots of the equation (2.8.21) become $z_1 = 1/(1 + hg_1^2)$ and $z_2 = (\lambda + 2\mu)/\mu$. The wave corresponding to (2.8.20), with $z = z_1$, reduces to the longitudinal wave of the classical thermoelasticity, and the wave corresponding to (2.8.20), with $z = z_2$, reduces to the ordinary transverse wave. We now discuss two special cases in which the equation (2.8.17) simplifies.

i) *Plane waves propagating in the direction of the x_3-axis.* In this case $l_3 = 1$, and the equation (2.8.17) becomes

$$(\xi^2 a_{54} - \chi^2 a_{11})\big[(b\xi^2 - i\chi)(\xi^2 a_{33} - \chi^2 a_{11}) - i\xi^2 g_3^2 h a_{11}\chi\big] = 0\,. \tag{2.8.22}$$

The solution

$$\xi^2 = \chi^2 a_{11}/a_{54}\,, \tag{2.8.23}$$

corresponds to a transverse wave propagated with velocity

$$(a_{54}/\varrho)^{1/2} = \lambda_1\lambda_3[\lambda_3(\Phi + \lambda_1^2\Psi)/\varrho_0]^{1/2}\,. \tag{2.8.24}$$

This velocity coincides with that given by (2.8.11) when $l_3 = 1$. The remaining roots of (2.8.22) are

$$\xi^2 = \chi^2(hg_3^2 + a_{33}/a_{11})^{-1} + O(\chi^3)\,,$$

and

$$\xi^2 = i(1 + ha_{11}g_3^2/a_{33})\chi/b + O(\chi^2)\,.$$

ii) *Plane waves propagating in planes parallel to the x_1Ox_2-plane.* In this case $l_3 = 0$, and the equation (2.8.17) has a solution given by

$$\xi^2 = \chi^2 a_{11}/a_{45}\,. \tag{2.8.25}$$

The corresponding wave velocity is

$$(a_{45}/\varrho)^{1/2} = \big[(\lambda_1^2\Phi + \lambda_1^4\Psi)/\varrho\big]^{1/2}\,. \tag{2.8.26}$$

We note that the transverse wave velocity given by (2.8.11) is, when $l_3 = 0$, equal to

$$\lambda_1^2[\lambda_3(\Phi + \lambda_3^2\Psi)/\varrho_0]^{1/2}\,.$$

The remaining roots of (2.8.17) when $l_3 = 0$, are

$$\xi^2 = \chi^2(1 + hg_1^2)^{-1} + O(\chi^3)\,, \qquad \xi^2 = i(1 + hg_1^2)\chi + O(\chi^2)\,.$$

b) **Rayleigh waves.** We now study the propagation of thermoelastic Rayleigh waves in a half space which has been subjected to large uniform extensions at constant temperature, given by (2.8.1). The result we give here is due to Flavin [131]. Thermoelastic Rayleigh waves within the classical theory have been studied by Chadwick [50], p.307.

We assume that the body B occupies the half space $x_3 \geq 0$, and the stress component t_{33} is zero. We then consider a thermoelastic deformation of B such that

$$u_1 = u_1(x_1,\, x_3,\, t)\,, \quad u_2 = 0, \quad u_3 = u_3(x_1,\, x_3,\, t)\,, \quad \theta = \theta(x_1,\, x_3,\, t)\,. \tag{2.8.27}$$

We assume that the body forces and the heat supply are absent. Substitution of (2.8.27) into the equations (2.8.4) yields the equations

$$\begin{aligned}
&a_{11}u_{1,11} + a_{54}u_{1,33} + (d_{13} + a_{45})u_{3,13} - \beta_1\theta_{,1} = \varrho\ddot{u}_1\,,\\
&(d_{13} + a_{45})u_{1,13} + a_{45}u_{3,11} + a_{33}u_{3,33} - \beta_3\theta_{,3} = \varrho\ddot{u}_3\,,\\
&k_1\theta_{,11} + k_3\theta_{,33} - m\dot{\theta} - T_0(\beta_1\dot{u}_{1,1} + \beta_3\dot{u}_{3,3}) = 0\,,
\end{aligned} \tag{2.8.28}$$

on $B \times (0, t_1)$. We assume that the boundary $x_3 = 0$ is a free surface so that

$$s_{3i} = 0 \qquad \text{on } x_3 = 0\,. \tag{2.8.29}$$

In addition we have the thermal boundary condition

$$\theta_{,3} + h_0\theta = 0 \qquad \text{on } x_3 = 0\,, \tag{2.8.30}$$

where h_0 is a given constant. The condition (2.8.30) corresponds to a thermally radiating boundary.

We seek solutions of the form

$$(u_1,\, u_3,\, \theta) = \text{Re}\,\left[(u_1^\circ,\, u_3^\circ,\, \theta^\circ)\exp\left\{[i(x_1 - ct) - qx_3]\omega/c\right\}\right]. \tag{2.8.31}$$

Clearly (2.8.31) represents a Rayleigh wave of frequency $\omega/(2\pi)$. The physical reasoning requires that

$$\text{Re}\,(q/c) > 0\,. \tag{2.8.32}$$

Since $t_{33} = 0$, we conclude from (2.8.2) and (2.8.27) that the boundary conditions (2.8.29) reduce to

$$a_{13}u_{1,1} + a_{33}u_{3,3} = \beta_1\theta\,, \qquad u_{1,1} + u_{3,3} = 0 \quad \text{on } x_3 = 0\,. \tag{2.8.33}$$

It follows from (2.8.28) and (2.8.31) that

$$u_1 = \mathrm{Re}\left\{i(v_1/\omega^*)\sum_{s=1}^{3}[(g_3b_4 - g_1b_1)q_s^2 - g_1(y^2 - b_2)]A_sE_s\right\},$$
$$u_3 = \mathrm{Re}\left\{i(v_1/\omega^*)\sum_{s=1}^{3}[g_3(b_2q_s^2 - 1 + y^2) + g_1b_4]q_sA_sE_s\right\}, \tag{2.8.34}$$
$$\theta = \mathrm{Re}\left\{(T_0/hy)\chi\sum_{s=1}^{3}A_sL(q_s)E_s\right\},$$

where we have used the notations (2.8.14), (2.8.16), and

$$\begin{aligned} &E_s = \exp\left\{[i(x_1 - ct) - q_sx_3]\omega/c\right\}, \\ &L(q) = (b_2q^2 - 1 + y^2)(b_1q^2 - b_3 + y^2) + b_4^2q^2, \\ &b_1 = a_{33}/a_{11}, \qquad b_2 = a_{54}/a_{11}, \qquad b_3 = a_{45}/a_{11}, \\ &b_4 = (d_{13} + a_{45})/a_{11}, \qquad y = c/v_1. \end{aligned} \tag{2.8.35}$$

In the relations (2.8.34), A_s are arbitrary constants and q_j are the roots of the equation

$$\begin{aligned} &[b_2q^2 - (1 + hg_1^2) + y^2][(b_1 + hg_3^2)q^2 - b_3 + y^2]y^2 + \\ &\qquad + (b_4 + hg_1g_3)^2q^2y^2 - i\chi(bq^2 - 1)L(q) = 0. \end{aligned} \tag{2.8.36}$$

In view of (2.8.30), (2.8.33) and (2.8.34), the boundary conditions on $x_3 = 0$ reduce to

$$p_{ij}A_j = 0, \tag{2.8.37}$$

where

$$\begin{aligned} p_{1s} &= b_2q_s^2[g_3(y^2 - b_3 + b_4) - g_1b_1] + (y^2 - b_3)[g_3(y^2 - 1) + \\ &\quad + g_1(b_4 - b_2)], \\ p_{2s} &= [g_3(b_4 - b_2) - g_1b_1]q_s^3 - [g_1(y^2 - b_3 + b_4) + \\ &\quad + g_3(y^2 - 1)]q_s, \\ p_{3s} &= [1 - \chi q_s/(jy)]L(q_s), \qquad j = h_0v_1/\omega^*. \end{aligned} \tag{2.8.38}$$

For non-zero solutions of (2.8.37) we have

$$\det(p_{ij}) = 0. \tag{2.8.39}$$

The equations (2.8.36) and (2.8.39) determine y and q_j in terms of χ. We now suppose that $\chi \ll 1$ and take $y^2 = O(1)$ as $\chi \to 0$. Then $q_3^2 = O(\chi^{-1})$, and the remaining roots of the equations (2.8.36) satisfy

$$\begin{aligned} b_2(b_1 + hg_3^2)(q_1^2 + q_2^2) &= -(b_4 + hg_1g_3)^2 + (b_1 + hg_3^2)(1 + \\ &\quad + hg_1^2 - y^2) + b_2(b_3 - y^2), \\ b_2(b_1 + hg_3^2)q_1^2q_2^2 &= (1 + hg_1^2 - y^2)(b_3 - y^2), \end{aligned} \tag{2.8.40}$$

where we have omitted the terms of order $O(\chi)$. Using (2.8.38) and (2.8.40) we can write the equation (2.8.39) in the form

$$(q_1 - q_2)G(y)[F(y) + O(\chi^{1/2})] = 0\,, \tag{2.8.41}$$

where

$$\begin{aligned} F(y) &= (y^2 - b_3 + b_2)(b_1 + hg_3^2)b_2q_1q_2 - \\ &\quad - \left[(1 + hg_1^2 - y^2)(b_1 + hg_3^2) - (b_4 + hg_1g_3 - b_2)^2\right](b_3 - y)^2, \\ G(y) &= g_3\left[g_3b_4 - g_1(b_1 - b_2)\right]y^2 + (g_3 - g_1b_4)(g_1b_1 - g_3b_4) - g_1g_3b_2b_3\,. \end{aligned}$$

If $q_1 = q_2$ or $G(y) = 0$, then $u_1 = u_3 = 0$ and $\theta = 0$. In view of (2.8.40) the equation (2.8.41) reduces to

$$\begin{aligned} &(b_3 - y^2)\left[(1 + hg_1^2 - y^2)(b_1 + hg_3^2) - (b_4 + hg_1g_3 - b_2)^2\right]^2 - \\ &\quad - b_2(b_1 + hg_3^2)(1 + hg_1^2 - y^2)(y^2 - b_3 + b_4)^2 + O(\chi^{1/2}) = 0\,. \end{aligned} \tag{2.8.42}$$

If we consider wave motion under isentropic or isothermal conditions, then we can show that the values of y^2 differ from the corresponding values derived from the equation (2.8.42) by terms of order $O(\chi^{1/2})$ only.

In [58], Chadwick has established interesting results concerning the propagation of plane harmonic waves in anisotropic bodies.

2.9 Thermoelastostatics

a) **The body force analogy.** In this section we consider the equilibrium theory of thermoelastic bodies. We assume that the primary state is a configuration of equilibrium at constant temperature. In the absence of time dependence, the fundamental system of field equations describing the behavior of a thermoelastic body, referred to the primary state, consists of the equations of equilibrium

$$s_{ji,j} + \varrho F_i = 0\,, \tag{2.9.1}$$

the equilibrium energy equation

$$\varphi_{i,i} + \varrho S = 0\,, \tag{2.9.2}$$

and the constitutive equations

$$\begin{aligned} s_{ji} &= d_{ijrs}u_{r,s} - \beta_{ij}\theta\,, \\ \varphi_i &= k_{ij}\theta_{,j}\,, \end{aligned} \tag{2.9.3}$$

on B. We consider the boundary conditions

$$\begin{aligned} &\boldsymbol{u} = \tilde{\boldsymbol{u}} \quad \text{on } \bar{S}_1\,, \qquad &\boldsymbol{s} = \tilde{\boldsymbol{p}} \quad \text{on } S_2\,, \\ &\theta = \tilde{\theta} \quad \text{on } \bar{S}_3\,, \qquad &q = \tilde{\varphi} \quad \text{on } S_4\,. \end{aligned} \tag{2.9.4}$$

We assume that: (i) d_{ijrs}, β_{ij} and k_{ij} are continuously differentiable on $\bar{B}$ and $d_{ijrs} = d_{rsij}$, $\beta_{ij} = \beta_{ji}$; (ii) ϱ is continuous and strictly positive on $\bar{B}$; (iii) $\boldsymbol{F}$ and S are continuous on $\bar{B}$; (iv) $\tilde{\boldsymbol{u}}$ is continuous on $\bar{S}_1$, and $\tilde{\theta}$ is continuous on $\bar{S}_3$; (v) $\tilde{\boldsymbol{p}}$ is piecewise regular on S_2, and $\tilde{\varphi}$ is positive regular on S_4.

Substitution of (2.9.3) into the equations (2.9.1), (2.9.2) and (2.9.4) yields the equations

$$(d_{ijrs}u_{r,s})_{,j} - (\beta_{ij}\theta)_{,j} + \varrho F_i = 0\,, \tag{2.9.5}$$

$$(k_{ij}\theta_{,j})_{,i} = -\varrho S\,, \tag{2.9.6}$$

on B, and the boundary conditions

$$u_i = \tilde{u}_i \ \text{ on } \bar{S}_1\,, \qquad (d_{ijrs}u_{r,s} - \beta_{ij}\theta)n_j = \tilde{p}_i \ \text{ on } S_2\,, \tag{2.9.7}$$

$$\theta = \tilde{\theta} \ \text{ on } \bar{S}_3\,, \qquad k_{ij}\theta_{,j}n_i = \tilde{\varphi} \ \text{ on } S_4\,. \tag{2.9.8}$$

We note that the above system is uncoupled in the sense that the temperature field can be found by solving the heat flow problem (2.9.6), (2.9.8). In this chapter, we shall treat the temperature field as having already been so determined. The ordered array $\Lambda = (F_i,\ \tilde{u}_i,\ \tilde{p}_i,\ \theta)$ is called external data system on $\bar{B}$ for the mixed problem of thermoelastostatics. By a solution of the mixed problem of thermoelastostatics we mean an admissible displacement field $\boldsymbol{u}$ on $\bar{B}$ that satisfies the equations (2.9.5) and the boundary conditions (2.9.7).

It follows from (2.9.5) and (2.9.7) that $\boldsymbol{u}$ is a solution of the mixed problem of thermoelastostatics corresponding to the external data system $\Lambda = (F_i,\ \tilde{u}_i,\ \tilde{p}_i,\ \theta)$ on $\bar{B}$ if and only if $\boldsymbol{u}$ is a solution of the mixed problem of elastostatics corresponding to the external data system $L = (F_i - (1/\varrho)(\beta_{ij}\theta)_{,j}\,,\ \tilde{u}_i,\ \tilde{p}_i + \beta_{ij}\theta n_j)$ on $\bar{B}$. Following the classical thermoelastostatics we refer to this result as *the body force analogy* (see, for example, Carlson [46], p.316). Thus, for every theorem in elastostatics, there is an associated result in the equilibrium theory of linear thermoelasticity.

b) **Thermoelastic stresses in homogeneously deformed bodies.** We assume throughout this section that the body B_0 is homogeneous and isotropic. We suppose that the primary state is obtained from B_0 by the isothermal homogeneous deformation (2.8.1). In addition, we assume throughout this section that the body forces F_i and the heat supply S are zero. We first establish a solution of the displacement-temperature equations of equilibrium. Then, this solution is used to solve a problem concerning with a penny-shaped crack in an infinite medium. The results we give here are due to England and Green [120].

The special expressions assumed by the relations (2.9.3) have already been given in (2.8.2). In absence of body forces and heat supply, the equations (2.9.5) and (2.9.6) reduces to

$$\begin{gathered}
a_{11}u_{1,11}+\frac{1}{2}(a_{11}-d_{12})u_{1,22}+a_{54}u_{1,33}+\frac{1}{2}(a_{11}+d_{12})u_{2,12}+\\
+(d_{13}+a_{45})u_{3,13}-\beta_1\theta_{,1}=0\,,\\
\frac{1}{2}(a_{11}+d_{12})u_{1,12}+\frac{1}{2}(a_{11}-d_{12})u_{2,11}+a_{11}u_{2,22}+a_{45}u_{2,33}+\\
+(d_{13}+a_{45})u_{3,23}-\beta_1\theta_{,2}=0\,,\\
(d_{13}+a_{45})u_{\alpha,\alpha 3}+a_{45}\Delta_0 u_3+a_{33}u_{3,33}-\beta_3\theta_{,3}=0\,,
\end{gathered} \tag{2.9.9}$$

and

$$k_1\Delta_0\theta+k_3\theta_{,33}=0\,, \tag{2.9.10}$$

respectively. The constants $a_{rs}(r,s=1,2,\dots,6)$, β_1, β_3, d_{12} and d_{13} are given in Section 2.8. Since θ is assumed to be known we first seek a particular solution of the equations (2.9.9) which expresses u_i in terms of θ. The general solution is then found by adding to this particular solution the general solution of (2.9.9) when $\theta=0$.

We seek a particular solution in the form (England and Green [120])

$$u_\alpha=r_1G_{,\alpha}\,,\qquad u_3=r_3G_{,3}\,, \tag{2.9.11}$$

where G is a class C^2 field on B. The equations (2.9.9) are satisfied if

$$\begin{gathered}
r_1a_{11}\Delta_0G+[r_1a_{54}+r_3(d_{13}+a_{45})]G_{,33}-\beta_1\theta=0\,,\\
[r_1(d_{13}+a_{45})+r_3a_{45}]\Delta_0G+r_3a_{33}G_{,33}-\beta_3\theta=0\,.
\end{gathered} \tag{2.9.12}$$

Recalling (2.9.10) we assume that G is any particular solution of the equations

$$k_1\Delta_0 G + k_3 G_{,33} = 0\,, \qquad G_{,33} = \theta\,. \tag{2.9.13}$$

We suppose that $k_1 \neq 0$, $k_2 \neq 0$. Equations (2.9.12) and (2.9.13) are compatible if

$$\begin{aligned} (a_{54} - a_{11}b)r_1 + (d_{13} + a_{45})r_3 &= \beta_1\,, \\ -(d_{13} + a_{54})br_1 + (a_{33} - a_{45}b)r_3 &= \beta_3\,, \end{aligned}$$

where $b = k_3/k_1$. We obtain

$$\begin{aligned} a_{11}a_{45}(m_1 - b)(m_2 - b)r_1 &= \beta_1(a_{33} - a_{45}b) - \beta_3(d_{13} + a_{45})\,, \\ a_{11}a_{45}(m_1 - b)(m_2 - b)r_3 &= \beta_3(a_{54} - a_{11}b) - \beta_1 b(d_{13} + a_{45})\,, \end{aligned} \tag{2.9.14}$$

where m_α have been defined in Section 2.8. We assume that a_{11} and a_{45} are non-zero. The particular solution is degenerate when $m_1 = b$ or $m_2 = b$. In this case there is a particular solution of the form

$$u_\alpha = d_1 x_3 G_{,\alpha 3}\,, \qquad u_3 = d_2 x_3 G_{,33} + d_3 G_{,3}\,, \tag{2.9.15}$$

where G is any particular integral of (2.9.15) and d_i are given by

$$\begin{aligned} D_1 d_1 &= k_1\big[(d_{13} + a_{45})k_1\beta_3 - (k_1 a_{33} - k_3 a_{45})\beta_1\big]\,, \\ D_1 d_2 &= k_1\big[(k_3 a_{11} - k_1 a_{54})\beta_3 - (d_{13} + a_{45})k_3\beta_1\big]\,, \\ (d_{13} + a_{45})D_1 d_3 &= \beta_1(a_{11}k_3 - a_{54}k_1)(k_1 a_{33} + k_3 a_{45}) - \\ &\quad - k_1\beta_3(d_{13} + a_{45})(k_1 a_{54} + k_3 a_{11})\,, \\ D_1 &= 2(k_3^2 a_{11}a_{45} - k_1^2 a_{33}a_{54})\,. \end{aligned} \tag{2.9.16}$$

When the equation (2.8.12) has equal roots the particular solution (2.9.15) degenerates. In this case there is a particular solution of the form

$$u_\alpha = e_1 x_3^2 G_{,\alpha 3}\,, \qquad u_3 = e_2 x_3^2 G_{,33} + e_3 x_3 G_{,3} + e_4 G_{,3}\,, \tag{2.9.17}$$

where G is any particular integral of (2.9.13) and

$$\begin{aligned} &8a_{33}a_{54}de_1 = -(d_{13} + a_{45})(\beta_3 d - \beta_1 b)\,, \qquad d = a_{54}/a_{45}\,, \\ &e_2 = de_1\,, \qquad 4a_{33}a_{54}dk_1 e_3 = (k_1 a_{54} + k_3 a_{11})(\beta_3 d - \beta_1 b)\,, \\ &4(d_{13} + a_{45})a_{54}k_3 e_4 = 4a_{54}k_3\beta_1 - (d_{13} + a_{45})dk_1(\beta_3 d - \beta_1 b)\,. \end{aligned}$$

Application. In what follows we study a problem concerning with a penny-shaped crack in an infinite medium. We deal only with non-degenerate case, since the solutions in the other cases may be found from this by a limiting process. The result we give here is due to England and Green [120]. We consider an infinite homogeneous and isotropic medium which is subject to the isothermal deformation (2.8.1) in which the stress component t_{33} is zero. We assume that in B_0 there is a penny-shaped crack of radius a_0 in the plane $X_3 = 0$ which becomes a crack of radius a in the same plane, where $a = \lambda_1 a_0$.

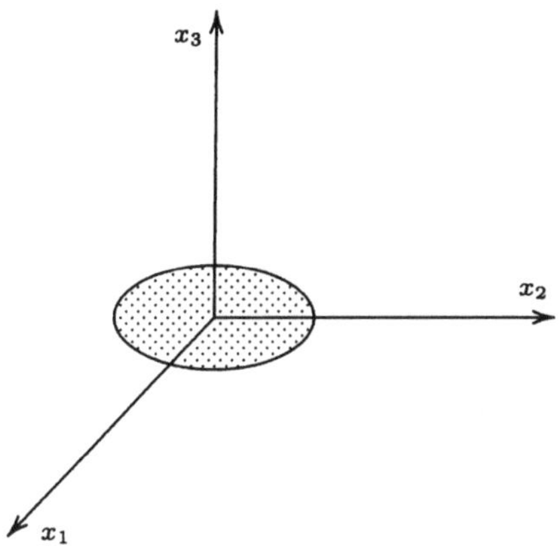

Fig. 2

We suppose that the equal distributions of temperature are applied to both surfaces of the crack. This problem is equivalent to the problem for the half-space $x_3 \geq 0$, which is obtained from the unstrained body by the finite deformation (2.8.1), and then subjected to the following conditions for $x_3 = 0$,

$$\begin{aligned} &s_{33} = 0\,, \quad 0 \leq r \leq a\,, \qquad u_3 = 0\,, \quad a < r\,, \\ &s_{3\alpha} = 0\,, \quad r \geq 0\,, \end{aligned} \tag{2.9.18}$$

$$\theta = f(r)\,, \quad 0 \leq r \leq a\,, \qquad \theta_{,3} = 0\,, \quad r > a\,, \tag{2.9.19}$$

where f is a prescribed function, and $r = (x_\alpha x_\alpha)^{1/2}$. We impose the further condition that s_{ij} and θ vanish at infinity.

The solution of the boundary-value problem (2.9.10), (2.9.19) is (cf. Green and Zerna [152], p. 175)

$$\theta = \frac{1}{2}\int_{-a}^{a} [r^2 + (z + it)^2]^{-1/2} h(t)\, dt\,, \tag{2.9.20}$$

where $z = x_3(k_1/k_3)^{1/2}$, and h is an even function given by

$$h = \frac{2}{\pi}\frac{d}{dt}\int_0^t (t^2 - r^2)^{-1/2} r f(r)\, dr\,. \tag{2.9.21}$$

It follows from (2.9.13) and (2.9.20) that

$$\begin{aligned} G = \frac{k_3}{2k_1}\int_{-a}^{a} &\left[(z+it)\ln\left\{z+it+\left[r^2+(z+it)^2\right]^{1/2}\right\} - \right. \\ &\left. - \left\{r^2+(z+it)\right\}^{1/2}\right] h(t)\, dt\,. \end{aligned} \tag{2.9.22}$$

Clearly, the second-order derivatives of G vanish at infinity. We seek the displacement field in the form

$$\begin{aligned} u_\alpha &= (g_1+g_2)_{,\alpha} + r_1 G_{,\alpha}\,, \\ u_3 &= (h_1 g_1 + h_2 g_2)_{,3} + r_3 G_{,3}\,, \end{aligned} \tag{2.9.23}$$

where g_α are class C^2 fields that satisfy

$$\left(\Delta_0 + m_\alpha \frac{\partial^2}{\partial x_3^2}\right) g_\alpha = 0 \qquad (\alpha \text{ not summed})\,, \tag{2.9.24}$$

and G satisfies the equations (2.9.13). It follows from (2.8.2), (2.8.3) and (2.9.23) that

$$\begin{aligned} s_{33} &= (c_{33}h_1 - c_{13}m_1) g_{1,33} + (c_{33}h_2 - c_{13}m_2) g_{2,33} + \\ &\quad + (c_{33}r_3 + c_{13}r_1 b) G_{,33} - \beta_3\theta\,, \\ s_{3\alpha} &= c_{44}\left[(1+h_1) g_{1,3\alpha} + (1+h_2) g_{2,3\alpha} + (r_1+r_3) G_{,3\alpha}\right]. \end{aligned} \tag{2.9.25}$$

The boundary conditions (2.9.18) are satisfied if

$$\begin{aligned} &(1+h_1) g_{1,3} + (1+h_2) g_{2,3} + (r_1+r_3) G_{,3} = 0\,, \qquad r \geq 0\,, \\ &h_1 g_{1,3} + h_2 g_{2,3} + r_3 G_{,3} = 0\,, \qquad r > a\,, \\ &(c_{33}h_1 - c_{13}m_1) g_{1,33} + (c_{33}h_2 - c_{13}m_2) g_{2,33} + (c_{33}r_3 + \\ &+ c_{13}r_1 b) G_{,33} = \beta_3\theta\,, \qquad 0 \leq r \leq a\,, \end{aligned} \tag{2.9.26}$$

for $x_3 = 0$. In view of equations (2.9.24) and (2.9.13) we put

$$\begin{aligned} g_1 &= d_1 G(x_1,\, x_2,\, v_1 x_3) + C_1 F(x_1,\, x_2,\, t_1 x_3)\,, \\ g_2 &= d_2 G(x_1,\, x_2,\, v_2 x_3) - C_2 F(x_1,\, x_2,\, t_2 x_3)\,, \end{aligned} \tag{2.9.27}$$

where d_α are constants and

$$t_\alpha = m_\alpha^{-1/2}, \qquad v_\alpha = t_\alpha (k_3/k_1)^{1/2}, \qquad C_\alpha^{-1} = t_\alpha (1 + h_\alpha). \tag{2.9.28}$$

The function F satisfies the equation

$$F_{,ii} = 0. \tag{2.9.29}$$

We choose the constants d_α so that

$$\begin{aligned} &(1 + h_1)v_1 d_1 + (1 + h_2)v_2 d_2 = -(r_1 + r_3), \\ &h_1 v_1 d_1 + h_2 v_2 d_2 = -r_3. \end{aligned} \tag{2.9.30}$$

Thus

$$d_1 = \frac{h_2 r_1 - r_3}{v_1(h_1 - h_2)}, \qquad d_2 = \frac{r_3 - h_1 r_1}{v_2(h_1 - h_2)}.$$

It follows from (2.9.27) and (2.9.30) that the boundary conditions (2.9.26) reduce to

$$F_{,3} = 0, \quad r > a, \qquad F_{,33} = Af, \quad 0 \le r \le a, \tag{2.9.31}$$

on $x_3 = 0$, where

$$\begin{aligned} &A = (\beta_3 - c_{33} r_3 - c_{13} r_1 b - A_1 d_1 v_1^2 - A_2 d_2 v_2^2)/(A_1 C_1 t_1^2 - A_2 C_2 t_2^2), \\ &A_\alpha = c_{33} h_\alpha - c_{13} m_\alpha. \end{aligned} \tag{2.9.32}$$

The solution of the boundary-value problem (2.9.29), (2.9.31) is (cf. Green and Zerna [152], p. 174)

$$F = \frac{1}{2i} \int_{-a}^{a} g(t) \ln \left\{ x_3 + it + \left[r^2 + (x_3 + it)^2 \right]^{1/2} \right\} dt, \tag{2.9.33}$$

where g is an odd function defined by

$$g = 2A\pi^{-1} \int_0^t s f(s)(t^2 - s^2)^{-1/2}\, ds. \qquad t \ge 0. \tag{2.9.34}$$

The corresponding stresses may be found by using the relations (2.8.2). We note that all stress components vanish at infinity. The normal displacement at the surface of the crack is given by

$$u_3 = h_1 g_{1,3} + h_2 g_{2,3} + r_3 G_{,3}, \qquad 0 \le r \le a, \qquad x_3 = 0. \tag{2.9.35}$$

It follows from (2.9.27), (2.9.30) and (2.9.29) that

$$u_3 = (h_1 t_1 C_1 - h_2 t_2 C_2) F_{,3} \,, \qquad 0 \le r \le a \,, \qquad x_3 = 0 \,. \tag{2.9.36}$$

In the case when a constant temperature field is applied to the surface of the crack, so that $f = f_0$, where f_0 is a constant, the relation (2.9.34) yields

$$g = 2 A f_0 \pi^{-1} t \,. \tag{2.9.37}$$

It follows from (2.9.33), (2.9.36) and (2.9.37) that the normal displacement at the crack is given by

$$u_3 = -2 A f_0 \pi^{-1} (h_1 t_1 C_1 - h_2 t_2 C_2)(a^2 - r^2)^{1/2} \,, \qquad 0 \le r \le a \,.$$

The solution of the problem may now be completed by straightforward calculation, but details are left to the reader who may also refer to England and Green [120]. In the classical thermoelasticity, the problem of thermal stresses in an infinite elastic solid containing a penny-shaped crack was studied by Olesiak and Sneddon [287].

The thermoelastic crack problem has been intensively studied (see, e.g. Barber [15], Bui, Ehralcher and Nguyen [42], Clements [81], Fu [138], Sladek and Sladek [326] and the literature cited therein).

3 Thermoelastodynamics of Bodies without Initial Deformations

3.1 Basic equations

In what follows we assume that the primary state is identical with that of the initial body B_0. Moreover, we suppose that the primary state is a configuration of equilibrium at the constant temperature T_0 and without stresses. It follows from (2.1.77)–(2.1.79) or (2.2.14)–(2.2.17) that the basic system of field equations consists of the equations of motion

$$t_{ji,j} + f_i = \varrho \ddot{u}_i \,, \tag{3.1.1}$$

the energy equation

$$T_0 \dot{\eta} = q_{i,i} + S \,, \tag{3.1.2}$$

and the constitutive equations

$$\begin{aligned} t_{ij} &= C_{ijrs} u_{r,s} - \beta_{ij} \theta \,, \\ \eta &= \beta_{ij} u_{i,j} + a\theta \,, \\ q_i &= k_{ij} \theta_{,j} \,, \end{aligned} \tag{3.1.3}$$

on $B \times (0, t_1)$. Here we have used the notations $t_{ij} = s_{ij}$, $\eta = \gamma$, $q_i = \varphi_i$, $f_i = \varrho_0 F_i$, $\varrho = \varrho_0$ and we have denoted $\varrho_0 S$ by S. In the classical linear theory of thermoelasticity the Lagrangian coordinates are usually indicated by x_i.
The constitutive coefficients have the properties

$$C_{ijrs} = C_{jirs} = C_{rsij} \,, \quad \beta_{ij} = \beta_{ji} \,, \quad k_{ij}\xi_i\xi_j \geq 0 \qquad \text{for any } \xi_i \,. \tag{3.1.4}$$

The form of the free energy ψ may be obtained from (2.1.81),

$$\psi = \psi_0 - f\theta + \frac{1}{2}(C_{ijrs} e_{ij} e_{rs} - 2\beta_{ij} e_{ij}\theta - a\theta^2) \,, \tag{3.1.5}$$

where the coefficients ψ_0, f, C_{ijrs}, β_{ij} and a are, in general, functions of the coordinates x_i, and

$$e_{ij} = \frac{1}{2}(u_{i,j} + u_{j,i}) \,. \tag{3.1.6}$$

The constitutive coefficients are constants for homogeneous bodies.

It follows from (3.1.3) and (3.1.4) that the constitutive equations can be expressed in the form

$$\begin{aligned} t_{ij} &= C_{ijrs}e_{rs} - \beta_{ij}\theta\,, \\ \eta &= \beta_{ij}e_{ij} + a\theta\,, \qquad q_i = k_{ij}\theta_{,j}\,. \end{aligned} \tag{3.1.7}$$

k_{ij} is called *the conductivity tensor.*

When the body is isotropic, then from (2.1.80) we find that

$$\begin{aligned} C_{ijrs} &= \lambda\delta_{ij}\delta_{rs} + \mu(\delta_{ir}\delta_{js} + \delta_{is}\delta_{jr})\,, \\ \beta_{ij} &= \beta\delta_{ij}\,, \qquad k_{ij} = k\delta_{ij}\,, \end{aligned} \tag{3.1.8}$$

and the constitutive equations (3.1.7) become

$$\begin{aligned} t_{ij} &= \lambda e_{rr}\delta_{ij} + 2\mu e_{ij} - \beta\theta\delta_{ij}\,, \\ \eta &= \beta e_{rr} + a\theta\,, \\ q_i &= k\theta_{,i}\,, \end{aligned} \tag{3.1.9}$$

If $\mu \neq 0$ and $3\lambda + 2\mu \neq 0$, then $(3.9.1)_1$ can be inverted to yield

$$e_{ij} = \frac{1+\nu}{E}t_{ij} - \frac{\nu}{E}t_{rr}\delta_{ij} + \alpha\theta\delta_{ij}\,, \tag{3.1.10}$$

where

$$E = \frac{\mu(3\lambda + 2\mu)}{\lambda + \mu}\,, \qquad \nu = \frac{\lambda}{2(\lambda + \mu)}\,, \qquad a = \frac{\beta}{3\lambda + 2\mu}\,.$$

E is known as Young's modulus, ν is known as Poisson's ratio and α is called the coefficient of thermal expansion.

The surface traction $\boldsymbol{t}$ and the surface heat flux q are defined at every regular point of $\partial B \times [0, t_1)$ by

$$t_i = t_{ji}n_j\,, \qquad q = q_i n_i\,. \tag{3.1.11}$$

To the field equations (3.1.1)–(3.1.3) we add the initial conditions (2.3.1) and the boundary conditions

$$\begin{aligned} u_i &= \tilde{u}_i \quad \text{on } \bar{S}_1 \times [0,\, t_1)\,, \qquad t_{ji}n_j = \tilde{t}_i \quad \text{on } S_2 \times [0,\, t_1)\,, \\ \theta &= \tilde{\theta} \quad \text{on } \bar{S}_3 \times [0,\, t_1)\,, \qquad q_i n_i = \tilde{q} \quad \text{on } S_3 \times [0,\, t_1)\,. \end{aligned} \tag{3.1.12}$$

a) **Special results for homogeneous and isotropic bodies.** We now assume that the body is homogeneous and isotropic. From (3.1.6), (3.1.9), (3.1.1) and (3.1.2) we obtain the field equations expressed in terms of the displacement vector field and temperature variation

$$\begin{aligned} &\mu\Delta\boldsymbol{u} + (\lambda+\mu)\text{grad div }\boldsymbol{u} - \beta\text{grad }\theta - \varrho\ddot{\boldsymbol{u}} = -\boldsymbol{f}\,, \\ &k\Delta\theta - T_0\beta\text{div }\dot{\boldsymbol{u}} - \mathrm{a}\mathrm{T}_0\dot{\theta} = -\mathrm{S}\,, \end{aligned} \tag{3.1.13}$$

where Δ is the Laplacian. We introduce the notations

$$\begin{aligned} &\Box_\alpha = \Delta - \frac{1}{c_\alpha^2}\frac{\partial^2}{\partial t^2}\,, \qquad \bigcirc_1 = \frac{k}{c}\Delta - \frac{\partial}{\partial t}\,, \\ &\bigcirc_2 = \bigcirc_1 - \frac{\lambda+2\mu}{\lambda+\mu}\varepsilon\frac{\partial}{\partial t}\,, \qquad c_1 = \sqrt{\frac{\lambda+2\mu}{\varrho}}\,, \\ &c_2 = \sqrt{\frac{\mu}{\varrho}}\,, \qquad c = aT_0\,, \qquad \varepsilon = \frac{\beta^2 T_0}{c\varrho c_1^2}\,. \end{aligned} \tag{3.1.14}$$

The equations (3.1.13) can be expressed in the form

$$\begin{aligned} &c_2^2\Box_2\,\boldsymbol{u} + (c_1^2 - c_2^2)\text{grad div }\boldsymbol{u} - \frac{\beta}{\varrho}\text{grad }\theta = -\frac{1}{\varrho}\boldsymbol{f}\,, \\ &\bigcirc_1\,\theta - \frac{1}{c}T_0\beta\text{div }\dot{\boldsymbol{u}} = -\frac{1}{c}S\,. \end{aligned} \tag{3.1.15}$$

The basic equations of the approximate *uncoupled theory* are

$$\mu\Delta\boldsymbol{u} + (\lambda+\mu)\text{grad div }\boldsymbol{u} - \beta\text{grad }\theta - \varrho\ddot{\boldsymbol{u}} = -\boldsymbol{f}\,,$$

$$k\Delta\theta - aT_0\dot{\theta} = -S\,.$$

The next theorem provides a solution of Galerkin type of the field equations.

Theorem 3.1.1. *Let*

$$\begin{aligned} \boldsymbol{u} =& (\lambda+2\mu)\Big[\bigcirc_1\Box_1 - \varepsilon\Delta\frac{\partial}{\partial t}\Big]\boldsymbol{\Phi} - (\lambda+\mu)\bigcirc_2\,\text{grad div }\boldsymbol{\Phi} + \\ &+ \frac{\beta}{\lambda+2\mu}\text{grad }\Psi\,, \\ \theta =& \frac{\beta T_0\mu}{c}\Box_2\frac{\partial}{\partial t}\text{div }\boldsymbol{\Phi} + \Box_1\,\Psi\,, \end{aligned} \tag{3.1.16}$$

where the fields Φ_i *of class* $C^{6,5}$ *and* Ψ *of class* $C^{4,3}$ *on* $B \times (0, t_1)$ *satisfy*

$$\begin{aligned} &\Box_2 \left(\Box_1 \Box_1 - \varepsilon \Delta \frac{\partial}{\partial t} \right) \Phi_i = -\frac{1}{\mu(\lambda + 2\mu)} f_i , \\ &\left(\Box_1 \Box_1 - \varepsilon \Delta \frac{\partial}{\partial t} \right) \Psi = -\frac{1}{c} S . \end{aligned} \tag{3.1.17}$$

Then $\boldsymbol{u}$ *and* θ *satisfy the equations (3.1.15).*
Proof. We note that the following identities

$$\begin{aligned} &c_2^2(c_1^2 - c_2^2) \Box_2 \Box_2 + (c_1^2 - c_2^2)^2 \Box_2 \Delta + \\ &+ \varepsilon c_1^2 c_2^2 \Box_2 \frac{\partial}{\partial t} = c_1^2(c_1^2 - c_2^2) \left(\Box_1 \Box_1 - \varepsilon \Delta \frac{\partial}{\partial t} \right) , \\ &c_2^2 c_1^{-2} \Box_2 + (c_1^2 - c_2^2) c_1^{-2} \Delta = \Box_1 , \end{aligned} \tag{3.1.18}$$

hold. Now, we substitute $\boldsymbol{u}$ and θ given by (3.1.16) in the left-hand side of (3.1.15). In view of (3.1.18) and (3.1.17) we obtain

$$\begin{aligned} &c_2^2 \Box_2 \boldsymbol{u} + (c_1^2 - c_2^2) \text{grad div } \boldsymbol{u} - \frac{\beta}{\varrho} \text{grad } \theta = \\ &= \varrho c_1^2 c_2^2 \Box_2 \left(\Box_1 \Box_1 - \varepsilon \Delta \frac{\partial}{\partial t} \right) \boldsymbol{\Phi} - \Big[\varrho c_1^2 (c_2^2 - c_1^2) \left(\Box_1 \Box_1 - \varepsilon \Delta \frac{\partial}{\partial t} \right) + \\ &+ \varrho c_2^2 (c_1^2 - c_2^2) \Box_2 \Box_2 + \varrho (c_1^2 - c_2^2)^2 \Box_2 \Delta + \varepsilon \varrho c_1^2 c_2^2 \Box_2 \frac{\partial}{\partial t} \Big] \text{grad div } \boldsymbol{\Phi} + \\ &+ \frac{\beta}{\varrho} \left[c_2^2 c_1^{-2} \Box_2 + (c_1^2 - c_2^2) c_1^{-2} \Delta - \Box_1 \right] \text{grad} \Psi = -\frac{1}{\varrho} \boldsymbol{f} . \end{aligned}$$

Similarly,

$$\begin{aligned} &\Box_1 \theta - \frac{1}{c} T_0 \beta \text{div } \dot{\boldsymbol{u}} = \left(\Box_1 \Box_1 - \varepsilon \Delta \frac{\partial}{\partial t} \right) \Psi + \Big[\frac{1}{\beta} \varepsilon \varrho^2 c_1^2 c_2^2 \Box_2 \Box_1 + \\ &+ \frac{1}{c} T_0 \beta \varrho (c_1^2 - c_2^2) \Box_2 \Delta - \frac{1}{c} T_0 \beta \varrho c_1^2 \left(\Box_1 \Box_1 - \varepsilon \Delta \frac{\partial}{\partial t} \right) \Big] \text{div } \frac{\partial}{\partial t} \boldsymbol{\Phi} = -\frac{1}{c} S . \; \Box \end{aligned}$$

Theorem 3.1.1 was first established by Nowacki [278]. In Section 3.2 we shall use this theorem to study the effect of concentrated loads in an unbounded domain.

Assume that $\boldsymbol{f}$ is of class $C^{2,1}$ on $B \times (0, t_1)$ and of class $C^{0,0}$ on $\bar{B} \times [0, t_1)$. Then we have Helmholtz resolution

$$\boldsymbol{f} = \operatorname{grad} F + \operatorname{curl} \boldsymbol{G}, \tag{3.1.19}$$

where F and G_i are fields of class $C^{2,1}$ on $B \times (0, t_1)$ and $\operatorname{div} \boldsymbol{G} = 0$.

Theorem 3.1.2. *Let*

$$\begin{aligned} \boldsymbol{u} &= \operatorname{grad} \varphi + \operatorname{curl} \boldsymbol{g}, \\ \theta &= \frac{1}{\beta}\left[(\lambda + 2\mu)\square_1 \varphi + F\right], \end{aligned} \tag{3.1.20}$$

where the fields φ *of class* $C^{4,3}$ *and* $\boldsymbol{g}$ *of class* $C^{3,2}$ *on* $B \times (0, t_1)$ *satisfy*

$$\begin{aligned} \left(\square_1 \Box_1 - \varepsilon \frac{\partial}{\partial t} \Delta\right)\varphi &= -\frac{\beta}{c(\lambda + 2\mu)} S - \frac{1}{\lambda + 2\mu} \Box_1 F, \\ \Box_2 \boldsymbol{g} &= -\frac{1}{\mu} \boldsymbol{G}. \end{aligned} \tag{3.1.21}$$

Then $\boldsymbol{u}$ *and* θ *satisfy the equations (3.1.13).*

Proof. We substitute the Helmholtz decomposition $\boldsymbol{u} = \operatorname{grad} \varphi + \operatorname{curl} \boldsymbol{g}$ into the equations $(3.1.13)_1$ and get

$$\operatorname{grad} \left[(\lambda + 2\mu)\Delta\varphi - \varrho\ddot{\varphi} - \beta\theta + F\right] + \operatorname{curl} \left[\mu\Delta\boldsymbol{g} - \varrho\ddot{\boldsymbol{g}} + \boldsymbol{G}\right] = 0.$$

This equation is satisfied in view of $(3.1.20)_2$ and $(3.1.21)_2$. If we substitute (3.1.20) into the equation $(3.1.13)_2$ we obtain

$$\left(\square_1 \Box_1 - \varepsilon \frac{\partial}{\partial t} \Delta\right)\varphi = -\frac{\beta}{c(\lambda + 2\mu)} S - \frac{1}{\lambda + 2\mu} \Box_1 F,$$

which is satisfied on the basis of (3.1.21). □

This result is due to Deresiewicz [106] and Zorski [371]. The completeness of solution (3.1.20) has been established in [46], p.330.

b) **Curvilinear coordinates.** In the solution of many boundary value problems in thermoelasticity, it is convenient to employ orthogonal curvilinear coordinates. We shall list here the commonly used formulae for cylindrical coordinates and spherical polar coordinates.

We note that in the linear theory the stress tensors τ_{ij}, S^{ij}, π^{ij} and p^{ij} coincide, and that $Q^i = q^i$. The relations (1.6.12) become

$$2\gamma_{ij} = u_{i|j} + u_{j|i}\,. \tag{3.1.22}$$

The form of the free energy in curvilinear coordinates is

$$\psi = \psi_0 - f\theta + \frac{1}{2}\left(C^{ijrs}\gamma_{ij}\gamma_{rs} - 2B^{ij}\gamma_{ij}\theta - a\theta^2\right).$$

By using the results from Section 1.6 we obtain the following constitutive equations

$$\tau^{ij} = C^{ijrs}\gamma_{rs} - B^{ij}\theta\,,$$

$$\eta = B^{ij}\gamma_{ij} + a\theta\,,$$

$$q^i = K^{ij}\theta_{|j}\,.$$

The equations of motion become

$$\tau^{ji}|_j + \varrho\hat{f}^i = \varrho\hat{a}^i\,, \tag{3.1.23}$$

and the energy equation can be expressed in the form

$$T_0\dot{\eta} = q^i|_i + S\,. \tag{3.1.24}$$

When the body is isotropic the constitutive equations become

$$\begin{aligned} \tau^{ij} &= \lambda\gamma^r_r g^{ij} + 2\mu\gamma^{ij} - \beta\theta g^{ij}\,, \\ \eta &= \beta\gamma^r_r + a\theta\,, \\ q^i &= kg^{ir}\theta_{|r}\,, \end{aligned} \tag{3.1.25}$$

where $\gamma^i_j = g^{ir}\gamma_{rj}\,,\quad \gamma^{ij} = g^{rj}\gamma^i_r\,.$

Let us consider *the cylindrical coordinates* (r, φ, z) such that $X_1 = r\cos\varphi$, $X_2 = r\sin\varphi$, $X_3 = z$. Then

$$\begin{aligned} &g_{11} = g_{33} = 1\,,\quad g_{22} = r^2\,,\quad g^{11} = g^{33} = 1\,,\quad g^{22} = \frac{1}{r^2}\,, \\ &g_{ij} = g^{ij} = 0\ (i \neq j)\,,\quad g = r^2\,,\quad \boldsymbol{g}^1 = \boldsymbol{g}_1\,,\quad \boldsymbol{g}^3 = \boldsymbol{g}_3\,, \\ &\boldsymbol{g}^2 = \frac{1}{r^2}\boldsymbol{g}_2\,,\quad \Gamma_{122} = \Gamma_{212} = r\,,\quad \Gamma_{221} = -r\,, \end{aligned} \tag{3.1.26}$$

while the other Christoffel symbols Γ_{ijk} are zero. We denote

$$\boldsymbol{\varepsilon}_r = \boldsymbol{g}_1 , \qquad \boldsymbol{\varepsilon}_\varphi = \frac{1}{r}\boldsymbol{g}_2 , \qquad \boldsymbol{\varepsilon}_z = \boldsymbol{g}_3 ,$$

$$\boldsymbol{u} = u^i\boldsymbol{g}_i = u_i\boldsymbol{g}^i = u_r\boldsymbol{\varepsilon}_r + u_\varphi\boldsymbol{\varepsilon}_\varphi + u_z\boldsymbol{\varepsilon}_z .$$

Here, u_r u_φ and u_z are the physical components of the displacement vector. We have

$$u^1 = u_1 = u_r , \qquad u^2 = r^{-2}u_2 = r^{-1}u_\varphi , \qquad u^3 = u_3 = u_z . \tag{3.1.27}$$

The physical components of the acceleration vector field are given by

$$a_r = \ddot{u}_r , \qquad a_\varphi = r^{-1}\ddot{u}_\varphi , \qquad a_z = \ddot{u}_z . \tag{3.1.28}$$

It follows from (3.1.22) and (3.1.26) that

$$\gamma_{11} = \frac{\partial u_r}{\partial r} , \qquad \gamma_{22} = r\frac{\partial u_\varphi}{\partial \varphi} + r u_r , \qquad \gamma_{33} = \frac{\partial u_z}{\partial z} ,$$

$$\gamma_{12} = \frac{1}{2}\Big(\frac{\partial u_r}{\partial \varphi} + r\frac{\partial u_\varphi}{\partial r} - u_\varphi\Big) , \qquad \gamma_{13} = \frac{1}{2}\Big(\frac{\partial u_r}{\partial z} + \frac{\partial u_z}{\partial r}\Big) ,$$

$$\gamma_{23} = \frac{1}{2}\Big(r\frac{\partial u_\varphi}{\partial z} + \frac{\partial u_z}{\partial \varphi}\Big) .$$

The physical components of the tensor γ_{ij} are

$$\varepsilon_{rr} = \gamma_{11} , \qquad \varepsilon_{\varphi\varphi} = r^{-2}\gamma_{22} , \qquad \varepsilon_{zz} = \gamma_{33} ,$$

$$\varepsilon_{r\varphi} = r^{-1}\gamma_{12} , \qquad \varepsilon_{\varphi z} = r^{-1}\gamma_{23} , \qquad \varepsilon_{rz} = \gamma_{13} .$$

Thus, we obtain the following geometrical relations

$$\begin{gathered} \varepsilon_{rr} = \frac{\partial u_r}{\partial r} , \quad \varepsilon_{\varphi\varphi} = \frac{1}{r}\Big(\frac{\partial u_\varphi}{\partial \varphi} + u_r\Big) , \quad \varepsilon_{zz} = \frac{\partial u_z}{\partial z} , \\ \varepsilon_{r\varphi} = \frac{1}{2}\Big(\frac{1}{r}\frac{\partial u_r}{\partial \varphi} + \frac{\partial u_\varphi}{\partial r} - \frac{1}{r}u_\varphi\Big) , \\ \varepsilon_{\varphi z} = \frac{1}{2}\Big(\frac{\partial u_\varphi}{\partial z} + \frac{1}{r}\frac{\partial u_z}{\partial \varphi}\Big) , \quad \varepsilon_{rz} = \frac{1}{2}\Big(\frac{\partial u_r}{\partial z} + \frac{\partial u_z}{\partial r}\Big) . \end{gathered} \tag{3.1.29}$$

We define E^i by

$$E^i = \tau^{ji}|_j \,.$$

In view of (3.1.26) we obtain

$$E^1 = \frac{\partial \tau^{11}}{\partial r} + \frac{\partial \tau^{12}}{\partial \varphi} + \frac{\partial \tau^{13}}{\partial z} + \frac{1}{r}\left(\tau^{11} - r^2\tau^{22}\right),$$

$$E^2 = \frac{\partial \tau^{21}}{\partial r} + \frac{\partial \tau^{22}}{\partial \varphi} + \frac{\partial \tau^{23}}{\partial z} + \frac{3}{r}\tau^{12},$$

$$E^3 = \frac{\partial \tau^{13}}{\partial r} + \frac{\partial \tau^{23}}{\partial \varphi} + \frac{\partial \tau^{33}}{\partial z} + \frac{1}{r}\tau^{13}.$$

The physical components of the stress tensor τ^{ij} are

$$\begin{aligned} \sigma_{rr} &= \tau^{11}, & \sigma_{\varphi\varphi} &= r^2\tau^{22}, & \sigma_{zz} &= \tau^{33}, \\ \sigma_{r\varphi} &= r\tau^{12}, & \sigma_{\varphi z} &= r\tau^{23}, & \sigma_{rz} &= \tau^{13}, \end{aligned} \tag{3.1.30}$$

so that we can write

$$\begin{aligned} E^1 &= \frac{\partial \sigma_{rr}}{\partial r} + \frac{1}{r}\frac{\partial \sigma_{r\varphi}}{\partial \varphi} + \frac{\partial \sigma_{rz}}{\partial z} + \frac{1}{r}(\sigma_{rr} - \sigma_{\varphi\varphi}), \\ E^2 &= \frac{1}{r}\left(\frac{\partial \sigma_{r\varphi}}{\partial r} + \frac{1}{r}\frac{\partial \sigma_{\varphi\varphi}}{\partial \varphi} + \frac{\partial \sigma_{\varphi z}}{\partial z} + \frac{2}{r}\sigma_{r\varphi}\right), \\ E^3 &= \frac{\partial \sigma_{rz}}{\partial r} + \frac{1}{r}\frac{\partial \sigma_{\varphi z}}{\partial \varphi} + \frac{\partial \sigma_{zz}}{\partial z} + \frac{1}{r}\sigma_{rz}. \end{aligned} \tag{3.1.31}$$

We denote by f_r, f_φ and f_z the physical components of the body force. Then

$$\hat{f}^1 = f_r, \qquad \hat{f}^2 = r^{-1}f_\varphi, \qquad \hat{f}^3 = f_z. \tag{3.1.32}$$

It follows from (3.1.23), (3.1.28), (3.1.31) and (3.1.32) that the equations of motion can be expressed in the form

$$\begin{aligned} &\frac{\partial \sigma_{rr}}{\partial r} + \frac{1}{r}\frac{\partial \sigma_{r\varphi}}{\partial \varphi} + \frac{\partial \sigma_{rz}}{\partial z} + \frac{1}{r}(\sigma_{rr} - \sigma_{\varphi\varphi}) + f_r = \varrho\frac{\partial^2 u_r}{\partial t^2}, \\ &\frac{\partial \sigma_{r\varphi}}{\partial r} + \frac{1}{r}\frac{\partial \sigma_{\varphi\varphi}}{\partial \varphi} + \frac{\partial \sigma_{\varphi z}}{\partial z} + \frac{2}{r}\sigma_{r\varphi} + f_\varphi = \varrho\frac{\partial^2 u_\varphi}{\partial t^2}, \\ &\frac{\partial \sigma_{rz}}{\partial r} + \frac{1}{r}\frac{\partial \sigma_{\varphi z}}{\partial \varphi} + \frac{\partial \sigma_{zz}}{\partial z} + \frac{1}{r}\sigma_{rz} + f_z = \varrho\frac{\partial^2 u_z}{\partial t^2}. \end{aligned} \tag{3.1.33}$$

From (3.1.22), (3.1.25)–(3.1.27) we obtain

$$\begin{aligned}
&\boldsymbol{q} = q^i \boldsymbol{g}_i = k\text{grad}\ \theta = k\Big(\frac{\partial\theta}{\partial r}\boldsymbol{\varepsilon}_r + \frac{1}{r}\frac{\partial\theta}{\partial\varphi}\boldsymbol{\varepsilon}_\varphi + \frac{\partial\theta}{\partial z}\boldsymbol{\varepsilon}_z\Big)\,,\\
&\gamma_s^s = u^i|_i = \text{div}\ \boldsymbol{u} = \frac{1}{r}\frac{\partial}{\partial r}(ru_r) + \frac{1}{r}\frac{\partial u_\varphi}{\partial\varphi} + \frac{\partial u_z}{\partial z}\,,\\
&q^i|_i = k\text{div grad}\ \theta = k\Delta\theta = \\
&\quad = k\left[\frac{1}{r}\frac{\partial}{\partial r}\Big(r\frac{\partial\theta}{\partial r}\Big) + \frac{1}{r^2}\frac{\partial^2\theta}{\partial\varphi^2} + \frac{\partial^2\theta}{\partial z^2}\right].
\end{aligned} \tag{3.1.34}$$

These relations permit us to write the other equations of the linear theory of thermoelasticity for homogeneous and isotropic bodies. We note that

$$\begin{aligned}
&\sigma_{rr} = \lambda e + 2\mu\varepsilon_{rr} - \beta\theta\,, \qquad \sigma_{\varphi\varphi} = \lambda e + 2\mu\varepsilon_{\varphi\varphi} - \beta\theta\,,\\
&\sigma_{zz} = \lambda e + 2\mu\varepsilon_{zz} - \beta\theta\,, \qquad \sigma_{r\varphi} = 2\mu\varepsilon_{r\varphi}\,,\\
&\sigma_{\varphi z} = 2\mu\varepsilon_{\varphi z}\,, \qquad\qquad\qquad \sigma_{rz} = 2\mu\varepsilon_{rz}\,,
\end{aligned} \tag{3.1.35}$$

where $e = \varepsilon_{rr} + \varepsilon_{\varphi\varphi} + \varepsilon_{zz}$.

We now consider the spherical coordinates $(R,\ \varphi,\ \psi)$ such that

$$X_1 = R\sin\varphi\cos\psi\,, \quad X_2 = R\sin\varphi\sin\psi\,, \quad X_3 = R\cos\varphi\,.$$

We denote by u_R, u_φ and u_ψ the physical components of the displacement field. Let ε_{RR}, $\varepsilon_{\varphi\varphi}$, $\varepsilon_{\psi\psi}$, $\varepsilon_{\varphi\psi}$, $\varepsilon_{\psi R}$, $\varepsilon_{\varphi R}$ be the physical components of the strain tensor γ_{ij}. In this case we obtain

$$\begin{aligned}
&\varepsilon_{RR} = \frac{\partial u_R}{\partial R}\,, \qquad \varepsilon_{\varphi\varphi} = \frac{1}{R}\frac{\partial u_\varphi}{\partial\varphi} + \frac{1}{R}u_R\,,\\
&\varepsilon_{\psi\psi} = \frac{1}{R\sin\varphi}\frac{\partial u_\psi}{\partial\psi} + \frac{1}{R}u_R + \frac{1}{R}u_\varphi\cot\varphi\,,\\
&\varepsilon_{\varphi\psi} = \frac{1}{2}\Big(\frac{1}{R}\frac{\partial u_\psi}{\partial\varphi} - \frac{1}{R}u_\psi\cot\varphi + \frac{1}{R\sin\varphi}\frac{\partial u_\varphi}{\partial\psi}\Big)\,,\\
&\varepsilon_{\psi R} = \frac{1}{2}\Big(\frac{1}{R\sin\varphi}\frac{\partial u_R}{\partial\psi} - \frac{1}{R}u_\psi + \frac{\partial u_\psi}{\partial R}\Big)\,,\\
&\varepsilon_{\varphi R} = \frac{1}{2}\Big(\frac{1}{R}\frac{\partial u_R}{\partial\varphi} - \frac{1}{R}u_\varphi + \frac{\partial u_\varphi}{\partial R}\Big)\,.
\end{aligned} \tag{3.1.36}$$

The equations of motion become

$$\frac{\partial \sigma_{RR}}{\partial R} + \frac{1}{R\sin\varphi}\frac{\partial \sigma_{R\psi}}{\partial \psi} + \frac{1}{R}\frac{\partial \sigma_{\varphi R}}{\partial \varphi} +$$

$$+ \frac{1}{R}(2\sigma_{RR} - \sigma_{\varphi\varphi} - \sigma_{\psi\psi} + \sigma_{\varphi R}\cot\varphi) + f_R = \varrho\frac{\partial^2 u_R}{\partial t^2},$$

$$\frac{\partial \sigma_{\psi R}}{\partial R} + \frac{1}{R\sin\varphi}\frac{\partial \sigma_{\varphi\psi}}{\partial \psi} + \frac{1}{R}\frac{\partial \sigma_{\varphi\varphi}}{\partial \varphi} +$$

$$\frac{1}{R}\left[3\sigma_{\varphi R} + (\sigma_{\varphi\varphi} - \sigma_{\psi\psi})\cot\varphi\right] + f_\varphi = \varrho\frac{\partial^2 u_\varphi}{\partial t^2}, \tag{3.1.37}$$

$$\frac{\partial \sigma_{\psi R}}{\partial R} + \frac{1}{R\sin\varphi}\frac{\partial \sigma_{\psi\psi}}{\partial \psi} + \frac{1}{R}\frac{\partial \sigma_{\varphi\psi}}{\partial \varphi} + \frac{1}{R}(3\sigma_{\psi R} +$$

$$+ 2\sigma_{\varphi\psi}\cot\varphi) + f_\psi = \varrho\frac{\partial^2 u_\psi}{\partial t^2},$$

where σ_{RR}, $\sigma_{\varphi\varphi}$, $\sigma_{\psi\psi}$, $\sigma_{\varphi\psi}$, $\sigma_{\psi R}$, $\sigma_{\varphi R}$ are the physical components of the stress tensor, and f_R, f_φ, f_ψ are the physical components of the body force. From (3.1.25) we obtain

$$\sigma_{RR} = \lambda e + 2\mu\varepsilon_{RR} - \beta\theta, \qquad \sigma_{\varphi\varphi} = \lambda e + 2\mu\varepsilon_{\varphi\varphi} - \beta\theta,$$

$$\sigma_{\psi\psi} = \lambda e + 2\mu\varepsilon_{\psi\psi} - \beta\theta, \qquad \sigma_{\varphi\psi} = 2\mu\varepsilon_{\varphi\psi}, \tag{3.1.38}$$

$$\sigma_{\psi R} = 2\mu\varepsilon_{\psi R}, \qquad \sigma_{\varphi R} = 2\mu\varepsilon_{\varphi R},$$

where $e = \varepsilon_{RR} + \varepsilon_{\varphi\varphi} + \varepsilon_{\psi\psi}$. We also note that

$$\boldsymbol{q} = k\text{grad}\,\theta = k\Big(\frac{\partial\theta}{\partial R}\boldsymbol{\varepsilon}_R + \frac{1}{R}\frac{\partial\theta}{\partial\varphi}\boldsymbol{\varepsilon}_\varphi + \frac{1}{R\sin\varphi}\frac{\partial\theta}{\partial\psi}\boldsymbol{\varepsilon}_\psi\Big),$$

$$\text{div}\,\boldsymbol{u} = \frac{1}{R^2}\frac{\partial}{\partial R}(R^2 u_R) + \frac{1}{R\sin\varphi}\frac{\partial}{\partial\varphi}(u_\varphi \sin\varphi) +$$

$$+ \frac{1}{R\sin\varphi}\frac{\partial u_\psi}{\partial\psi}, \tag{3.1.39}$$

$$\Delta\theta = \text{div grad}\,\theta = \frac{1}{R^2}\frac{\partial}{\partial R}\Big(R^2\frac{\partial\theta}{\partial R}\Big) + \frac{1}{R^2\sin\varphi}\frac{\partial}{\partial\varphi}\Big(\frac{\partial\theta}{\partial\varphi}\sin\varphi\Big) +$$

$$+ \frac{1}{R^2\sin^2\varphi}\frac{\partial^2\theta}{\partial\psi^2}.$$

3.2 Harmonic waves

a) **Planes waves.** We briefly consider harmonic plane wave solutions for homogeneous and isotropic bodies with zero body force and heat supply. It is easy to see that transverse waves are independent of thermal effects. We may, without loss of generality, take x_1-axis along the direction of propagation and choose $u_1 = u_1(x_1, t),\ u_2 = u_3 = 0,\ \theta = \theta(x_1, t)$. If we assume that

$$u_1 = \text{Re}\left[u_1^*(x_1;\,\omega)\exp\left(-i\omega t\right)\right],$$
$$\theta = \text{Re}\left[\theta^*(x_1;\,\omega)\exp\left(-i\omega t\right)\right], \tag{3.2.1}$$

then from (3.1.15) we obtain the following system for the functions u_1^* and θ^*

$$\left(\frac{d^2}{dx_1^2} + \sigma^2\right)u_1^* - \frac{\beta}{\varrho c_1^2}\frac{d\theta^*}{dx_1} = 0\,,$$
$$\left(\frac{d^2}{dx_1^2} + q\right)\theta^* + \frac{iT_0\beta\omega}{k}\frac{du_1^*}{dx_1} = 0\,, \tag{3.2.2}$$

where

$$\sigma^2 = \frac{\omega^2}{c_1^2}\,, \qquad q = \frac{i\omega c}{k}\,. \tag{3.2.3}$$

In what follows we assume that ω is a given real constant.

First we consider the case of a material with $\beta = 0$. Then, from (3.2.2) we obtain the solution $\hat{u}_1^*,\ \hat{\theta}^*$, where

$$\hat{u}_1^* = C_1\exp\left(i\lambda_1 x_1\right) + C_2\exp\left(-i\lambda_1 x_1\right),$$
$$\hat{\theta}^* = D_1\exp\left(i\lambda_2 x_1\right) + D_2\exp\left(-i\lambda_2 x_1\right), \tag{3.2.4}$$

where C_α and D_α are arbitrary constant and

$$\lambda_1 = \sigma = \frac{\omega}{c_1}\,, \qquad \lambda_2 = \sqrt{q} = (1+i)\frac{\omega}{V}\,, \qquad V = \sqrt{\frac{2k\omega}{c}}\,. \tag{3.2.5}$$

Thus the corresponding displacement and temperature fields are given by

$$\hat{u}_1 = \text{Re}\left\{C_1\exp\left[-i\omega(t - x_1/c_1)\right] + C_2\exp\left[-i\omega(t + x_1/c_1)\right]\right\},$$
$$\hat{\theta} = \text{Re}\left\{D_1\exp\left[-\omega x_1/V - i\omega(t - x_1/V)\right] + D_2\exp\left[\omega x_1/V - \right.\right. \tag{3.2.6}$$
$$\left.\left. - i\omega(t + x_1/V)\right]\right\}.$$

The equation $(3.2.6)_1$ represents elastic waves moving along the x_1-axis. The equation $(3.2.6)_2$ represents the thermal wave. It will be noted that the thermal wave is attenuated and, in the sense that its phase velocity $\hat{v}_2 = \omega/\mathrm{Re}\ \lambda_2 = V = \sqrt{2k\omega/c}$ is a function of the frequency, is subject to dispersion. The attenuation coefficient is $\hat{\alpha}_2 = \mathrm{Im}\ \lambda_2 = \omega/V$.

We now study the equations (3.2.2) in the qeneral case $\beta \neq 0$. From (3.2.2) we conclude that both u_1^* and θ^* satisfy the equation

$$\left[\left(\frac{d^2}{dx_1^2} + \sigma^2\right)\left(\frac{d^2}{dx_1^2} + q\right) + \varepsilon q \frac{d^2}{dx_1^2}\right] y = 0\,, \tag{3.2.7}$$

where ε is given by (3.1.14). We consider the algebraic equation

$$z^4 - (\sigma^2 + q + \varepsilon q)z^2 + q\sigma^2 = 0\,. \tag{3.2.8}$$

The roots of this equation are

$$k_{1,2}^2 = \frac{1}{2}\left\{\sigma^2 + q + \varepsilon q \pm \left[(\sigma^2 + q + \varepsilon q)^2 - 4q\sigma^2\right]^{1/2}\right\}. \tag{3.2.9}$$

For $\varepsilon = 0$ we have

$$k_1^2(0) = \sigma^2\,, \qquad k_2^2(0) = q\,. \tag{3.2.10}$$

We denote the roots of (3.2.8) such that

$$k_1(0) = \lambda_1\,, \qquad k_2(0) = \lambda_2\,. \tag{3.2.11}$$

The functions u_1^* and θ^* that satisfy the equation (3.2.2) are given by

$$\begin{aligned} u_1^* &= A_1 \exp(ik_1x_1) + A_2 \exp(-ik_1x_1) + \frac{ik_2\beta}{\varrho c_1^2(\sigma^2 - k_2^2)}[B_1 \exp(ik_2x_1) - \\ &\quad - B_2 \exp(-ik_2x_1)]\,, \\ \theta^* &= B_1 \exp(ik_2x_1) + B_2 \exp(-ik_2x_1) + \frac{\beta T_0 \omega k_1}{k(q - k_1^2)}[A_1 \exp(ik_1x_1) - \\ &\quad - A_2 \exp(-ik_1x_1)]\,, \end{aligned} \tag{3.2.12}$$

where A_α and B_α are arbitrary constants. If we introduce the notation

$$k_\beta = \frac{\omega}{v_\beta} + i\alpha_\beta\,, \tag{3.2.13}$$

then u_1 and θ can be expressed in the form

$$\begin{aligned} u_1 = &\operatorname{Re}\Big\{A_1 \exp\big[-\alpha_1 x_1 - i\omega(t - x_1/v_1)\big] + A_2 \exp\big[\alpha_1 x_1 - i\omega(t + x_1/v_1)\big] + \\ &+ \frac{ik_2\beta}{\varrho c_1^2(\sigma^2 - k_2^2)}\{B_1 \exp\big[-\alpha_2 x_1 - i\omega(t - x_1/v_2)\big] - \\ &- B_2 \exp\big[\alpha_2 x_1 - i\omega(t + x_1/v_2)\big]\}\Big\}, \\ \theta = &\operatorname{Re}\Big\{B_1 \exp\big[-\alpha_2 x_1 - i\omega(t - x_1/v_2)\big] + B_2 \exp\big[\alpha_2 x_1 - \\ &+ i\omega(t + x_1/v_2)\big] + \frac{\beta T_0 \omega k_1}{k(q - k_1^2)}\{A_1 \exp\big[-\alpha_1 x_1 - i\omega(t - x_1/v_1)\big] - \\ &- A_2 \exp\big[\alpha_1 x_1 - i\omega(t + x_1/v_1)\big]\}\Big\}. \end{aligned} \tag{3.2.14}$$

We note that for $\varepsilon = 0$ $(\beta = 0)$ we have

$$v_1(0) = c_1\,, \qquad v_2(0) = \hat{v}_2\,, \qquad \alpha_1(0) = 0\,, \qquad \alpha_2(0) = \hat{\alpha}_2\,.$$

In equations (3.2.14) we recognize two types of harmonic waves. Those with exponents $(t \pm x_1/v_1)$ are called *quasi-elastic* waves while those with exponents $(t \pm x_1/v_2)$ are called *quasi-thermal* waves.

In view of (3.2.11) we conclude that k_1 corresponds to modified elastic waves, and k_2 corresponds to modified thermal waves.

We note that in the expressions of u_1 and θ appear both quasi-elastic and quasi-thermal waves. This fact characterizes the *coupling* effect.
It is convenient to introduce the notations

$$\xi_\alpha = \frac{c_1}{\omega^*}k_\alpha\,, \qquad \chi = \frac{\omega}{\omega^*}\,, \qquad \omega^* = \frac{cc_1^2}{k}\,. \tag{3.2.15}$$

From (3.2.9) we obtain

$$\xi_1 = \frac{1}{2}\sqrt{\chi}\Big\{\big[\chi + (1+i)\sqrt{2\chi} + i(1+\varepsilon)\big]^{1/2} + \big[\chi - (1+i)\sqrt{2\chi} + i(1+\varepsilon)\big]^{1/2}\Big\},$$

$$\xi_2 = \frac{1}{2}\sqrt{\chi}\Big\{\big[\chi + (1+i)\sqrt{2\chi} + i(1+\varepsilon)\big]^{1/2} - \big[\chi - (1+i)\sqrt{2\chi} + i(1+\varepsilon)\big]^{1/2}\Big\}.$$

In practical applications of the theory we have (cf. Chadwick [50], p.289)

$$\chi \ll 1\,.$$

For that reason it is convenient to use power series in χ to express approximate results for the phase velocities and attenuation constants. These are shown to be

$$
\begin{aligned}
v_1 &= c_1\sqrt{1+\varepsilon}\Big[1-\frac{\chi^2\varepsilon(4-3\varepsilon)}{8(1+\varepsilon)^4}+O(\chi^4)\Big],\\
\alpha_1 &= \frac{\omega^*}{c_1\sqrt{1+\varepsilon}}\Big[\frac{\chi^2\varepsilon}{2(1+\varepsilon)^2}+O(\chi^4)\Big],\\
v_2 &= c_1\sqrt{\frac{2\chi}{1+\varepsilon}}\Big[1-\frac{\chi\varepsilon}{2(1+\varepsilon)^2}+\frac{\chi^2\varepsilon(4+\varepsilon)}{8(1+\varepsilon)^4}+\\
&\quad+\frac{\chi^3\varepsilon(8-20\varepsilon+\varepsilon^2)}{16(1+\varepsilon)^6}+O(\chi^4)\Big],\\
d_2 &= \frac{\omega^*}{c_1}\sqrt{\frac{1}{2}\chi(1+\varepsilon)}\Big[1-\frac{\chi\varepsilon}{2(1+\varepsilon)^2}-\\
&\quad-\frac{\chi^2\varepsilon(4-\varepsilon)}{8(1+\varepsilon)^4}+\frac{\chi^3\varepsilon(8-12\varepsilon+\varepsilon^2)}{16(1+\varepsilon)^6}+O(\chi^4)\Big].
\end{aligned}
\tag{3.2.16}
$$

We note that, in particular, the equations (3.2.16) imply that $v_1 = c_1\sqrt{1+\varepsilon}$ as $\chi \to 0$.

For other results concerning the propagation of plane harmonic waves the reader is refered to [50], [19].

b) **Fundamental solutions**. We now use Theorem 3.1.1 to derive fundamental solutions of the system (3.1.15) in the theory of steady vibrations. The results we give here are due to Nowacki [278].

We assume that

$$
\boldsymbol{f} = \text{Re}\left[\boldsymbol{f}^*(x)\exp(-i\omega t)\right], \quad S = \text{Re}\left[S^*(x)\exp(-i\omega t)\right], \tag{3.2.17}
$$

where ω is a given frequency of vibration. If we seek for solutions to (3.1.15) of the form

$$
\boldsymbol{u} = \text{Re}\left[\boldsymbol{u}^*(x;\omega)\exp(-i\omega t)\right], \quad \theta = \text{Re}\left[\theta^*(x;\omega)\exp(-i\omega t)\right], \tag{3.2.18}
$$

then the field equations reduce to a differential system for the amplitudes $\boldsymbol{u}^*$ and θ^*.

If we introduce the notations

$$
\Phi_j = \text{Re}\left[\Phi_j^*(x;\omega)\exp(-i\omega t)\right], \quad \Psi = \text{Re}\left[\Psi^*(x;\omega)\exp(-i\omega t)\right],
$$

then from (3.1.17) we find that the functions Φ_j^* and Ψ^* satisfy the equations

$$\begin{aligned}(\Delta+\tau^2)\big[(\Delta+\sigma^2)(\Delta+q)+\varepsilon q\Delta\big]\Phi_j^* &= -\frac{c}{k\mu(\lambda+2\mu)}f_j^*\,,\\ \big[(\Delta+\sigma^2)(\Delta+q)+\varepsilon q\Delta\big]\Psi^* &= -\frac{1}{k}S^*\,,\end{aligned} \tag{3.2.19}$$

where we have used the notations (3.2.3) and

$$\tau^2=\frac{\omega^2}{c_2^2}\,. \tag{3.2.20}$$

We note that

$$(\Delta+\sigma^2)(\Delta+q)+\varepsilon q\Delta=(\Delta+k_1^2)(\Delta+k_2^2)\,, \tag{3.2.21}$$

where k_1 and k_2 are defined in (3.2.9), (3.2.11). Thus, the equations (3.2.19) become

$$\begin{aligned}(\Delta+\tau^2)(\Delta+k_1^2)(\Delta+k_2^2)\Phi_j^* &= -\frac{c}{k\mu(\lambda+2\mu)}f_j^*\,,\\ (\Delta+k_1^2)(\Delta+k_2^2)\Psi^* &= -\frac{1}{k}S^*\,.\end{aligned} \tag{3.2.22}$$

In what follows, we assume that $k_1^2\neq k_2^2\neq\tau^2\neq k_1^2$. From (3.1.16) we obtain

$$\begin{aligned}u_j^* &= \frac{1}{c}(\lambda+2\mu)k(\Delta+k_1^2)(\Delta+k_2^2)\Phi_j^*-\frac{1}{c}(\lambda+\mu)k(\Delta+\\ &\quad+q+\frac{\lambda+2\mu}{\lambda+\mu}q\varepsilon)\Phi_{s,sj}^*+\frac{\beta}{\lambda+2\mu}\Psi_{,j}^*\,,\\ \theta^* &= -\frac{1}{c}\beta T_0\mu i\omega(\Delta+\tau^2)\Phi_{s,s}^*+(\Delta+\sigma^2)\Psi^*\,.\end{aligned} \tag{3.2.23}$$

If

$$f_j^*=\delta_{1j}f^*\,,\qquad S^*=0\,, \tag{3.2.24}$$

then we take $\Phi_2^*=\Phi_3^*=\Psi^*=0$. From (3.2.22) we conclude that the function Φ_1^* satisfies the equation

$$(\Delta+\tau^2)(\Delta+k_1^2)(\Delta+k_2^2)g=-\frac{c}{k\mu(\lambda+2\mu)}f^*\,. \tag{3.2.25}$$

If the functions G_j^* satisfy the equations

$$\begin{aligned}(\Delta + k_1^2)G_1^* &= -\frac{c}{k\mu(\lambda+2\mu)}f^*,\\ (\Delta + k_2^2)G_2^* &= -\frac{c}{k\mu(\lambda+2\mu)}f^*,\\ (\Delta + \tau^2)G_3^* &= -\frac{c}{k\mu(\lambda+2\mu)}f^*,\end{aligned} \tag{3.2.26}$$

then the solution of equation (3.2.25) can be expressed as

$$g = a_1 G_1^* + a_2 G_2^* + a_3 G_3^*, \tag{3.2.27}$$

where

$$\begin{aligned}a_1 &= \frac{1}{(k_1^2 - k_2^2)(k_1^2 - \tau^2)},\\ a_2 &= \frac{1}{(k_2^2 - k_1^2)(k_2^2 - \tau^2)},\\ a_3 &= \frac{1}{(\tau^2 - k_1^2)(\tau^2 - k_2^2)}.\end{aligned} \tag{3.2.28}$$

Let us assume that $f^* = \delta(x-y)$, where δ is the Dirac delta and y is a fixed point of the body. Then, the solutions of the equations (3.2.26) that satisfy the radiation conditions at infinity are

$$\begin{aligned}G_\alpha^* &= \frac{c}{4\pi k\mu(\lambda+2\mu)r}\exp(ik_\alpha r),\\ G_3^* &= \frac{c}{4\pi k\mu(\lambda+2\mu)r}\exp(i\tau r),\end{aligned} \tag{3.2.29}$$

where $r = |x-y|$. From (3.2.27) and (3.2.29) we find that

$$\begin{aligned}g(x,y;\omega) = \frac{c}{4\pi k\mu(\lambda+2\mu)(k_1^2-k_2^2)r}\Big[\frac{1}{k_1^2-\tau^2}\exp(ik_1 r) -\\ -\frac{1}{k_2^2-\tau^2}\exp(ik_2 r) + \frac{(k_1^2-k_2^2)}{(\tau^2-k_1^2)(\tau^2-k_2^2)}\exp(i\tau r)\Big].\end{aligned} \tag{3.2.30}$$

The amplitudes corresponding to $\Phi_1^* = g(x,y;\omega)$, $\Phi_2^* = \Phi_3^* = \Psi^* = 0$ are denoted by $u_j^{*(1)}$, $\theta^{*(1)}$. If we take into account the relations

$$(k_\alpha^2 - q)(k_\alpha^2 - \sigma^2) = k_\alpha^2 q\varepsilon\,,$$

$$(k_1^2 - \tau^2)(k_2^2 - \tau^2) = -\frac{1}{\mu}\sigma^2(\lambda+\mu)\Big(q - \tau^2 + \frac{\lambda+2\mu}{\lambda+\mu}\varepsilon q\Big)\,,$$

$$q - k_\alpha^2 + \frac{\lambda+2\mu}{\lambda+\mu}\varepsilon q = \frac{\mu(k_\alpha^2 - q)(k_\alpha^2 - \tau^2)}{(\lambda+\mu)k_\alpha^2}\,, \qquad (\alpha = 1,2)\,,$$

then from (3.2.23) we obtain

$$\begin{aligned} u_j^{*(1)} &= \frac{1}{4\pi\varrho c_2^2 r}\delta_{1j}\exp(i\tau r) - \frac{1}{4\pi\varrho\omega^2}\frac{\partial^2 E(r;\omega)}{\partial x_1 \partial x_j}\,, \\ \theta^{*(1)} &= \frac{q\varepsilon}{4\pi\beta(k_1^2 - k_2^2)}\frac{\partial}{\partial x_1}\Big\{\frac{1}{r}\big[\exp(ik_1 r) - \exp(ik_2 r)\big]\Big\}\,, \end{aligned} \tag{3.2.31}$$

where

$$E(r;\omega) = \frac{1}{r}\Big[\frac{\sigma^2(k_1^2 - q)}{k_1^2(k_1^2 - k_2^2)}\exp(ik_1 r) - \frac{\sigma^2(k_2^2 - q)}{k_2^2(k_1^2 - k_2^2)}\exp(ik_2 r) - \exp(i\tau r)\Big]. \tag{3.2.32}$$

If $f_i^* = \delta_{is}\delta(x-y)$, $S^* = 0$, then we take $\Phi_i^* = \delta_{is} g(x,y;\omega)$, $\Psi^* = 0$ and from (3.2.23) we obtain

$$\begin{aligned} u_j^{*(s)} &= \frac{1}{4\pi\varrho c_2^2 r}\delta_{sj}\exp(i\tau r) - \frac{1}{4\pi\varrho\omega^2}\frac{\partial^2 E(r;\omega)}{\partial x_j \partial x_s}\,, \\ \theta^{*(s)} &= \frac{q\varepsilon}{4\pi\beta(k_1^2 - k_2^2)}\frac{\partial}{\partial x_s}\Big\{\frac{1}{r}\big[\exp(ik_1 r) - \exp(ik_2 r)\big]\Big\}\,. \end{aligned} \tag{3.2.33}$$

Next we suppose that $f_j^* = 0$, $S^* = \delta(x-y)$. In this case $\Phi_j^* = 0$, and the function Ψ^* satisfies the equation

$$(\Delta + k_1^2)(\Delta + k_2^2)\Psi^* = -\frac{1}{k}\delta(x-y)\,. \tag{3.2.34}$$

From (3.2.24) we find that

$$\Psi^* = -\frac{1}{4\pi k(k_1^2 - k_2^2)r}\big[\exp(ik_1 r) - \exp(ik_2 r)\big]\,.$$

From (3.2.23) we obtain the solution

$$u_j^{*(4)} = -\frac{\beta}{4\pi k\varrho c_1^2(k_1^2 - k_2^2)}\frac{\partial}{\partial x_j}\left\{\frac{1}{r}\left[\exp(ik_1r) - \exp(ik_2r)\right]\right\}.$$

$$\theta^{*(4)} = -\frac{1}{4\pi k(k_1^2 - k_2^2)r}\left[(k_1^2 - \sigma^2)\exp(ik_2r) - (k_1^2 - \sigma^2)\exp(ik_1r)\right]. \tag{3.2.35}$$

The functions $u_j^{*(s)}$, $\theta^{*(s)}$, $(s = 1,\ 2,\ 3,\ 4)$ given by (3.2.33) and (3.2.35) represent the fundamental solutions of the system of field equations for steady vibrations.

3.3 Potential theory

In this section we use the method of potentials to study boundary-value problems within the theory of steady vibrations. In [192], Ignaczak and Nowacki have reduced the boundary-value problems to singular integral equations. Kupradze *et al.* [228], p.529 have studied these equations and have established existence and uniqueness results.

a) **Preliminaries. Integral identities.** In view of (3.2.18) we can write

$$e_{rs} = \text{Re}\left[e_{rs}^*(x;\omega)\exp(-i\omega t)\right], \qquad t_{rs} = \text{Re}\left[t_{rs}^*(x;\omega)\exp(-i\omega t)\right],$$

$$q_j = \text{Re}\left[q_j^*(x;\omega)\exp(-i\omega t)\right], \qquad \eta = \text{Re}\left[\eta^*(x;\omega)\exp(-i\omega t)\right],$$

where

$$e_{ij}^* = \frac{1}{2}(u_{i,j}^* + u_{j,i}^*). \tag{3.3.1}$$

and

$$t_{ij}^* = \lambda e_{rr}^*\delta_{ij} + 2\mu e_{ij}^* - \beta\theta^*\delta_{ij},$$

$$\eta^* = \beta e_{rr}^* + a\theta^*, \tag{3.3.2}$$

$$q_i^* = k\theta_{,i}^*.$$

Throughout this section we assume that

$$\mu > 0, \qquad 3\lambda + 2\mu > 0, \qquad k > 0. \tag{3.3.3}$$

We introduce the notations

$$\mu^* = \mu/\varrho, \qquad \lambda^* = \lambda/\varrho, \qquad \gamma = \beta/\varrho,$$

$$\frac{T_0\beta}{k} = \zeta, \qquad \frac{c}{k} = \frac{1}{\kappa}, \qquad F_i = -\frac{1}{\varrho}f_i^*, \qquad Z = -\frac{1}{k}S^*. \tag{3.3.4}$$

From (3.1.13), (3.2.18) and (3.3.4) we obtain the basic set of differential equations of steady vibrations. In what follows we suppress the asterisk so that these equations can be expressed in the form

$$\mu\Delta u_j + (\lambda+\mu)u_{r,rj} + \omega^2 u_j - \gamma\theta_{,j} = F_j\,,$$
$$\Delta\theta + \frac{i\omega}{\kappa}\theta + i\omega\zeta u_{r,r} = Z\,. \tag{3.3.5}$$

The system (3.3.5) can be written in matrix form. Following [228], the vector $v = (v_1,\ v_2,\ \cdots,\ v_m)$ shall be considered as the column matrix

$$v = \begin{Vmatrix} v_1 \\ \vdots \\ v_n \end{Vmatrix}.$$

Thus, the product of the matrix $A = \|a_{ij}\|_{m\times m}$ and the vector $v = (v_1,\ v_2,\ \cdots,\ v_m)$ is a m dimensional vector. The vector v multiplied by the matrix A will denote the matrix product between the row matrix $\|v_1, v_2, \cdots, v_m\|$ and the matrix A. We introduce the matricial differential operator

$$A\Big(\frac{\partial}{\partial x},\,\omega\Big) = \Big\|A_{rs}\Big(\frac{\partial}{\partial x},\,\omega\Big)\Big\|_{4\times4}, \tag{3.3.6}$$

where

$$A_{ij}\Big(\frac{\partial}{\partial x},\,\omega\Big) = \mu\delta_{ij}\Delta + (\lambda+\mu)\frac{\partial^2}{\partial x_i\partial x_j} + \omega^2\delta_{ij}\,,$$
$$A_{44}\Big(\frac{\partial}{\partial x},\,\omega\Big) = \Delta + \frac{i\omega}{\kappa}, \qquad A_{j4}\Big(\frac{\partial}{\partial x},\,\omega\Big) = -\gamma\frac{\partial}{\partial x_j}, \tag{3.3.7}$$
$$A_{4j}\Big(\frac{\partial}{\partial x},\,\omega\Big) = i\omega\zeta\frac{\partial}{\partial x_j}.$$

We denote

$$U = (u_1, u_2, u_3, \theta)\,, \qquad F = (F_1, F_2, F_3, Z)\,. \tag{3.3.8}$$

The equations (3.3.5) can be written in the form

$$A\Big(\frac{\partial}{\partial x},\,\omega\Big)U = F\,. \tag{3.3.9}$$

Clearly, we may write

$$A\Big(\frac{\partial}{\partial x},\,\omega\Big) = A_0(\omega) + A_1\Big(\frac{\partial}{\partial x},\,\omega\Big) + A_2\Big(\frac{\partial}{\partial x}\Big),$$

where

$$A_0(\omega) = \|a_{rs}(\omega)\|_{4\times 4}, \qquad A_1\Big(\frac{\partial}{\partial x}, \omega\Big) = \Big\|B_{rs}\Big(\frac{\partial}{\partial x}, \omega\Big)\Big\|_{4\times 4},$$

$$A_2\Big(\frac{\partial}{\partial x}\Big) = \Big\|M_{rs}\Big(\frac{\partial}{\partial x}\Big)\Big\|_{4\times 4}, \qquad a_{ij}(\omega) = \omega^2\delta_{ij}, \qquad a_{i4} = a_{4i} = 0,$$

$$a_{44}(\omega) = \frac{i\omega}{\kappa}, \qquad B_{ij}\Big(\frac{\partial}{\partial x}, \omega\Big) = 0, \qquad B_{j4}\Big(\frac{\partial}{\partial x}, \omega\Big) = -\gamma\frac{\partial}{\partial x_j},$$

$$B_{4j}\Big(\frac{\partial}{\partial x}, \omega\Big) = i\omega\zeta\frac{\partial}{\partial x_j}, \qquad B_{44}\Big(\frac{\partial}{\partial x}, \omega\Big) = 0,$$

$$M_{ij}\Big(\frac{\partial}{\partial x}\Big) = \mu\delta_{ij}\Delta + (\lambda+\mu)\frac{\partial^2}{\partial x_i \partial x_j}, \qquad M_{i4} = M_{4i} = 0,$$

$$M_{44}\Big(\frac{\partial}{\partial x}\Big) = \Delta.$$

For any real vector $\xi = (\xi_1, \xi_2, \xi_3)$ with $\xi_1^2 + \xi_2^2 + \xi_3^2 = 1$ we have

$$\det\|M_{rs}(\xi)\|_{4\times 4} = (\lambda + 2\mu)\mu^2. \tag{3.3.10}$$

It follows from (3.3.3) and (3.3.10) that the system (3.3.5) is elliptic. Let $\tilde{A}$ be the adjoint operator of the operator A. Clearly,

$$\tilde{A}\Big(\frac{\partial}{\partial x}, \omega\Big) = A_0(\omega) - A_1^*\Big(\frac{\partial}{\partial x}, \omega\Big) + A_2\Big(\frac{\partial}{\partial x}\Big),$$

where

$$A_1^*\Big(\frac{\partial}{\partial x}, \omega\Big) = \Big\|C_{rs}\Big(\frac{\partial}{\partial x}, \omega\Big)\Big\|_{4\times 4}, \qquad C_{rs}\Big(\frac{\partial}{\partial x}, \omega\Big) = B_{rs}\Big(\frac{\partial}{\partial x}, \omega\Big).$$

The adjoint system is

$$\tilde{A}\Big(\frac{\partial}{\partial x}, \omega\Big)V = \chi, \tag{3.3.11}$$

where $V = (v_1, v_2, v_3, v_4)$ and $\chi = (\chi_1, \chi_2, \chi_3, \chi_4)$. The system (3.3.11) may be written in the form

$$\begin{aligned} \mu\Delta v_j + (\lambda+\mu)v_{r,rj} - i\omega\zeta v_{4,j} + \omega^2 v_j &= \chi_j, \\ \Delta v_4 + \gamma v_{j,j} + \frac{i\omega}{\kappa}v_4 &= \chi_4. \end{aligned} \tag{3.3.12}$$

Let D^+ be a finite region bounded by the closed surface S, and D^- the complementary of $D^+ \cup S$ to infinite space. In what follows letters in boldface stand for three-dimensional vectors. We say that $U = (u_1, u_2, u_3, u_4) \equiv (\boldsymbol{u}, u_4)$ is regular in D^+ if $u_s \in C_1(\bar{D}^+) \cap C^2(D^+)$ and $u_{s,rm}$ $(r,\ s,\ m = 1,\ 2,\ 3,\ 4)$ are integrable on D^+. Let $\Sigma(x;\ \delta)$ be the sphere with the centre x and radius δ. We say that $U = (\boldsymbol{u},\ u_4)$ is regular in D^- if $u_s \in C^1(\bar{D}^-) \cap \Sigma(0,\ \delta)$ for any δ.

Let $A_i\left(\frac{\partial}{\partial x}, \omega\right)$ be the row-matrix with the elements $A_{ij}\left(\frac{\partial}{\partial x}, \omega\right)$ $(i,\ j = 1,\ 2,\ 3,\ 4)$ given by (3.3.7). If we introduce the notations

$$L_r U = A_r\left(\frac{\partial}{\partial x}, \omega\right)U, \qquad (r = 1,\ 2,\ 3,\ 4),$$

then we can write the system (3.3.9) in the form

$$L_i U = F_i, \qquad L_4 U = Z. \tag{3.3.13}$$

Let $\tilde{A}_i\left(\frac{\partial}{\partial x}, \omega\right)$ be the row-matrix with the elements $\tilde{A}_{ij}\left(\frac{\partial}{\partial x}, \omega\right)$, $(i, j = 1,\ 2,\ 3,\ 4)$. If we denote

$$\tilde{L}_s V = \tilde{A}_s\left(\frac{\partial}{\partial x}, \omega\right)V, \qquad (r = 1,\ 2,\ 3,\ 4),$$

where $V = (\boldsymbol{v},\ v_4)$, then the system (3.3.12) becomes

$$\tilde{L}_i V = \chi_i, \qquad \tilde{L}_4 V = \chi_4. \tag{3.3.14}$$

Let $U = (\boldsymbol{u},\ u_4)$ and $V = (\boldsymbol{v},\ v_4)$ be regular vectors in D^+ which satisfy the equations (3.3.9) and (3.3.11), respectively. We introduce the notations

$$\begin{aligned}
&T_{ij} = \lambda u_{r,r}\delta_{ij} + \mu(u_{i,j} + u_{j,i}) - \gamma\theta\delta_{ij}, \\
&Q_i = \theta_{,i}, \qquad g = i\omega\zeta u_{r,r} + \frac{i\omega}{\kappa}\theta, \\
&\tilde{T}_{kj} = \lambda v_{r,r}\delta_{kj} + \mu(v_{k,j} + v_{j,k}) - i\omega\zeta v_4\delta_{kj}, \\
&\tilde{Q}_i = v_{4,i}, \qquad \tilde{g} = \gamma v_{k,k} + \frac{i\omega}{\kappa}v_4.
\end{aligned} \tag{3.3.15}$$

Then the equations (3.3.13) and (3.3.14) may be written in the forms

$$T_{ji,j} + \omega^2 u_i = F_i\,, \qquad Q_{i,i} + g = Z\,, \tag{3.3.16}$$

$$\tilde{T}_{ji,j} + \omega^2 v_i = \chi_i\,, \qquad \tilde{Q}_{i,i} + \tilde{g} = \chi_4\,, \tag{3.3.17}$$

respectively. It follows from (3.3.13) and (3.3.16) that

$$v_i L_i U + v_4 L_4 U = v_i(T_{ji,j} + \omega^2 u_i) + v_4(Q_{i,i} + g) =$$

$$= (v_i T_{ji})_{,j} + (v_4 Q_i)_{,i} - T_{ji} v_{i,j} - Q_i v_{4,i} + v_4 g + \omega^2 v_i u_i\,.$$

Applying the divergence theorem, we are led to

$$\begin{aligned} \int_{D^+} VA\Big(\frac{\partial}{\partial x}, \omega\Big) U\, dv &= \int_S \Big(v_i T_{ji} n_j + v_4 \frac{\partial \theta}{\partial n}\Big)\, da - \\ &- \int_{D^+} \big[E(\boldsymbol{u}, \boldsymbol{v}) - \gamma\theta v_{r,r} + \theta_{,i} v_{4,i} - i\omega\zeta u_{r,r} v_4 - \\ &- \frac{i\omega}{\kappa} v_4 \theta - \omega^2 v_i u_i\big]\, dv\,, \end{aligned} \tag{3.3.18}$$

where $\partial f/\partial n = f_{,i} n_i$ and

$$E(\boldsymbol{u}, \boldsymbol{v}) = \lambda u_{r,r} v_{j,j} + \frac{1}{2}\mu(u_{i,j} + u_{j,i})(v_{i,j} + v_{j,i})\,. \tag{3.3.19}$$

In a similar manner we find that

$$\begin{aligned} \int_{D^+} U\tilde{A}\Big(\frac{\partial}{\partial x}, \omega\Big) V\, dv &= \int_S \Big(u_i \tilde{T}_{ji} n_j + u_4 \frac{\partial v_4}{\partial n}\Big)\, da - \\ &- \int_{D^+} \big[E(\boldsymbol{u}, \boldsymbol{v}) - \gamma\theta v_{r,r} + \theta_{,i} v_{4,i} - i\omega\zeta u_{r,r} v_4 - \\ &- \frac{i\omega}{\kappa} v_4 \theta - \omega^2 v_i u_i\big]\, dv\,. \end{aligned} \tag{3.3.20}$$

Thus, from (3.3.18) and (3.3.20) we conclude that

$$\begin{aligned} &\int_{D^+} \Big[U\tilde{A}\Big(\frac{\partial}{\partial x}, \omega\Big) V - VA\Big(\frac{\partial}{\partial x}, \omega\Big) U\Big]\, dv = \\ &= \int_S \Big[U\tilde{R}\Big(\frac{\partial}{\partial x}, n_x\Big) V - VR\Big(\frac{\partial}{\partial x}, n_x\Big) U\Big]\, da\,, \end{aligned} \tag{3.3.21}$$

where

$$R\Big(\frac{\partial}{\partial x}, n_x\Big) = \Big\| R_{pq}\Big(\frac{\partial}{\partial x}, n_x\Big)\Big\|_{4\times 4}, \qquad \tilde{R}\Big(\frac{\partial}{\partial x}, n_x\Big) = \Big\| \tilde{R}_{pq}\Big(\frac{\partial}{\partial x}, n_x\Big)\Big\|_{4\times 4},$$

$$R_{ij}\Big(\frac{\partial}{\partial x}, n_x\Big) = \mu\delta_{ij}\frac{\partial}{\partial n} + \lambda n_i \frac{\partial}{\partial x_j} + \mu n_j \frac{\partial}{\partial x_i}, \qquad R_{i4} = -\gamma n_i\,, \qquad R_{4i} = 0\,,$$

$$R_{44}\Big(\frac{\partial}{\partial x}, n_x\Big) = \frac{\partial}{\partial n_x}, \qquad \tilde{R}_{ij}\Big(\frac{\partial}{\partial x}, n_x\Big) = R_{ij}\Big(\frac{\partial}{\partial x}, n_x\Big), \tag{3.3.22}$$

$$\tilde{R}_{j4}\Big(\frac{\partial}{\partial x}, n_x\Big) = -i\omega T_0 \zeta n_j\,, \qquad \tilde{R}_{4j} = 0\,, \qquad \tilde{R}_{44}\Big(\frac{\partial}{\partial x}, n_x\Big) = \frac{\partial}{\partial n_x}\,.$$

Let $R_i\Big(\frac{\partial}{\partial x}, n_x\Big)$ be the row-matrix with the elements $R_{ij}\Big(\frac{\partial}{\partial x}, n_x\Big)$, $(i,\, j = 1,\, 2,\, 3,\, 4)$. Then, from (3.3.15) and (3.3.22) we obtain

$$T_{ji}n_j = R_i\Big(\frac{\partial}{\partial x}, n_x\Big)U\,, \qquad Q_i n_i = R_4\Big(\frac{\partial}{\partial x}, n_x\Big)U\,. \tag{3.3.23}$$

Next, we establish other integral identities. In what follows we denote by $\bar{f}$ the complex conjugate of f. We assume that $U = (\boldsymbol{u},\, u_4)$ is a regular vector field in D^+ that satisfies the equation

$$A\Big(\frac{\partial}{\partial x}, \omega\Big)U = 0\,. \tag{3.3.24}$$

Then, the vector $\bar{U} = (\bar{u}_j,\, \bar{u}_4)$ satisfies the equations

$$\mu\Delta\bar{u}_j + (\lambda+\mu)\bar{u}_{r,rj} - \gamma\bar{u}_{4,j} + \omega^2\bar{u}_j = 0\,,$$

$$\Delta\bar{u}_4 - \frac{i\omega}{\kappa}\bar{u}_4 - i\omega\zeta\bar{u}_{s,s} = 0\,. \tag{3.3.25}$$

These equations can be written in the form

$$\bar{T}_{ji,j} + \omega^2\bar{u}_i = 0\,, \qquad \bar{Q}_{i,i} + \bar{g} = 0\,, \tag{3.3.26}$$

where

$$\bar{T}_{ij} = \lambda\bar{u}_{r,r}\delta_{ij} + \mu(\bar{u}_{i,j} + \bar{u}_{j,i}) - \gamma\bar{u}_4\delta_{ij}\,,$$

$$\bar{Q}_i = \bar{u}_{4,i}\,, \qquad \bar{g} = -i\omega\zeta\bar{u}_{r,r} - \frac{i\omega}{\kappa}\bar{u}_4\,. \tag{3.3.27}$$

We note that

$$\int_{D^+} u_i\bar{T}_{ji,j}dv = \int_S u_i\bar{T}_{ji}n_j da - \int_{D^+} \left[E(\boldsymbol{u},\, \bar{\boldsymbol{u}}) - \gamma\bar{u}_4 u_{r,r}\right]dv\,. \tag{3.3.28}$$

On the other hand, from (3.3.24) we get

$$\gamma u_{r,r} = -\frac{\gamma}{i\omega\zeta}\Delta u_4 - \frac{\gamma}{\eta\kappa}u_4\,,$$

so that

$$\begin{aligned}\int_{D^+} \gamma\bar{u}_4 u_{r,r}dv &= -\frac{\gamma}{i\omega\zeta}\int_{D^+} \bar{u}_4\Delta u_4 dv - \frac{\gamma}{\zeta\kappa}\int_{D^+} u_4\bar{u}_4 dv = \\ &= -\frac{\gamma}{i\omega\zeta}\int_S \bar{u}_4\frac{\partial u_4}{\partial n}\,da + \int_{D^+}\Big(\frac{\gamma}{i\omega\zeta}\bar{u}_{4,i}u_{4,i} - \frac{\gamma}{\zeta\kappa}u_4\bar{u}_4\Big)dv\,.\end{aligned} \tag{3.3.29}$$

It follows from (3.3.26), (3.3.27) and (3.3.29) that

$$\begin{aligned}&\int_S \Big(u_i\bar{T}_{ji}n_j - \frac{\gamma}{i\omega\zeta}\bar{u}_4\frac{\partial u_4}{\partial n}\Big)da = \\ &= \int_{D^+}\Big[E(\boldsymbol{u},\, \bar{\boldsymbol{u}}) - \omega^2 u_j\bar{u}_j + \frac{\gamma}{\zeta}\Big(\frac{1}{\kappa}u_4\bar{u}_4 - \frac{1}{i\omega}u_{4,i}\bar{u}_{4,i}\Big)\Big]dv\,.\end{aligned} \tag{3.3.30}$$

In a similar manner we find that

$$\begin{aligned}&\int_S \Big(\bar{u}_j T_{sj}n_s + \frac{\gamma}{i\omega\zeta}u_4\frac{\partial \bar{u}_4}{\partial n}\Big)da = \\ &= \int_{D^+}\Big[E(\boldsymbol{u},\, \bar{\boldsymbol{u}}) - \omega^2 u_j\bar{u}_j + \frac{\gamma}{\zeta}\Big(\frac{1}{\kappa}u_4\bar{u}_4 + \frac{1}{i\omega}u_{4,i}\bar{u}_{4,i}\Big)\Big]dv\,.\end{aligned} \tag{3.3.31}$$

By (3.3.30) and (3.3.31),

$$\begin{aligned}&\int_S \Big[(\bar{u}_i T_{ji} - u_i\bar{T}_{ji})n_j + \frac{\gamma}{i\omega\zeta}\Big(u_4\frac{\partial\bar{u}_4}{\partial n} + \bar{u}_4\frac{\partial u_4}{\partial n}\Big)\Big]da = \\ &= \frac{2\gamma}{i\omega\zeta}\int_{D^+} u_{4,i}\bar{u}_{4,i}dv\,.\end{aligned} \tag{3.3.32}$$

b) **A representation of solutions.** We consider the equation (3.3.24).

Theorem 3.3.1. *Let $U = (\boldsymbol{u},\, u_4)$ be a regular solution of the equation (3.3.24). Then*

$$U = (\boldsymbol{u}^{(1)} + \boldsymbol{u}^{(2)},\, u_4)\,, \tag{3.3.33}$$

where

$$(\Delta + k_1^2)(\Delta + k_2^2)\boldsymbol{u}^{(1)} = \boldsymbol{0}, \qquad \text{curl } \boldsymbol{u}^{(1)} = \boldsymbol{0},$$
$$(\Delta + \tau^2)\boldsymbol{u}^{(2)} = \boldsymbol{0}, \qquad \text{div } \boldsymbol{u}^{(2)} = 0, \tag{3.3.34}$$
$$(\Delta + k_1^2)(\Delta + k_2^2)u_4 = 0.$$

Proof. The equations (3.3.24) can be expressed in the form

$$\mu\Delta\boldsymbol{u} + (\lambda + \mu)\text{grad div } \boldsymbol{u} + \omega^2\boldsymbol{u} - \gamma\text{grad } u_4 = \boldsymbol{0},$$
$$\Delta u_4 + \frac{i\omega}{\kappa}u_4 + i\omega\zeta\text{div } \boldsymbol{u} = 0. \tag{3.3.35}$$

By

$$\text{curl curl } \boldsymbol{u} = \text{grad div } \boldsymbol{u} - \Delta\boldsymbol{u},$$

the first equation from (3.3.35) becomes

$$(\lambda + 2\mu)\text{grad div } \boldsymbol{u} - \mu\text{curl curl } \boldsymbol{u} + \omega^2\boldsymbol{u} - \gamma\text{grad } u_4 = \boldsymbol{0}.$$

Thus we can write

$$\boldsymbol{u} = \boldsymbol{u}^{(1)} + \boldsymbol{u}^{(2)}, \tag{3.3.36}$$

where

$$\boldsymbol{u}^{(1)} = -\frac{\lambda + 2\mu}{\omega^2}\text{grad div } \boldsymbol{u} + \frac{\gamma}{\omega^2}\text{grad } u_4,$$
$$\boldsymbol{u}^{(2)} = \frac{\mu}{\omega^2}\text{curl curl } \boldsymbol{u}. \tag{3.3.37}$$

Clearly,

$$\text{curl } \boldsymbol{u}^{(1)} = \boldsymbol{0}, \qquad \text{div } \boldsymbol{u}^{(2)} = 0. \tag{3.3.38}$$

From (3.3.35) we get

$$(\mu\Delta + \omega^2)\text{curl } \boldsymbol{u} = \boldsymbol{0}. \tag{3.3.39}$$

In view of (3.1.14), (3.2.20), (3.3.4) and (3.3.39) we obtain

$$(\Delta + \tau^2)\text{curl } \boldsymbol{u} = \boldsymbol{0},$$

so that from (3.3.37) we find that

$$(\Delta + \tau^2)\boldsymbol{u}^{(2)} = \boldsymbol{0}. \tag{3.3.40}$$

By (3.1.14), (3.2.3) and (3.3.4), from (3.3.5) we get

$$(\Delta+\sigma^2)\mathrm{div}\,\boldsymbol{u}-\frac{\gamma}{c_1^2}\Delta u_4=0\,,$$
$$(\Delta+q)u_4+i\omega\zeta\mathrm{div}\,\boldsymbol{u}=0\,. \tag{3.3.41}$$

It follows from (3.3.41) that

$$\big[(\Delta+\sigma^2)(\Delta+q)+\varepsilon q\Delta\big]\mathrm{div}\,\boldsymbol{u}=0\,,$$
$$\big[(\Delta+\sigma^2)(\Delta+q)+\varepsilon q\Delta\big]u_4=0\,. \tag{3.3.42}$$

In view of (3.2.21), (3.3.42) and (3.3.37) we conclude that

$$(\Delta+k_1^2)(\Delta+k_2^2)\boldsymbol{u}^{(1)}=\mathbf{0}\,,\qquad (\Delta+k_1^2)(\Delta+k_2^2)u_4=0\,,$$

and the proof is complete. □

c) **Radiation conditions. A uniqueness result.** We now study the regular solution in D^- of the equation (3.3.24). Let $U=(\boldsymbol{u}^{(1)}+\boldsymbol{u}^{(2)},u_4)$ be a regular solution in D^- of the equation (3.3.24). We say that U satisfies the radiation conditions if

$$\boldsymbol{u}^{(1)}=o(r^{-1})\,,\qquad \boldsymbol{u}^{(1)}_{,j}=O(r^{-2})\,,$$
$$u_4=o(r^{-1})\,,\qquad u_{4,j}=O(r^{-2})\,, \tag{3.3.43}$$
$$\boldsymbol{u}^{(2)}=O(r^{-1})\,,\qquad \frac{\partial\boldsymbol{u}^{(2)}}{\partial r}-i\tau\boldsymbol{u}^{(2)}=o(r^{-1})\,,$$

where $r^2=x_ix_i$. Here, $f=O(g)$ means that f/g is bounded, and $f=o(g)$ means that f/g tends to zero when g increases. We introduce the notation

$$T_i^{(e)}(\boldsymbol{u})=\big[\lambda u_{r,r}\delta_{ij}+\mu(u_{i,j}+u_{j,i})\big]n_j\,. \tag{3.3.44}$$

It is known from [227], p.44 that if (3.3.43) holds then we have

$$\boldsymbol{u}^{(2)}=O(r^{-1})\,,\qquad \boldsymbol{T}^{(e)}(\boldsymbol{u}^{(2)})-i\omega\sqrt{\mu}\boldsymbol{u}^{(2)}=o(r^{-1})\,. \tag{3.3.45}$$

Theorem 3.3.2. *Let $U = (\boldsymbol{u}^{(1)} + \boldsymbol{u}^{(2)}, u_4)$ be a regular solution in D^- of the equation (3.3.24) that satisfies the radiation conditions and one of the conditions*

$$\boldsymbol{u} = \boldsymbol{0}, \qquad u_4 = 0 \qquad \text{on } S, \tag{3.3.46}$$

$$T_{ji}n_j = 0, \qquad Q_i n_i = 0 \qquad \text{on } S. \tag{3.3.47}$$

Then $\boldsymbol{u}^{(1)} = \boldsymbol{0}, \quad \boldsymbol{u}^{(2)} = \boldsymbol{0}, \quad u_4 = 0$.

Proof. We consider the sphere $\Sigma(0;\, R)$ with sufficiently large R such that $D^+ \subset \Sigma(0;\, R)$. Let $\partial\Sigma$ be the boundary of $\Sigma(0;\, R)$ and $\Omega = D^- \cap \Sigma(0;\, R)$. If $U = (\boldsymbol{u},\, u_4)$ is a regular solution in D^- of the equation (3.3.24), that satisfies the radiation conditions and one of the conditions (3.3.46), (3.3.47), then by (3.3.32) we get

$$\begin{aligned} \frac{2\gamma}{i\omega\zeta}\int_\Omega u_{4,i}\bar{u}_{4,i}dv = \int_{\partial\Sigma}\Big[\bar{u}_j^{(2)}\left(T_j^{(e)}(\boldsymbol{u}^{(2)}) - i\omega\sqrt{\mu}u_j^{(2)}\right) - \\ - u_j^{(2)}\left(\bar{T}_j^{(e)}(\boldsymbol{u}^{(2)}) + i\omega\sqrt{\mu}\bar{u}_j^{(2)}\right) + 2i\omega\sqrt{\mu}u_j^{(2)}\bar{u}_j^{(2)}\Big]da + o(1). \end{aligned} \tag{3.3.48}$$

Clearly, (3.3.45) implies that

$$\bar{T}_j^{(e)}(\boldsymbol{u}^{(2)}) + i\omega\sqrt{\mu}u_j^{(2)} = o(r^{-1}), \tag{3.3.49}$$

Thus, if we let $R \to \infty$ and use (3.3.49), we conclude from (3.3.48) that

$$\frac{\gamma}{\omega\zeta}\int_{D^-} u_{4,i}\bar{u}_{4,i}dv + \omega\sqrt{\mu}\lim_{R\to\infty}\int_{\partial\Sigma}|\boldsymbol{u}^{(2)}|^2 da = 0. \tag{3.3.50}$$

Therefore,

$$\int_D u_{4,i}\bar{u}_{4,i}dv = 0, \qquad \lim_{R\to\infty}\int_{\partial\Sigma}|\boldsymbol{u}^{(2)}|^2 da = 0. \tag{3.3.51}$$

In view of Theorem 3.3.1,

$$(\Delta + \tau^2)\boldsymbol{u}^{(2)} = \boldsymbol{0}. \tag{3.3.52}$$

By a well-known theorem (see Kupradze [227], p.60) from $(3.3.51)_2$ and (3.3.52) we conclude that $\boldsymbol{u}^{(2)} = \boldsymbol{0}$. It follows from $(3.3.51)_1$ and (3.3.43) that $u_4 = 0$. By (3.3.34) we find that div $\boldsymbol{u}^{(1)} = 0$. From div $\boldsymbol{u}^{(1)} = 0$, curl $\boldsymbol{u}^{(1)} = \boldsymbol{0}$ and the conditions at infinity we conclude that $\boldsymbol{u}^{(1)} = \boldsymbol{0}$. □

d) **Potentials**. Let $\Gamma(x,\, y;\, \omega)$ be the matrix of fundamental solutions of the system (3.3.5),

$$\Gamma(x,\, y;\, \omega) = \|\Gamma_{rs}(x,\, y;\, \omega)\|_{4\times 4}.$$

Clearly, the fundamental solutions of the system (3.3.5) can be obtained easily from

(3.2.33) and (3.2.35).

We denote by $\Gamma^{(s)}$ $(s = 1, 2, 3, 4)$ the columns of the matrix Γ. We have

$$A_i\Big(\frac{\partial}{\partial x}, \omega\Big)\Gamma^{(k)} = \delta_{ik}\delta(x-y)\,,$$

$$A_4\Big(\frac{\partial}{\partial x}, \omega\Big)\Gamma^{(4)} = \delta(x-y)\,,$$

$$A_i\Big(\frac{\partial}{\partial x}, \omega\Big)\Gamma^{(4)} = 0\,, \qquad A_4\Big(\frac{\partial}{\partial x}, \omega\Big)\Gamma^{(i)} = 0\,.$$

Thus, for $x \neq y$, we can write

$$A\Big(\frac{\partial}{\partial x}, \omega\Big)\Gamma(x, y; \omega) = 0\,, \qquad x \neq y\,. \tag{3.3.53}$$

Let $\widetilde{\Gamma}(x, y; \omega)$ be the matrix of fundamental solutions of the system (3.3.12). Then we have

$$\widetilde{A}\Big(\frac{\partial}{\partial x}, \omega\Big)\widetilde{\Gamma}(x, y; \omega) = 0\,, \qquad \text{for} \quad x \neq y\,. \tag{3.3.54}$$

Moreover,

$$\widetilde{\Gamma}^*(y, x; \omega) = \Gamma(x, y; \omega)\,. \tag{3.3.55}$$

If $U = (u_1, u_2, u_3, u_4)$ is a regular vector field in D^+ and $V = \widetilde{\Gamma}^{(k)}(x, y; \omega)$, $(k = 1, 2, 3, 4)$, $y \in D^+$, then from (3.3.21) we obtain

$$\begin{aligned} u_k(y) = \int_S \Big[U(x)\widetilde{R}\Big(\frac{\partial}{\partial x}, n_x\Big)\widetilde{\Gamma}^{(k)}(x, y; \omega) - \\ - \widetilde{\Gamma}^{(k)}(x, y; \omega)R\Big(\frac{\partial}{\partial x}, n_x\Big)U(x)\Big]da_x + \\ + \int_{D^+} \widetilde{\Gamma}^{(k)}(x, y; \omega)A\Big(\frac{\partial}{\partial x}, \omega\Big)U(x)dv_x\,, \quad (k = 1, 2, 3, 4)\,. \end{aligned} \tag{3.3.56}$$

This relation can be written in the form

$$\begin{aligned} U(y) = \int_S \Big\{\Big[\widetilde{R}\Big(\frac{\partial}{\partial x}, n_x\Big)\widetilde{\Gamma}(x, y; \omega)\Big]^* U(x) - \\ - \widetilde{\Gamma}^*(x, y; \omega)R\Big(\frac{\partial}{\partial x}, n_x\Big)U(x)\Big\}da_x + \\ + \int_{D^+} \widetilde{\Gamma}^*(x, y; \omega)A\Big(\frac{\partial}{\partial x}, \omega\Big)U(x)dv_x\,. \end{aligned} \tag{3.3.57}$$

It follows from (3.3.57) and (3.3.55) that

$$U(x) = \int_S \Big[\Lambda(x,\, y;\, \omega)U(y) - \Gamma(x,\, y;\, \omega)R\Big(\frac{\partial}{\partial y},\, n_y\Big)U(y)\Big]da_y + \\ + \int_{D^+} \Gamma(x,\, y;\, \omega)A\Big(\frac{\partial}{\partial y},\, \omega\Big)U(y)dv_y\,, \qquad x \in D^+\,, \tag{3.3.58}$$

where

$$\Lambda(x,\, y;\, \omega) = \Big[\widetilde{R}\Big(\frac{\partial}{\partial y},\, n_y\Big)\Gamma^*(x,\, y;\, \omega)\Big]^*\,. \tag{3.3.59}$$

Theorem 3.3.3. *Each column of the matrix $\Lambda(x,\, y;\, \omega)$ satisfies at x the equation (3.3.24) for $x \neq y$.*
Proof. By (3.3.59),

$$\Lambda_{ij} = \widetilde{R}_{js}\Big(\frac{\partial}{\partial y},\, n_y\Big)\Gamma_{is}(x,\, y;\, \omega) \qquad (i, j = 1,\, 2,\, 3,\, 4)\,,$$

where Λ_{ij} are the elements of the matrix $\Lambda(x,\, y;\, \omega)$. In view of (3.3.53) we have

$$\Big[A\Big(\frac{\partial}{\partial x},\, \omega\Big)\Lambda(x,\, y;\, \omega)\Big]_{kj} = A_{km}\Big(\frac{\partial}{\partial x},\, \omega\Big)\Lambda_{mj} = \\ = A_{km}\Big(\frac{\partial}{\partial x},\, \omega\Big)\widetilde{R}_{js}\Big(\frac{\partial}{\partial y},\, n_y\Big)\Gamma_{ms}(x,\, y;\, \omega) = \\ = \widetilde{R}_{js}\Big(\frac{\partial}{\partial y},\, n_y\Big)A_{km}\Big(\frac{\partial}{\partial x},\, \omega\Big)\Gamma_{ms}(x,\, y;\, \omega) = 0\,. \quad \square$$

If U is a regular vector field in D^- that satisfies the radiation conditions, then we can establish a representation of U analogous with (3.3.58).

We introduce the potential of a single-layer

$$V(x; \psi) = \int_S \Gamma(x,\, y;\, \omega)\psi(y)da_y\,, \tag{3.3.60}$$

the potential of a double-layer

$$W(x; \eta) = \int_S \Lambda(x,\, y;\, \omega)\eta(y)da_y\,, \tag{3.3.61}$$

and the potential of mass

$$U(x; \varrho) = \int_{D^+} \Gamma(x,\, y;\, \omega)\varrho(y)dv_y\,. \tag{3.3.62}$$

We assume that ψ is Hölder continuous on S, η is Hölder continuously differentiable on S, and ϱ is Hölder continuous on $\bar{D}^+$. As in classical theory, we have

$$A\Big(\frac{\partial}{\partial x}, \omega\Big)V(x;\psi) = 0\,, \qquad A\Big(\frac{\partial}{\partial x}, \omega\Big)W(x;\eta) = 0\,, \tag{3.3.63}$$

and

$$A\Big(\frac{\partial}{\partial x}, \omega\Big)U(x;\varrho) = \varrho\,, \qquad x \in D^+\,. \tag{3.3.64}$$

Moreover, we can prove (see, for example, [227], p.33)

Theorem 3.3.4. *The potential of a single-layer is continuous throughout.*

Theorem 3.3.5. *The potential of a double-layer has finite limits when the point x tends to $z \in S$ both from within and from without, and these limits are respectively equal to*

$$\begin{aligned} W^+(z;\eta) &= -\frac{1}{2}\eta(z) + \int_S \Lambda(z,\, y)\eta(y)da_y\,,\\ W^-(z;\eta) &= \frac{1}{2}\eta(z) + \int_S \Lambda(z,\, y)\eta(y)da_y\,, \end{aligned} \tag{3.3.65}$$

the integrals being conceived in the sense of Cauchy's principal value.

Theorem 3.3.6. *$R(\partial/\partial x,\, n_x)V(x;\,\psi)$ has finite limits as the point x tends to $z \in S$ from within and from without and these limits are*

$$\begin{aligned} \Big[R\Big(\frac{\partial}{\partial z}, n_z\Big)V(z;\,\psi)\Big]^+ &= \frac{1}{2}\psi(z) + \int_S \Big[R\Big(\frac{\partial}{\partial z}, n_z\Big)\Gamma(z,\, y;\, \omega)\Big]\psi(y)da_y\,,\\ \Big[R\Big(\frac{\partial}{\partial z}, n_z\Big)V(z;\,\psi)\Big]^- &= -\frac{1}{2}\psi(z) + \int_S \Big[R\Big(\frac{\partial}{\partial z}, n_z\Big)\Gamma(z,\, y;\, \omega)\Big]\psi(y)da_y\,, \end{aligned} \tag{3.3.66}$$

respectively.

In what follows we restrict our attention to the equation (3.3.24). We consider the following boundary value problems.

Interior problems. To find a regular solution in D^+ of the equation (3.3.24) satisfying one of the conditions

$$(I_1) \qquad U^+ = G_1 \qquad \text{on } S\,,$$

$$(I_2) \qquad \Big[R\Big(\frac{\partial}{\partial z}, n_z\Big)U(z)\Big]^+ = G_2(z)\,, \qquad z \in S\,,$$

where G_1 and G_2 are prescribed vector fields.

Exterior problems. To find a regular solution in D^- of the equation (3.3.24) that satisfies the radiation conditions and one of the conditions

$$(E_1) \qquad U^- = G_3 \qquad \text{on } S,$$

$$(E_2) \qquad \left[R\left(\frac{\partial}{\partial z}, n_z\right)U(z)\right]^- = G_4(z), \qquad z \in S,$$

where G_3 and G_4 are given.

We assume that G_1 and G_3 are Hölder continuously differentiable on S, and G_2 and G_4 are Hölder continuous on S. We denote by (I_α^o) and (E_α^o) the homogeneous problems corresponding to (I_α) and (E_α), $(\alpha = 1, 2)$, respectively.

We seek the solutions of the problems (I_1) and (E_1) in the form of a double-layer potential and the solutions of the problems (I_2) and (E_2) in the form of a single-layer potential. In view of Theorems 3.3.5 and 3.3.6 we obtain for unknown densities the following equations

$$(I_1) \qquad -\frac{1}{2}\eta(z) + \int_S \Lambda(z, y; \omega)\eta(y)da_y = G_1(z),$$

$$(I_2) \qquad \frac{1}{2}\psi(z) + \int_S \left[R\left(\frac{\partial}{\partial z}, n_z\right)\Gamma(z, y; \omega)\right]\psi(y)da_y = G_2(z),$$

$$(E_1) \qquad \frac{1}{2}\eta(z) + \int_S \Lambda(z, y; \omega)\eta(y)da_y = G_3(z),$$

$$(E_2) \qquad -\frac{1}{2}\psi(z) + \int_S \left[R\left(\frac{\partial}{\partial z}, n_z\right)\Gamma(z, y; \omega)\right]\psi(y)da_y = G_4(z), \quad z \in S.$$

These equations are two-dimensional singular integral equations. They can only be understood in the sense of principal values. Let us show that the Fredholm theory is applicable. The integral operator of the problem (I_1) may be written in the form

$$H\eta = -\eta(z) + \int_S K(z, y; \omega)\eta(y)da_y, \tag{3.3.67}$$

where

$$K(z, y; \omega) = 2\Lambda(z, y; \omega).$$

It follows from (3.2.33), (3.2.35) and (3.3.59) that

$$K(z, y) = K_1(z, y) + K_2(z, y),$$

where

$$K_1(z, y) = \|k_{rs}(z, y)\|_{4\times 4},$$

$$k_{ij}(z, y; \omega) = \frac{\mu}{2\pi(\lambda + 2\mu)r^3}\left[(z_j - y_j)\cos(n_y, z_i) - (z_i - y_i)\cos(n_y, z_j)\right],$$

$$k_{4j} = k_{j4} = k_{44} = 0, \qquad K_2(z, y; \omega) = O\big(|z - y|^{\alpha - 2}\big), \qquad \alpha > 0.$$

Next, we determine the characteristic matrix of the operator H introducing a local coordinate system $(\zeta_1, \zeta_2, \zeta_3)$ at each point $z \in S$ with the ζ_3-axis directed along the outside normal to S at z. Let $\chi = \|\chi_{ij}\|_{4\times 4}$ be the characteristic matrix of H and $\sigma = \|\sigma_{ij}\|_{4\times 4}$ be the symbolic matrix of the operator H. We find that

$$\chi_{11} = \chi_{22} = \chi_{33} = \chi_{44} = -1, \qquad \chi_{ij} = -\chi_{ji} \quad (i \neq j),$$

$$\chi_{i4} = \chi_{4i} = 0, \qquad \sigma_{11} = \sigma_{22} = \sigma_{33} = \sigma_{44} = -1,$$

$$\sigma_{rs} = 2\pi i \chi_{rs} \qquad (r \neq s), \qquad \sigma_{i4} = \sigma_{4i} = 0,$$

and

$$\det \sigma = 1 - \frac{\mu^2}{(\lambda + 2\mu)^2}.$$

In what follows we assume that

$$\frac{\mu}{\lambda + 2\mu} \neq \pm 1,$$

so that the integral operator corresponding to the equation (I_1) is of the normal type. Since H is of the normal type and the symbolic matrix is antisymmetric, then the index of this operator is zero (cf. Mikhlin [252], p. 195). Thus we may conclude that the operator H is a Fredholm operator. In a similar way we can prove that the operators corresponding to the remaining equations are Fredholm operators in the space of Hölder continuous functions.

e) **Existence and uniqueness results.** In Section 3.3.c we have established a uniqueness result in the case of exterior problems for any value of ω^2. For bounded regions uniqueness does not occur because of the existence of a discrete spectrum of eigenfrequencies.

First interior problem. We assume that $U = (\boldsymbol{u}^{(1)} + \boldsymbol{u}^{(2)}, u_4)$ is a regular solution in D^+ of the equation (3.3.24) that satisfies the condition

$$U = 0 \quad \text{on } S: \tag{3.3.68}$$

From (3.3.32) we obtain

$$\int_{D^+} u_{4,i}\bar{u}_{4,i}dv = 0\,,$$

so that $u_4 = 0$. By (3.3.35) we obtain div $\boldsymbol{u} = 0$ and (3.3.37) implies that $\boldsymbol{u}^{(1)} = \boldsymbol{0}$ on D^+. In view of (3.3.36) we have $\boldsymbol{u} = \boldsymbol{u}^{(2)}$. It follows from (3.3.34) and (3.3.68) that

$$(\Delta + \tau^{(2)})\boldsymbol{u}^{(2)} = \boldsymbol{0}\,, \qquad \text{div}\ \boldsymbol{u}^{(2)} = 0 \quad \text{on } D^+\,, \qquad \boldsymbol{u}^{(2)} = \boldsymbol{0} \quad \text{on } S\,. \tag{3.3.69}$$

Thus, we conclude that for any ω^2 different from the characteristic values of the problem (3.6.69) we obtain $U = 0$ on D^+.

Second interior problem. Let $U = (\boldsymbol{u}^{(1)} + \boldsymbol{u}^{(2)}, u_4)$ be a regular solution in D^+ of the equation (3.3.24) that satisfies the condition

$$R\Big(\frac{\partial}{\partial x}, n_x\Big)U = 0 \qquad \text{on } S\,. \tag{3.3.70}$$

Now, from (3.3.23), (3.3.70) and (3.3.32) we obtain $u_4 = c = \text{const}$. By (3.3.34) and (3.3.35) we find that div $\boldsymbol{u}^{(1)} = -c/\zeta\kappa$. Thus, from (3.3.37) we obtain $\boldsymbol{u}^{(1)} = \boldsymbol{0}$, and from (3.3.35) we find that $c = 0$. It follows from (3.3.22) that

$$R_i\Big(\frac{\partial}{\partial x}, n_x\Big)U = 2\mu\frac{\partial u_i^{(2)}}{\partial n} + \mu\big[\boldsymbol{n}\times \text{curl}\ \boldsymbol{u}^{(2)}\big]_i\,, \qquad R_4\Big(\frac{\partial}{\partial x}, n_x\Big)U = 0\,,$$

so that we obtain for $\boldsymbol{u}^{(2)}$ the following boundary-value problem

$$\begin{aligned} &(\Delta + \tau^2)\boldsymbol{u}^{(2)} = \boldsymbol{0}\,, \qquad \text{div}\ \boldsymbol{u}^{(2)} = 0 \qquad \text{on } D^+\,,\\ &2\frac{\partial}{\partial n}\boldsymbol{u}^{(2)} + \boldsymbol{n}\times\text{curl}\ \boldsymbol{u}^{(2)} = \boldsymbol{0} \qquad \text{on } S\,. \end{aligned} \tag{3.3.71}$$

We conclude that for any ω^2 different from the characteristic values of the problem (3.3.71) the regular solution in D^+ of the boundary-value problem (3.3.24), (3.3.70) is equal to zero.

Remark 3.3.1. If $U = (\boldsymbol{u},\, u_4)$ is a solution of one of the interior problems then $u_4 = 0$ and $\boldsymbol{u}$ is a solution of one of the following problems of classical elasticity

$$\mu\Delta\boldsymbol{u} + (\lambda+\mu)\text{grad div}\ \boldsymbol{u} + \omega^2\boldsymbol{u} = 0 \qquad \text{on } D^+$$

$$\boldsymbol{u} = \boldsymbol{0} \qquad \text{on } S\,, \qquad \text{or} \qquad \lambda u_{r,r}n_i + \mu(u_{i,j} + u_{j,i})n_j = 0 \qquad \text{on } S\,.$$

Remark 3.3.2. Let $U = (\boldsymbol{u}, u_4)$ and $U' = (\boldsymbol{u}', u_4')$ be vector fields that satisfy the equations

$$A\Big(\frac{\partial}{\partial x}, \omega\Big)U = 0, \qquad \widetilde{A}\Big(\frac{\partial}{\partial x}, \omega\Big)U' = 0.$$

It follows from (3.3.5) and (3.3.12) that

$$\boldsymbol{u} = \boldsymbol{u}', \qquad u_4 = \frac{i\omega\zeta}{\gamma}u_4'.$$

In order to establish an existence theorem, we now study the integral equation (E_1).

Theorem 3.3.7. *The necessary and sufficient condition for equation*

$$\frac{1}{2}\eta(z) + \int_S \Lambda(z, y; \omega)\eta(y)da_y = 0, \tag{3.3.72}$$

to have a non-trivial solution is that the parameter ω^2 equals one of the proper frequencies of the problem (I_2^o). If ω^2 is a ν-fold proper frequency for the problem (I_2^o), then the integral equation (3.3.72) has ν linearly independent solutions, and they coincide with the boundary values of the solutions for problem (I_2^o).

Proof. We assume that (3.3.72) has a non-trivial solution. Let us prove that ω^2 is proper frequency of the problem (I_2^o). We assume the contrary. Then, according to the Fredholm theory, the associate equation

$$\frac{1}{2}\eta(z) + \int_S \widetilde{R}\Big(\frac{\partial}{\partial z}, n_z\Big)\widetilde{\Gamma}(z, y; \omega)\eta(y)da_y = 0,$$

has the non-trivial solution η. Then the single-layer potential

$$\widetilde{V}(x; \eta) = \int_S \widetilde{\Gamma}(x, y; \omega)\eta(y)da_y,$$

is a solution of the problem

$$\widetilde{A}\Big(\frac{\partial}{\partial x}, \omega\Big)\widetilde{V}(x; \eta) = 0, \quad x \in D^+, \qquad \Big[\widetilde{R}\Big(\frac{\partial}{\partial x}, n_x\Big)\widetilde{V}\Big]^+ = 0 \quad \text{on } S. \tag{3.3.73}$$

Let $\widetilde{V} = (\boldsymbol{v}^*, v_4)$. From (3.3.73) and Remark 3.3.1 it follows that $v_4^* = 0$. If we consider the problem

$$A\Big(\frac{\partial}{\partial x}, \omega\Big)U = 0 \quad \text{on } D^+, \qquad \Big[R\Big(\frac{\partial}{\partial x}, n_x\Big)U\Big]^+ = 0 \quad \text{on } S, \tag{3.3.74}$$

then from (3.3.73) and Remark 3.3.2 we find that $U = (\boldsymbol{v}^*, 0) = \widetilde{V}$. Since ω^2 is different from the proper frequencies of the problem (3.3.74) we conclude that $\widetilde{V} = 0$ on D^+. We note that $\widetilde{V}$ satisfies the radiation conditions. From the continuity of the single-layer potentials and Theorem 3.3.2 we conclude that $\widetilde{V} = 0$ on D^-. By (3.3.66) we obtain $\eta = 0$. This contradiction proves that the condition of the theorem is necessary. The sufficiency of the condition of the theorem can be established as in classical elasticity [227, p. 159]. □

Theorem 3.3.6 enable us to establish an existence theorem for the first exterior problem (E_1).

Thus, if ω^2 is not a proper frequency of problem (I_2^o), in view of Theorem 3.3.6 we conclude that the integral equation (E_1^o) has only the trivial solution. According to Fredholm's first theorem, the integral equation (E_1) has a solution which is unique. The solution of the non-homogeneous problem (E_1) can be expressed uniquely in the form of a double-layer potential. Let ω^2 be a proper frequency of the problem (I_2^o). We seek the solution of the problem (E_1) in the form

$$U(x) = W(x;\, \eta) + V(x;\, h)\,, \tag{3.3.75}$$

where η and h are unknown densities. From the boundary condition we obtain the integral equation

$$\frac{1}{2}\eta(z) + \int_S \Lambda(z,\, y)\eta(y)da_y = G_3(z) - V(z;\, h)\,. \tag{3.3.76}$$

According to Theorem 3.3.6, the corresponding homogeneous equation has ν linearly independent solutions $\{\eta_s\}$ $(s = 1,\, 2,\, \cdots, \nu)$. The equation (3.3.76) can be solved only when the conditions

$$\int_S V(z;\, h)\varphi_s(z)da_z = C_s \qquad (s = 1,\, 2,\, \cdots, \nu)\,, \tag{3.3.77}$$

are satisfied, where $\{\varphi_s\}$ $(s = 1,\, 2,\, \cdots, \nu)$ are linearly independent solutions of the associate homogeneous equation

$$\frac{1}{2}\varphi(z) + \int_S \widetilde{R}\Big(\frac{\partial}{\partial z},\, n_z\Big)\Gamma^*(y,\, z;\, \omega)\varphi(y)da_y = 0\,, \tag{3.3.78}$$

and

$$C_s = \int_S G_3(z)\varphi_s(z)da\,. \tag{3.3.79}$$

The conditions (3.3.77) can be written in the form

$$\int_S \int_S \Gamma(z,\, y;\, \omega) h(y) \varphi_s(z) da_y da_z =$$

$$= \int_S \int_S \Gamma^*(z,\, y;\, \omega) \varphi_s(z) h(y) da_z da_y = \tag{3.3.80}$$

$$= \int_S \int_S \widetilde{\Gamma}(y,\, z;\, \omega) \varphi_s(z) h(y) da_z da_y = C_s \qquad (s = 1,\, 2,\, \cdots, \nu)\,.$$

We consider the potential of a single-layer

$$\widetilde{V}(x;\, \varphi_s) = \int_S \widetilde{\Gamma}(x,\, z;\, \omega) \varphi_s(z) da_z\,. \tag{3.3.81}$$

Since φ_s is solution of (3.3.78), the potential (3.3.81) is solution of the problem

$$\widetilde{A}\Big(\frac{\partial}{\partial x},\, \omega\Big)\widetilde{V} = 0 \qquad \text{on } D^+\,, \qquad \Big[\widetilde{R}\widetilde{V}\Big]^+ = 0 \qquad \text{on } S\,.$$

If $\widetilde{V} = (\boldsymbol{v},\, v_4)$, then according to Remark 3.3.1 we have $v_4 = 0$. Thus, the problem (I_2^o) coincides with $(\widetilde{I}_2^o)$. The conditions (3.3.80) take the form

$$\int_S \widetilde{V}(z;\, \varphi_s) h(z) da = C_s \qquad (s = 1,\, 2,\, \cdots, \nu)\,. \tag{3.3.82}$$

According to Theorem 3.3.6, $\widetilde{V}(z;\, \varphi_s) = \eta_s(z)$ and (3.3.82) becomes

$$\int_S \eta_s(z) h(z) da = C_s \qquad (s = 1,\, 2,\, \cdots, \nu)\,. \tag{3.3.83}$$

We assume that the system $\{\eta_s\}$ is made orthonormal. We choose h of the form

$$h(z) = \sum_{s=1}^{\nu} C_s \bar{\eta}_s(z)\,.$$

Then, the conditions (3.3.83) are satisfied. Thus we have proved the following result

Theorem 3.3.8. *The external non-homogeneous problem (E_1) has a unique solution for any ω^2. The solution can be expressed as a potential of a double-layer if ω^2 is different from the proper frequencies of problem (I_2^o), and as a linear combination of a double-layer potential and some single-layer potentials if ω^2 is one of the proper frequencies of the problem (I_2^o).*

The results established in [227] can be extended to obtain an existence theorem for the dynamic problem (E_2).

3.4 Transient waves

The study of specific problems of thermoelastodynamics has been the subject of many investigations (see, e.g. Bargmann [19], Nowacki [279], Day [102], Chandrasekharaiah [60], and the literature cited therein). The level of difficulty of these problems can be ascertained by the fact that even in the case of a single spatial variable the solutions are valid for short times only (Hetnarski [164]) and are usually based either on a few terms of the power series expansion with respect to the coupling parameter ε defined by (3.1.14) (Hetnarski [165], [166], Lessen [238]) or on asymptotic expansions governing short-and long-time responses (Boley and Tolins [29]). In this section we study a half-space problem and the problem of a concentrated heat supply.

a) **A half-space problem.** We consider a homogeneous and isotropic material which occupies the half-space $x_1 > 0$.

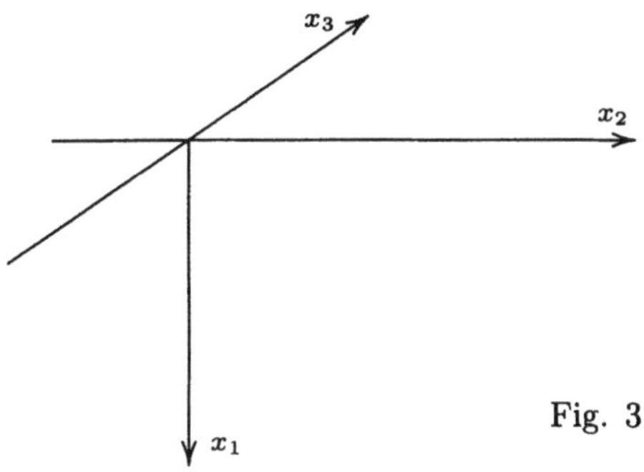

Fig. 3

We assume that the initial data, the body forces and the heat supply are zero. We suppose that the boundary $x_1 = 0$ is free of stresses and is subject to a thermal field independent of $\boldsymbol{x}$. We consider the following conditions at infinity

$$\lim_{|x|\to\infty} u_i = 0\,, \qquad \lim_{|\boldsymbol{x}|\to\infty} u_{i,j} = 0\,, \qquad \lim_{|\boldsymbol{x}|\to\infty} \theta = 0\,, \qquad \lim_{|x|\to\infty} \theta_{,i} = 0\,. \tag{3.4.1}$$

The data of the problem suggest that

$$u_1 = u_1(x_1, t)\,, \qquad \theta = \theta(x_1, t)\,, \qquad u_2 = u_3 = 0\,. \tag{3.4.2}$$

The equations (3.1.13) reduce to

$$(\lambda+2\mu)\frac{\partial^2 u_1}{\partial x_1^2}-\varrho\frac{\partial^2 u_1}{\partial t^2}-\beta\frac{\partial\theta}{\partial x_1}=0\,,$$
$$k\frac{\partial^2\theta}{\partial x^2}-c\frac{\partial\theta}{\partial t}-\beta T_0\frac{\partial^2 u_1}{\partial x_1\partial t}=0\,. \tag{3.4.3}$$

The non-zero components of the stress tensor are given by

$$t_{11}=(\lambda+2\mu)\frac{\partial u_1}{\partial x_1}-\beta\theta\,,\qquad t_{22}=t_{33}=\lambda\frac{\partial u_1}{\partial x_1}-\beta\theta\,. \tag{3.4.4}$$

Introduction of the dimensionless quantities

$$x=\frac{c_1 c}{k}x_1\,,\quad \tau=\frac{c_1^2 c}{k}t\,,\quad u=\frac{c_1 c}{k}u_1\,,\quad T=\frac{\theta}{T_0}\,, \tag{3.4.5}$$

allows equations (3.4.3) to be written in the form

$$\Big(\frac{\partial^2}{\partial x^2}-\frac{\partial^2}{\partial\tau^2}\Big)u-A\frac{\partial T}{\partial x}=0\,,$$
$$\Big(\frac{\partial^2}{\partial x^2}-\frac{\partial}{\partial\tau}\Big)T-\frac{\beta}{c}\frac{\partial^2 u}{\partial x\partial\tau}=0\,, \tag{3.4.6}$$

where

$$A=\frac{\beta T_0}{\lambda+2\mu}\,. \tag{3.4.7}$$

We consider the following boundary conditions

$$t_{11}(0,\tau)=0\,,\qquad \theta(0,\tau)=T^*H(\tau)\,, \tag{3.4.8}$$

where T^* is a given constant, and H is the Heaviside unit step function, i.e., $H(\tau)=0$ for $\tau\le 0$ and $H(\tau)=1$ for $\tau>0$. The initial conditions are

$$u(x,0)=0\,,\qquad \dot u(x,0)=0\,,\qquad T(x,0)=0\,. \tag{3.4.9}$$

(*i*) *The uncoupled theory.* In the context of the uncoupled theory the above boundary-initial-value problem was first studied by Danilovskaya [97]. In the uncoupled theory the field equations reduce to

$$\Big(\frac{\partial^2}{\partial x^2}-\frac{\partial^2}{\partial\tau^2}\Big)u-A\frac{\partial T}{\partial x}=0\,,$$
$$\Big(\frac{\partial^2}{\partial x^2}-\frac{\partial}{\partial\tau}\Big)T=0\,. \tag{3.4.10}$$

In this section we will designate by $\bar f$ the Laplace transform with respect to τ

of the function $f(x, \tau)$,

$$\bar{f}(x, p) = \mathscr{L}[f(x, \tau)] = \int_0^\infty \exp(-p\tau) f(x, \tau)\, d\tau\,. \qquad (3.4.11)$$

In view of (3.4.9), from (3.4.10) we obtain the equations

$$\begin{aligned} &\left(\frac{d^2}{dx^2} - p^2\right)\bar{u} - A\frac{\partial \bar{T}}{\partial x} = 0\,, \\ &\left(\frac{d^2}{dx^2} - p\right)\bar{T} = 0\,. \end{aligned} \qquad (3.4.12)$$

From (3.4.8) we get

$$\bar{t}_{11}(0, p) = 0\,, \qquad \bar{T}(0, p) = \frac{T^*}{pT_0}\,. \qquad (3.4.13)$$

The solution of the equation $(3.4.12)_2$ which satisfies the conditions (3.4.1) is

$$\bar{T} = C_1 \exp(-x\sqrt{p})\,,$$

where C_1 is an arbitrary constant. From (3.4.13) we can determine the constant C_1. We find that

$$\bar{T} = \frac{T^*}{pT_0} \exp(-x\sqrt{p})\,. \qquad (3.4.14)$$

We note that (see e.g. Carslaw and Jaeger [47])

$$\mathscr{L}^{-1}\left(\frac{\exp(-\sqrt{p}x)}{p}\right) = \operatorname{erfc} \frac{x}{2\sqrt{\tau}}\,,$$

where

$$\operatorname{erfc} y = \frac{2}{\sqrt{\pi}} \int_y^\infty \exp(-\xi^2)\, d\xi\,.$$

Thus, we get

$$\theta(x, \tau) = T^* \operatorname{erfc} \frac{x}{2\sqrt{\tau}}\,. \qquad (3.4.15)$$

It follows from (3.4.12) and (3.4.14) that

$$\left(\frac{d^2}{dx^2} - p^2\right)\bar{u} + \frac{AT^*}{T_0 p} \exp(-\sqrt{p}x) = 0\,. \qquad (3.4.16)$$

The solution of the equation (3.4.16) which satisfies the conditions (3.4.1) is given by

$$\bar{u} = C_2 \exp(-px) - \frac{AT^*}{T_0(1-p)p\sqrt{p}} \exp(-x\sqrt{p}), \tag{3.4.17}$$

where C_2 is an arbitrary constant. It follows from (3.4.4), (3.4.14) and (3.4.17) that

$$\begin{aligned} \bar{t}_{11} =& (\lambda + 2\mu)\Big[-pC_2 \exp(-px) + \frac{AT^*}{T_0(1-p)p\sqrt{p}} \exp(-x\sqrt{p})\Big] - \\ &- \frac{\beta T^*}{p} \exp(-x\sqrt{p}). \end{aligned} \tag{3.4.18}$$

By (3.4.13) and (3.4.18) we get

$$C_2 = \frac{AT^*}{T_0(1-p)p}.$$

Thus, we conclude that

$$\begin{aligned} \bar{t}_{11} &= -\frac{\beta T^*}{1-p}\left[\exp(-px) - \exp(-\sqrt{p}x)\right], \\ \bar{u} &= \frac{\beta T^*}{(\lambda+2\mu)(1-p)p}\left[\exp(-px) - \frac{1}{\sqrt{p}}\exp(-\sqrt{p}x)\right]. \end{aligned} \tag{3.4.19}$$

If we take into account the relations (cf. [47])

$$\begin{aligned} \mathscr{L}^{-1}\left(\frac{\exp(-\sqrt{p}x)}{p-1}\right) &= H(\tau - x)\exp(\tau - x), \\ \mathscr{L}^{-1}\left(\frac{\exp(-\sqrt{p}x)}{p-1}\right) &= \frac{1}{2}\exp(\tau)\Big\{\exp(-x)\operatorname{erfc}\left(\frac{x}{2\sqrt{\tau}} - \sqrt{\tau}\right) + \\ &\quad + \exp(x)\operatorname{erfc}\left(\frac{x}{2\sqrt{\tau}} + \sqrt{\tau}\right)\Big\}, \end{aligned}$$

then we find that

$$t_{11} = -\beta T^*[S(x,\tau) - H(\tau - x)]\exp(\tau - x), \tag{3.4.20}$$

where

$$S(x,\tau)=\frac{1}{2}\exp(\tau)\Big\{\exp(-x)\text{erfc}\left(\frac{x}{2\sqrt{\tau}}-\sqrt{\tau}\right)+ \\ +\exp(x)\text{erfc}\left(\frac{x}{2\sqrt{\tau}}+\sqrt{\tau}\right)\Big\}. \tag{3.4.21}$$

The term in (3.4.20) which contains $H(\tau-x)$ corresponds to an elastic wave. Clearly, $x=\tau$ (or $x_1=c_1t$) is the wavefront. The jump of the stress t_{11} at the wavefront is

$$[t_{11}]_{x_1=c_1t}=\beta T^*. \tag{3.4.22}$$

We note that this jump is constant.

(*ii*) *The coupled theory.* The coupled counterpart of Danilovskaya's problem has been considered by many authors (see, e.g. Hetnarski [164–166], Boley and Tolins [29], Nowacki [279], Bargmann [99], Brun [39], Molinari [256], Bahar and Hetnarski [10, 11] and the literature cited therein). In [10, 11] the authors have studied the problem by avoiding the use of the thermoelastic potential. Taking the Laplace transform of the equations (3.4.6), we get

$$\left(\frac{d^2}{dx^2}-p^2\right)\bar{u}-A\frac{d\bar{T}}{dx}=0, \\ \left(\frac{d^2}{dx^2}-p\right)\bar{T}-\frac{\beta}{c}p\frac{d\bar{u}}{dx}=0. \tag{3.4.23}$$

From (3.4.23) we conclude that $\bar{u}$ and $\bar{T}$ satisfy the equation

$$\left[\frac{d^4}{dx^4}-p(p+1+\varepsilon)\frac{d^2}{dx^2}+p^3\right]y=0. \tag{3.4.24}$$

The roots of the characteristic equation

$$k^4-p(1+p+\varepsilon)k^2+p^3=0,$$

are denoted by $\pm k_1$ and $\pm k_2$, for k_1 and k_2 positive. We have

$$k_{1,2}^2=\frac{p}{2}\{1+p+\varepsilon\pm\sqrt{p^2-2p(1-\varepsilon)+(1+\varepsilon)^2}\}. \tag{3.4.25}$$

It follows from (3.4.1) and (3.4.24) that

$$\bar{u}=B_1\exp(-k_1x)+B_2\exp(-k_2x), \\ \bar{T}=D_1\exp(-k_1x)+D_2\exp(-k_2x), \tag{3.4.26}$$

where B_α and D_α are arbitrary constants. The functions (3.4.26) satisfy the equations (3.4.23) only if

$$D_1 = -\frac{1}{Ak_1}(k_1^2 - p^2)B_1\,, \qquad D_2 = -\frac{1}{Ak_2}(k_2^2 - p^2)B_2\,.$$

Thus, we find that

$$\begin{aligned} \bar{T} &= -\frac{1}{A}\Big[\frac{k_1^2 - p^2}{k_1}B_1 \exp(-k_1 x) + \frac{k_2^2 - p^2}{k_2}B_2 \exp(-k_2 x)\Big]\,, \\ \bar{t}_{11} &= -\varrho c_1^2 p^2 \Big[\frac{1}{k_1}B_1 \exp(-k_1 x) + \frac{1}{k_2}B_2 \exp(-k_2 x)\Big]\,. \end{aligned} \tag{3.4.27}$$

From (3.4.13) and (3.4.27) we obtain

$$B_2 = -\frac{k_2}{k_1}B_1\,, \qquad B_1 = -\frac{Ak_1 T^*}{pT_0(k_1^2 - k_2^2)}\,. \tag{3.4.28}$$

By (3.4.27) and (3.4.28),

$$\begin{aligned} \bar{T} &= -\frac{T^*}{p(k_1^2 - k_2^2)}\left[(k_1^2 - p^2)\exp(-k_1 x) - (k_2^2 - p^2)\exp(-k_2 x)\right], \\ \bar{t}_{11} &= \frac{\beta T^* p}{k_1^2 - k_2^2}\left[\exp(-k_1 x) - \exp(-k_2 x)\right]. \end{aligned} \tag{3.4.29}$$

The problem of finding of the functions T and t_{11} is difficult.

According to Abel's theorem (see, e.g. [359]),

$$\lim_{\tau \to 0} f(\tau) = \lim_{p \to \infty} p\bar{f}(p)\,.$$

This fact permits the inversion of the Laplace transforms for a short time. It follows from (3.4.25) that

$$k_1 = p + \frac{\varepsilon}{2} + \frac{\varepsilon(4 - \varepsilon)}{8p} + O\Big(\frac{1}{p^2}\Big)\,,$$

$$k_2 = \sqrt{p} - \frac{\varepsilon}{2\sqrt{p}} + \frac{\varepsilon(3\varepsilon - 4)}{8p\sqrt{p}} + O\Big(\frac{1}{p^{5/2}}\Big)\,,$$

$$
\begin{aligned}
\frac{1}{k_1^2-k_2^2} &= \frac{1}{p^2}+\frac{1-\varepsilon}{p^3}+\frac{1-4\varepsilon+\varepsilon^2}{p^4}\,,\\
\frac{k_1^2-p^2}{k_1^2-k_2^2} &= \frac{\varepsilon}{p}+\frac{\varepsilon(2-\varepsilon)}{p^2}+\frac{\varepsilon(3+\varepsilon^2-6\varepsilon)}{p^3}+\frac{2\varepsilon(1+\varepsilon^2-3\varepsilon)}{p^4}\,,\\
\frac{k_2^2-p^2}{k_1^2-k_2^2} &= -1+\frac{\varepsilon}{p}+\frac{\varepsilon(2-\varepsilon)}{p^2}+\frac{1-6\varepsilon+3\varepsilon^2}{p^3}-\frac{2\varepsilon(1+\varepsilon^2-3\varepsilon)}{p^4}\,,\\
\frac{k_1}{k_1^2-k_2^2} &= \frac{1}{p}+\frac{2-\varepsilon}{2p^2}+\frac{8-24\varepsilon+3\varepsilon^2}{8p^3}+\frac{\varepsilon(5\varepsilon^2-21\varepsilon)}{8p^4}\,,\\
\frac{k_2}{k_1^2-k_2^2} &= \frac{1}{p^{3/2}}+\frac{2-3\varepsilon}{2p}+\frac{8-40\varepsilon+15\varepsilon^2}{8p^{7/2}}\,.
\end{aligned}
\tag{3.4.30}
$$

If we assume that

$$
\exp\left\{-\frac{\varepsilon(4-\varepsilon)}{8p}x\right\}\approx 1-\frac{\varepsilon(4-\varepsilon)}{8p}x\,,\qquad \exp\left(\frac{\varepsilon x}{2\sqrt{p}}\right)\approx 1+\frac{\varepsilon x}{2\sqrt{p}}\,,
$$

$$
\exp\left\{\frac{\varepsilon(4-3\varepsilon)}{8p\sqrt{p}}x\right\}\approx 1+\frac{\varepsilon(4-3\varepsilon)}{8p\sqrt{p}}x\,,
$$

then we obtain

$$
\begin{aligned}
\frac{t_{11}}{\beta T^*} =& \exp\left(-\frac{\varepsilon}{2}x\right)\exp(-px)\left[\frac{1}{p}-\frac{\varepsilon(4-\varepsilon)x-8(1-\varepsilon)}{8p^2}+O\Big(\frac{1}{p^3}\Big)\right]-\\
&-\exp(-\sqrt{p}x)\left[\frac{1}{p}+\frac{1-\varepsilon}{p^2}+\frac{\varepsilon x}{2p\sqrt{p}}+O\Big(\frac{1}{p^{5/2}}\Big)\right],\\
\frac{\bar\theta}{T^*} =& \exp\left(-\frac{\varepsilon}{2}x\right)\left[\frac{\exp(-px)}{p^2}\varepsilon+O\Big(\frac{1}{p^3}\Big)\right]-\\
&-\exp(-\sqrt{p}x)\left[-\frac{1}{p}-\frac{\varepsilon x}{2p\sqrt{p}}+\frac{\varepsilon}{p^2}+O\Big(\frac{1}{p^{5/2}}\Big)\right].
\end{aligned}
\tag{3.4.31}
$$

By using the relations (cf. Carslaw and Jaeger [47], p.380)

$$
\mathscr{L}^{-1}\Big(\frac{\exp(-px)}{p}\Big)=H(\tau-x)\,,\qquad \mathscr{L}^{-1}\Big(\frac{\exp(-\sqrt{p}x)}{p}\Big)=\operatorname{erfc}\frac{x}{2\sqrt{\tau}}\,,
$$

$$\mathscr{L}^{-1}\left(\frac{\exp(-px)}{p^2}\right) = (\tau - x)H(\tau - x)\,,$$

$$\mathscr{L}^{-1}\left(\frac{\exp(-\sqrt{p}x)}{p^2}\right) = \left(\tau + \frac{1}{2}x^2\right)\operatorname{erfc}\frac{x}{2\sqrt{\tau}} - x\left(\frac{\tau}{\pi}\right)^{1/2}\exp\left(-\frac{x^2}{4\tau}\right), \tag{3.4.32}$$

$$\mathscr{L}^{-1}\left(\frac{\exp(-\sqrt{p}x)}{p\sqrt{p}}\right) = 2\left(\frac{\tau}{\pi}\right)^{1/2}\exp\left(-\frac{x^2}{4\tau}\right) - x\operatorname{erfc}\frac{x}{2\sqrt{\tau}}\,,$$

we find

$$\begin{aligned}
t_{11} = {} & \beta T^*\exp\left(-\frac{\varepsilon}{2}x\right)\left\{1 - \frac{1}{8}\left[\varepsilon(4-\varepsilon)x - 8(1-\varepsilon)\right](\tau - x)\right\}H(\tau - x) - \\
& - \beta T^*\left\{\operatorname{erfc}\frac{x}{2\sqrt{\tau}} + (1-\varepsilon)\left[\left(\tau + \frac{1}{2}x^2\right)\operatorname{erfc}\frac{x}{2\sqrt{\tau}} - x\left(\frac{\tau}{\pi}\right)^{1/2}\exp\left(-\frac{x^2}{4\tau}\right)\right] + \right.\\
& \left. + \frac{\varepsilon x}{2}\left[2\left(\frac{\tau}{\pi}\right)^{1/2}\exp\left(-\frac{x^2}{4\tau}\right) - x\operatorname{erfc}\frac{x}{2\sqrt{\tau}}\right]\right\}, \\
\theta = {} & T^*\left\{\operatorname{erfc}\frac{x}{2\sqrt{\tau}} + \varepsilon\exp\left(-\frac{\varepsilon x}{2}\right)(\tau - x)H(\tau - x) + \right.\\
& + \frac{\varepsilon x}{2}\left[2\left(\frac{\tau}{\pi}\right)^{1/2}\exp\left(-\frac{x^2}{4\tau}\right) - x\operatorname{erfc}\frac{x}{2\sqrt{\tau}}\right] - \\
& \left. - \varepsilon\left[\left(\tau + \frac{1}{2}x^2\right)\operatorname{erfc}\frac{x}{2\sqrt{\tau}} - x\left(\frac{\tau}{\pi}\right)^{1/2}\exp\left(-\frac{x^2}{4\tau}\right)\right]\right\}.
\end{aligned} \tag{3.4.33}$$

The jump of t_{11} at the wave front $x = \tau$ is now given by

$$[t_{11}]_{x=\tau} = \beta T^*\exp\left(-\frac{1}{2}\varepsilon x\right). \tag{3.4.34}$$

This jump decays exponentially with distance from the boundary. From (3.4.33) we also obtain

$$[\dot{\theta}]_{x=\tau} = \varepsilon T^*\exp\left(-\frac{1}{2}\varepsilon x\right).$$

Although the jumps are obtained from the short time solutions, it is to be noted that their magnitudes are exact and hold for any time.

Finite-element solutions of Danilovskaya's problem have been obtained in [273], [286].

b) **Concentrated loads**. Next we illustrate the method of Lessen [238] who studied the action of a concentrated heat source. We consider a homogeneous and isotropic material which occupies the three-dimensional space. We assume that the initial data and body forces are zero. The conditions at infinity are given by (3.4.1). We seek the solution in the form (3.1.20). Then , from (3.1.21), we find that

$$\left[\left(\Delta - \frac{p^2}{c_1^2}\right)\left(\Delta - \frac{cp}{k}\right) - \frac{\varepsilon pc}{k}\Delta\right]\bar{\varphi} = -\frac{\beta}{k(\lambda+2\mu)}\bar{S}, \qquad \left(\Delta - \frac{p^2}{c_2^2}\right)\bar{g}_i = 0. \tag{3.4.35}$$

We can write the equation $(3.4.35)_1$ in the form

$$(\Delta - \zeta_1^2)(\Delta - \zeta_2^2)\bar{\varphi} = -\frac{\beta}{k(\lambda+2\mu)}\bar{S}, \tag{3.4.36}$$

where

$$\zeta_{1,2}^2 = \frac{1}{2}\left\{\frac{p^2}{c_1^2} + \frac{pc}{k}(1+\varepsilon) \pm \left[\left(\frac{p^2}{c_2^2} + \frac{pc}{k}(1+\varepsilon)\right)^2 - \frac{4cp^3}{kc_1^3}\right]^{1/2}\right\}. \tag{3.4.37}$$

Lessen [238] suggested the following expansion

$$\bar{\varphi} = \bar{\varphi}_0 + \varepsilon\bar{\varphi}_1 + \varepsilon^2\bar{\varphi}_2 + \dots$$

It follows from (3.4.35) that the functions $\bar{\varphi}_0$ and $\bar{\varphi}_1$ satisfy the equations

$$\left(\Delta - \frac{p^2}{c_1^2}\right)\left(\Delta - \frac{cp}{k}\right)\bar{\varphi}_0 = -\frac{\beta}{k(\lambda+2\mu)}\bar{S}, \qquad \left(\Delta - \frac{p^2}{c_1^2}\right)\left(\Delta - \frac{cp}{k}\right)\bar{\varphi}_1 = \frac{pc}{k}\Delta\bar{\varphi}_0. \tag{3.4.38}$$

Clearly, the function $\bar{\varphi}_0$ corresponds to a problem of uncoupled theory.

We assume that

$$S = Q\delta(x-y)\delta(t),$$

where Q is a given constant. We have $g_i = 0$, and from (3.4.38) we obtain by the procedure from Section 3.2 that

$$\bar{\varphi}_0 = \frac{\beta c_1^2 Q}{4\pi k(\lambda+2\mu)rp\left(p^2 - \frac{c_1^2 c}{k}\right)}\left[\exp\left(-\frac{r}{c_1}p\right) - \exp\left(-r\sqrt{\frac{pc}{k}}\right)\right], \qquad r = |\boldsymbol{x} - \boldsymbol{y}|. \tag{3.4.39}$$

We denote by θ_0 the temperature variation corresponding to $\varphi = \varphi_0,\ g_i = 0$. From (3.1.20) and (3.4.39) we get

$$\bar{\theta}_0 = \frac{cQ}{4\pi kr} \exp\left(-r\sqrt{\frac{pc}{k}}\right). \tag{3.4.40}$$

If we introduce the notations

$$\xi = \frac{cc_1}{k} r, \qquad \nu = \frac{cc_1}{k} t,$$

then we can express the original functions in the form

$$\begin{aligned} \varphi_0(\xi,\,\nu) &= \frac{\beta Q}{4\pi c(\lambda+2\mu)} \left[E_1(\xi,\,\nu) - E_2(\xi,\,\nu) H(\nu-\xi)\right], \\ \theta_0(\xi,\,\nu) &= \frac{c_1^2 c^3 \beta Q}{8k^3(\pi\nu)^{3/2}} \exp\left(-\frac{\xi^2}{4\nu}\right). \end{aligned} \tag{3.4.41}$$

Now we can determine the function $\bar{\varphi}_1$, and so on. The solution of this problem has been studied by Hetnarski [166]. In [165], Hetnarski studied the action of heat source $S = Q\delta(x-y)H(\tau)$. The generation of thermoelastic disturbances by the action of body forces has been studied in various papers (see, e.g. Lockett and Sneddon [243], Soós [334], Nowacki [279], Dragos [110]). We note that Dragos [110] has derived the fundamental solutions directly from the field equations, avoiding the representation of the solution by potentials.

A detailed treatment of the thermomechanical interaction in one-dimensional linear thermoelastodynamics is presented in the book of Day [102], p.1.

3.5 Propagation of singular surfaces

In this section we apply the theory from Section 1.7 to study the propagation of singular surfaces in homogeneous and isotropic thermoelastic bodies. The results we give here are due to Chadwick and Powdrill [51], Mc Carthy [249], p.153, Chen [64], p.333.

In the absence of body forces and heat supply, the field equations (3.1.15) reduce to

$$\begin{aligned} \frac{\partial^2 u_i}{\partial t^2} &= c_2^2 \frac{\partial^2 u_i}{\partial x_j \partial x_j} + (c_1^2 - c_2^2) \frac{\partial^2 u_j}{\partial x_i \partial x_j} - \frac{\beta}{\varrho} \frac{\partial \theta}{\partial x_i}, \\ c\frac{\partial \theta}{\partial t} + \beta T_0 \frac{\partial^2 u_j}{\partial x_j \partial t} &= k \frac{\partial^2 \theta}{\partial x_j \partial x_j}. \end{aligned} \tag{3.5.1}$$

In what follows we denote $\Omega = B \times I$. We say that (u_i, θ) is a *strict solution* of (3.5.1) on Ω if u_i and θ are functions of class C^2 on Ω which satisfy the equations (3.5.1) on Ω.

a) **Weak thermoelastic waves.** The singular surface of u_i and θ of order $r \geq 2$ are called weak singular surfaces. Let σ be a singular hypersurface of order r. If $r > 2$, then the equations (3.5.1) hold on Ω. If $r = 2$, then (3.5.1) hold on $\Omega \backslash \sigma$. Thus, for $r \geq 2$ the equations obtained by applying to each term in (3.5.1) the operator $\partial^{r-2}/\partial x_m \dots \partial x_p$ hold on $\Omega \backslash \sigma$. Forming the jump across σ of each term in the resulting equations, we obtain

$$\begin{aligned} \left[\frac{\partial^r u_i}{\partial x_m \dots \partial x_p \partial t^2}\right] = c_2^2 \left[\frac{\partial^r u_i}{\partial x_m \dots \partial x_p \partial x_j \partial x_j}\right] + \\ + \left(c_1^2 - c_2^2\right) \left[\frac{\partial^r u_j}{\partial x_m \dots \partial x_p \partial x_i \partial x_j}\right], \\ \beta T_0 \left[\frac{\partial^r u_j}{\partial x_m \dots \partial x_p \partial x_j \partial t}\right] = k \left[\frac{\partial^r \theta}{\partial x_m \dots \partial x_p \partial x_j \partial x_j}\right]. \end{aligned} \tag{3.5.2}$$

Substituting from (1.7.58) for the jumps in equations (3.5.2) then multiplying each term by $n_m \dots n_p$ and summing on the repeated indices, we arrive at

$$\begin{aligned} &\left(u_n^2 - c_2^2\right)\lambda_i - \left(c_1^2 - c_2^2\right)\lambda_j n_j n_i = 0\,, \\ &k\xi + \beta T_0 u_n n_i \lambda_i = 0\,, \end{aligned} \tag{3.5.3}$$

where

$$\lambda_i = \left[\frac{\partial^r u_i}{\partial n^r}\right], \qquad \xi = \left[\frac{\partial^r \theta}{\partial n^r}\right].$$

It follows from (1.7.58) that λ_i and ξ determine the jumps across σ of all r^{th} partial derivatives of u_i and θ and so, by hypothesis, they cannot all be zero. Thus we conclude that the determinant of coefficients of the system (3.5.3) must be zero. This condition reduces to

$$\left(u_n^2 - c_1^2\right)\left(u_n^2 - c_2^2\right)^2 = 0\,.$$

We therefore conclude that within the linear thermoelasticity the weak singular hypersurfaces propagate with one of the constant speed c_1, c_2. these hypersurfaces are called *weak thermoelastic waves*.

If we use (1.7.58), then from

$$v_i = \frac{\partial u_i}{\partial t}\,, \qquad e_{ij} = \frac{1}{2}\left(\frac{\partial u_i}{\partial x_j} + \frac{\partial u_j}{\partial x_i}\right), \qquad \omega_{ij} = \frac{1}{2}\left(\frac{\partial u_i}{\partial x_j} - \frac{\partial u_j}{\partial x_i}\right),$$

we obtain

$$\left[\frac{\partial^{r-1}v_i}{\partial t^{r-1}}\right]=(-u_n)^r\lambda_i\,,\qquad \left[\frac{\partial^{r-1}e_{ij}}{\partial t^{r-1}}\right]=(-u_n)^{r-1}\lambda_s n_s\,,$$
$$\left[\frac{\partial^{r-1}\omega_{ij}}{\partial t^{r-1}}\right]=-\frac{1}{2}(-u_n)^{r-1}(\lambda_j n_i-\lambda_i n_j)\,. \tag{3.5.4}$$

If the wave speed is c_1, equation $(3.5.3)_1$ yields

$$\lambda_i=\lambda n_i\,,\qquad \text{where } \lambda=\lambda_s n_s\,,$$

so that

$$\left[\frac{\partial^{r-1}v_i}{\partial t^{r-1}}\right]=(-u_n)^r\lambda n_i\neq 0\,,$$
$$\left[\frac{\partial^{r-1}e_{ii}}{\partial t^{r-1}}\right]=(-u_n)^{r-1}\lambda\neq 0\,,\qquad \left[\frac{\partial^{r-1}\omega_{ij}}{\partial t^{r-1}}\right]=0\,. \tag{3.5.5}$$

If the wave propagate with speed c_2, then from $(3.5.3)_1$ we obtain $\lambda=0$ so that we have

$$\left[\frac{\partial^{r-1}v_i}{\partial t^{r-1}}\right]n_i=0\,,\qquad \left[\frac{\partial^{r-1}e_{jj}}{\partial t^{r-1}}\right]=0\,,\qquad \left[\frac{\partial^{r-1}\omega_{ij}}{\partial t^{r-1}}\right]\neq 0\,. \tag{3.5.6}$$

In view of (3.5.5) and (3.5.6), we say that a weak thermoelastic wave is *dilatational* if it propagates with the speed c_1 and *rotational* if its speed of propagation is c_2. From $(3.5.3)_2$ we find that

$$\xi=-\frac{1}{k}\beta T_0\lambda c_1\,,\qquad \text{for a dilatational weak wave}\,,$$
$$\xi=0\,,\qquad \text{for a rotational weak wave}\,. \tag{3.5.7}$$

It follows that the jumps across a weak thermoelastic wave of order r in the rth partial derivatives of the u_i and θ are completely determined by the singularity vector $\boldsymbol{\lambda}$. The magnitude of this vector is a measure of the strength of wave.

In order to study the strengths of weak waves we apply to each term of (3.5.1) the operator $\partial^{r-1}/\partial x_m\ldots\partial x_p$, and form jumps across the wave. Then we use the compatibility conditions (1.7.59), (1.7.60) for the functions u_i and θ. Next we multiply the resulting equations by $n_m \dots n_p$ and sum on the repeated indices. Thus, with the aid of (1.7.19), we obtain

$$\begin{aligned}
&(u_n^2 - c_2^2)\mu_i - (c_1^2 - c_2^2)\{\mu_j n_j n_i + a^{\alpha\beta}\lambda_{j;\alpha}\varphi_{j;\beta} n_i + \\
&+ a^{\alpha\beta}\varphi_{i;\beta}(\lambda_{j;\alpha} n_j - \lambda_j a^{\nu\varrho} b_{\alpha\nu}\varphi_{j;\varrho})\} - \lambda_i \frac{\delta u_n}{\delta t} - \\
&- 2u_n \frac{\delta \lambda_i}{\delta t} + 2Hc_2^2\lambda_i + \frac{\beta}{\varrho}\xi n_i = 0\,, \\
&\beta T_0\big(u_n \mu_j n_j - n_j \frac{\delta \lambda_j}{\delta t} + u_n a^{\alpha\beta}\lambda_{j;\alpha}\varphi_{j;\beta} + \\
&+ a^{\alpha\beta} u_{n;\alpha}\lambda_j \varphi_{j;\beta}\big) + k\gamma + (cu_n - 2kH)\xi = 0\,,
\end{aligned} \tag{3.5.8}$$

where

$$\mu_i = \left[\frac{\partial^{r+1} u_i}{\partial n^{r+1}}\right], \qquad \gamma = \left[\frac{\partial^{r+1}\theta}{\partial n^{r+1}}\right].$$

For weak thermoelastic waves u_n is constant. Thus, by (1.7.41) we find that

$$\frac{\delta\lambda}{\delta t} = \frac{\delta}{\delta t}(\lambda_j n_j) = n_j \frac{\delta \lambda_j}{\delta t}\,. \tag{3.5.9}$$

By (1.7.9) and (1.7.23),

$$\begin{aligned}
&a^{\alpha\beta}\lambda_{j;\alpha}\varphi_{j;\beta} = a^{\alpha\beta}(\lambda_j\varphi_{j;\beta})_{;\alpha} - 2H\lambda\,, \\
&\lambda_{j;\alpha} n_j - \lambda_j a^{\nu\varrho} b_{\alpha\nu}\varphi_{j;\varrho} = \lambda_{;\alpha}\,.
\end{aligned} \tag{3.5.10}$$

With the aid of relations (3.5.9) and (3.5.10), the equations (3.5.8) can be expressed in the form

$$\begin{aligned}
&(u_n^2 - c_2^2)\mu_i - (c_1^2 - c_2^2)\{\mu_j n_j n_i + a^{\alpha\beta}(\lambda_j\varphi_{j;\beta})_{;\alpha} n_i + a^{\alpha\beta}\lambda_{;\alpha}\varphi_{i;\beta}\} - \\
&- 2u_n \frac{\delta\lambda_i}{\delta t} + 2H\{c_2^2\lambda_i + (c_1^2 - c_2^2)\lambda n_i\}\frac{\beta}{\varrho}\xi n_i = 0\,, \\
&\beta T_0 u_n \mu_j n_j + k\gamma - \beta T_0\Big\{\frac{\delta\lambda}{\delta t} - u_n a^{\alpha\beta}(\lambda_j\varphi_{j;\beta})_{;\alpha} + 2Hu_n\lambda\Big\} + \\
&+ (cu_n - 2kH)\xi = 0\,.
\end{aligned} \tag{3.5.11}$$

If we multiply $(3.5.11)_2$ by n_i and sum on the index i, then we get

$$(u_n^2 - c_1^2)\mu_j n_j - (c_1^2 - c_2^2)a^{\alpha\beta}(\lambda_j \varphi_{j;\beta})_{;\alpha} - 2u_n \frac{\delta\lambda}{\delta t} + 2Hc_1^2\lambda + \frac{\beta}{\varrho}\xi = 0 . \qquad (3.5.12)$$

In the case of *dilatational weak waves* we have

$$u_n = c_1 , \qquad \lambda_i = \lambda n_i , \qquad \xi = -\frac{1}{k}\beta T_0 c_1 \lambda ,$$

and equation (3.5.12) yields the *transport equation*

$$\frac{1}{c_1}\frac{\delta\lambda}{\delta t} = \frac{d\lambda}{dn} = \lambda\Big(H - \frac{1}{2c_1}\varepsilon\omega^*\Big) , \qquad (3.5.13)$$

where n is the distance measured along the normal to the wave surface, and ε and ω^* are given by (3.1.14) and (3.2.15), respectively. In view of (3.5.13), equations (3.5.11) reduce to

$$\begin{aligned} &\mu_i - \mu_j n_j n_i = a^{\alpha\beta}\lambda_{;\alpha}\varphi_{i;\beta} , \\ &\beta T_0 c_1 \mu_j n_j + k\gamma = \beta T_0 c_1 \lambda\Big[H + \frac{\omega^*}{c_1}\Big(T_0 - \frac{1}{2}\varepsilon\Big)\Big] . \end{aligned} \qquad (3.5.14)$$

We assume that at time $t = t_0$ the mean and Gaussian curvatures of the wavefront $\sigma(t_0)$ are H_0 and K_0, respectively. Then if the distance n is measured from $\sigma(t_0)$, the mean curvature of $\sigma(t)$ is given by (cf. Thomas [350], p.108)

$$H = \frac{H_0 - nK_0}{1 - 2nH_0 + n^2K_0} . \qquad (3.5.15)$$

It follows from (3.5.13) and (3.5.15) that

$$\lambda = \lambda_0(1 - 2nH_0 + n^2K_0)^{-1/2}\exp\Big(-\frac{\varepsilon\omega^* n}{2c_1}\Big) , \qquad (3.5.16)$$

where $\lambda = \lambda_0$ on $\sigma(t_0)$. In view of $(3.5.7)_1$ and (3.5.16) we conclude that the jumps of the rth partial derivatives of u_i and θ across a dilatational weak thermoelastic wavefront of order r are completely determined by the magnitude of the singularity vector on the initial wavefront $\sigma(t_0)$.

For *rotational weak waves*, we have $u_n = c_2$, $\lambda = 0$, $\xi = 0$, and the equation (3.5.12) reduces to

$$\mu_j n_j = -a^{\alpha\beta}(\lambda_s\varphi_{s;\beta})_{;\alpha} . \qquad (3.5.17)$$

From (3.5.11) we obtain the transport equation

$$\frac{1}{c_2}\frac{\delta\lambda_i}{\delta t} = \frac{d\lambda_i}{dn} = \lambda_i H\,, \tag{3.5.18}$$

and $\gamma = 0$. It follows from (3.5.15) and (3.5.18) that

$$\lambda_i = \lambda_i^0(1 - 2nH_0 + n^2K_0)^{-1/2}\,, \tag{3.5.19}$$

where $\lambda_i = \lambda_i^o$ on $\sigma(t_0)$. Thus, the jumps of the rth partial derivatives of the u_i across a rotational weak thermoelastic wavefront of order r are determined by their distributions on an initial wavefront. We note that the partial derivatives of orders r and $r+1$ of the temperature are continuous on Ω.

b) **Strong thermoelastic waves.** The singular surfaces of u_i and θ of orders $r = 0$ and $r = 1$ are called strong singular surfaces.

We first present a new approach to derive dynamical compatibility conditions within the linear theory of thermoelastodynamics. Let $\mathscr{M}_4(\Omega)$ be the set of all regular regions contained in Ω. If $\omega \subset \mathscr{M}_4(\Omega)$, then we denote by ω^0 the interior of ω. Let M be an open region in E^3. A function $f : E^3 \to R$ will be called a test function on M if it is of class C^∞ on E^3 and is zero in $E^3\backslash M$.

We assume that the functions u_i, θ, e_{ij}, η, t_{ij}, q_i and v_i satisfy the equations

$$\begin{gathered} v_i = \frac{\partial u_i}{\partial t}\,, \qquad e_{ij} = \frac{1}{2}(u_{i,j} + u_{j,i})\,, \qquad t_{ij} = \lambda e_{rr}\delta_{ij} + 2\mu e_{ij}\,, \\ \eta = \beta e_{rr} + a\theta\,, \qquad q_i = k\theta_{,i}\,, \end{gathered} \tag{3.5.20}$$

and, for all $\omega \subset \mathscr{M}_4(\Omega)$ and for all test functions ζ_i and ζ on ω^0 the following equations

$$\int_\omega \Big(\varrho v_i\frac{\partial \zeta_i}{\partial t} - t_{ij}\frac{\partial \zeta_j}{\partial x_i}\Big)dv = 0\,, \qquad \int_\omega \Big(T_0\eta\frac{\partial \zeta}{\partial t} - q_i\frac{\partial \zeta}{\partial x_i}\Big)dv = 0\,, \tag{3.5.21}$$

hold. Then we say that (u_i, θ) is a weak solution of the field equations (3.5.1) of the linear thermoelastodynamics on Ω. We can prove that if (u_i, θ) is a weak solution on Ω and u_i and θ are of class C^2 on Ω, then (u_i, θ) is a strict solution of (3.5.1) on Ω.

Let (u_i, θ) be a weak solutions of the equations (3.5.1) on Ω. Let σ be a strong singular hypersurface, and let ω be an open subregion of Ω which intersects σ. We assume that u_i and θ are of class C^2 on Ω^+ and Ω^-. We introduce the notations $\omega^+ = \omega \cap \Omega^+$, $\omega^- = \omega \cap \Omega^-$. Since (u_i, θ) is a strict solution of (3.5.1) in Ω^+ and in Ω^-, then

$$t_{ji,j} = \varrho\dot{v}_i\,, \qquad T_0\dot{\eta} = q_{i,i}\,,$$

hold in ω^+ and in ω^-. Thus, we have

$$\int_P \Big\{ \frac{\partial}{\partial t}(\varrho v_i \zeta_i) - \frac{\partial}{\partial x_j}(t_{ji}\zeta_i) \Big\} dv = 0 \,,$$
$$\int_P \Big\{ \frac{\partial}{\partial t}(T_0 \eta \zeta) - \frac{\partial}{\partial x_i}(\zeta q_i) \Big\} dv = 0 \,, \tag{3.5.22}$$

where $P = \omega^+ \cup \omega^-$. Let $(\eta_0, \eta_1, \eta_2, \eta_3)$ be the components of the unit normal at $\partial\omega$. In view of (1.7.31), we find that

$$\eta_0 = -\alpha u_n \,, \qquad \eta_i = \alpha n_i \,,$$

where α is a normalizing function. Since the test functions vanish on $\partial\omega$, from (3.5.23) we get

$$\int_Q \zeta_i \{ \varrho u_n [v_i] + [t_{ji}] n_j \} da = 0 \,,$$
$$\int_Q \zeta \{ T_0 u_n [\eta] + [q_i] n_i \} da = 0 \,, \tag{3.5.23}$$

where $Q = \sigma \cap \omega$. In view of the continuity of the integrands on σ, we obtain the dynamical compatibility conditions in the linear theory

$$\varrho u_n [v_i] + [t_{ji}] n_j = 0 \,, \qquad T_0 u_n [\eta] + [q_i] n_i = 0 \qquad \text{on } \sigma \,. \tag{3.5.24}$$

By (3.5.20), the conditions (3.5.24) can be expressed in the form

$$u_n \Big[\frac{\partial v_i}{\partial t} \Big] + (c_1^2 - 2c_2^2) \Big[\frac{\partial u_j}{\partial x_j} \Big] n_i + c_2^2 \Big[\frac{\partial u_i}{\partial x_j} + \frac{\partial u_j}{\partial x_i} \Big] n_j - \frac{\beta}{\varrho} [\theta] n_i = 0 \,,$$
$$\beta T_0 u_n \Big[\frac{\partial u_j}{\partial x_j} \Big] + c u_n [\theta] + k \Big[\frac{\partial \theta}{\partial x_j} \Big] n_j = 0 \,, \qquad \text{on } \sigma \,. \tag{3.5.25}$$

In what follows we assume that the functions u_i are continuous on Ω. The compatibility conditions for first partial derivatives of the functions u_i are given by (1.7.53), so that

$$\Big[\frac{\partial u_i}{\partial x_j} \Big] = \lambda_i n_j \,, \qquad \Big[\frac{\partial u_i}{\partial t} \Big] = -u_n \lambda_i \,, \qquad \lambda_i = \Big[\frac{\partial u_i}{\partial n} \Big] \,. \tag{3.5.26}$$

It follows from (1.7.51), (3.5.25) and (3.5.26) that

$$
\begin{aligned}
&(u_n^2 - c_2^2)\lambda_i - (c_1^2 - c_2^2)\lambda n_i + \frac{\beta}{\varrho}[\theta]n_i = 0\,, \\
&\beta T_0 u_n \lambda + k\xi + c u_n[\theta] = 0\,,
\end{aligned}
\tag{3.5.27}
$$

where

$$
\xi = \left[\frac{\partial \theta}{\partial n}\right], \qquad \lambda = \lambda_i n_i\,.
$$

Next we form the jump across σ of each term in (3.5.1) and apply the compatibility conditions (1.7.55) to the terms in u_i and conditions (1.7.51), (1.7.56) to the terms in θ. We obtain

$$
\begin{aligned}
&(u_n^2 - c_2^2)\mu_i - (c_1^2 - c_2^2)\{\mu_j n_j n_i + a^{\alpha\beta}(\lambda_j \varphi_{j;\beta})_{;\alpha} n_i + \\
&+ a^{\alpha\beta}\lambda_{;\alpha}\varphi_{i;\beta} - 2u_n \frac{\delta \lambda_i}{\delta t} - \lambda_i \frac{\delta u_n}{\delta t} + 2H\{c_2^2 \lambda_i + \\
&+ (c_1^2 - c_2^2)\lambda n_i\} + \frac{\beta}{\varrho}\{\xi n_i + a^{\alpha\beta}[\theta]_{;\alpha}\varphi_{i;\beta}\} = 0\,, \\
&\beta T_0 u_n \mu_j n_j + k\gamma - \beta T_0 \Big\{\frac{\delta \lambda}{\delta t} - u_n a^{\alpha\beta}(\lambda_j \varphi_{j;\beta})_{;\alpha} + 2u_n H \lambda\Big\} + \\
&+ (c u_n - 2kH)\xi - c\frac{\delta[\theta]}{\delta t} + k a^{\alpha\beta}[\theta]_{;\alpha\beta} + \\
&+ \beta T_0 \lambda_j \Big[a^{\alpha\beta}\varphi_{j;\beta} u_{n;\alpha} + \frac{\delta n_j}{\delta t}\Big] = 0\,,
\end{aligned}
\tag{3.5.28}
$$

where

$$
\mu_i = \left[\frac{\partial^2 u_i}{\partial n^2}\right], \qquad \gamma = \left[\frac{\partial^2 \theta}{\partial n^2}\right].
$$

The equations (3.5.27) and (3.5.28) are the basic results for the discussion of strong singular hypersurfaces. They differ from the equations (3.5.3) and (3.5.11) by the presence of terms in $[\theta]$ and $\delta u_n/\delta t$.

In what follows we consider only the case when σ is a singular hypersurface of order 1 of the functions u_i and θ. Then $[\theta] = 0$ on σ, and the equations (3.5.3) and (3.5.27) are formally identical. It follows that singular hypersurfaces of this type are waves propagating with one of the speeds c_1, c_2. Thus $\delta u_n/\delta t = 0$ and the equations (3.5.11) and (3.5.28) coincide. It follows that the decay laws (3.5.16) and (3.5.19) extend respectively to dilatational and rotational strong thermoelastic waves of order 1. If $u_n = c_1$, then from $(3.5.27)_1$ we obtain $\lambda_i = \lambda n_i$. In this case we have

$$[v_i] = -u_n \lambda n_i \neq 0\,, \qquad [e_{ss}] = \lambda \neq 0\,, \qquad [\omega_{ij}] = 0\,.$$

We note that the discontinuity of velocity is directed normally to $\sigma(t)$. We refer to dilatational strong thermoelastic waves of order 1 as *shock waves.*

If $u_n = c_2$, then from $(3.5.27)_1$ we obtain $\lambda = 0$. It follows from (3.5.6) that

$$[v_s]n_s = 0\,, \qquad [e_{rr}] = 0\,, \qquad [\omega_{ij}] \neq 0\,.$$

We note that the velocity discontinuity is directed tangentially to $\sigma(t)$.

3.6 Exercises

1. *Investigate the propagation of the thermoelastic Rayleigh waves in a homogeneous and isotropic half-space with zero body forces and heat supply.*

Solution. We consider a thermoelastic body which occupies the half space $x_3 > 0$. We assume that body is free to exchange heat by radiation into an atmosphere $x_3 < 0$, maintained at the reference temperature. Moreover, we suppose that the boundary $x_3 = 0$ is a free surface. Thus, we have the conditions

$$t_{i3} = 0\,, \qquad \frac{\partial \theta}{\partial x_3} + k\theta = 0 \qquad \text{on } x_3 = 0\,, \tag{3.6.1}$$

where h is a given constant. We assume a displacement field of the form

$$u_1 = \varphi_{,1} - \psi_{,3}\,, \qquad u_2 = 0\,, \qquad u_3 = \varphi_{,3} + \psi_{,1}\,, \tag{3.6.2}$$

where φ and ψ are functions of x_1 and x_3 only. The equations (3.1.13) reduce to

$$\begin{aligned} &c_1^2 \Delta_2 \varphi - \ddot{\varphi} - \frac{\beta}{\varrho}\theta = 0\,, \qquad c_2^2 \Delta_2 \psi - \ddot{\psi} = 0\,, \\ &k\Delta_2 \theta - \beta T_0 \Delta_2 \dot{\varphi} - c\dot{\theta} = 0\,, \end{aligned} \tag{3.6.3}$$

where

$$\Delta_2 = \frac{\partial^2}{\partial x_1^2} + \frac{\partial^2}{\partial x_3^2}\,.$$

We seek solutions of the form

$$\{\varphi,\, \psi,\, \theta\} = \text{Re}\,\left\{\left[F(x_3),\, G(x_3),\, H(x_3)\right] \exp\left[i(px_1 - \omega t)\right]\right\}. \tag{3.6.4}$$

Substituting (3.6.4) into (3.6.3), we see that the functions F, G and H satisfy the equations

$$F'' - \left(p^2 - \frac{\omega^2}{c_1^2}\right)F = \frac{\beta}{\varrho c_1^2}H\,, \qquad G'' - \left(p^2 - \frac{\omega^2}{c_2^2}\right)G = 0\,,$$
$$H'' - \left(p^2 - \frac{i\omega c}{k}\right)H = -\frac{i\omega\beta T_0}{k}(F'' - p^2 F)\,. \tag{3.6.5}$$

In addition to the conditions (3.6.1) we have the additional condition that the solutions will tend to zero as $x_3 \to \infty$. We therefore take solutions of the form

$$\begin{aligned}
F =& A\exp\left(-x_3\sqrt{p^2 - z_1^2}\right) + B\exp\left(-x_3\sqrt{p^2 - z_2^2}\right),\\
G =& C\exp\left(-x_3\sqrt{p^2 - z_3^2}\right),\\
H =& \frac{\varrho c_1^2}{\beta}\left[A\left(\frac{\omega^2}{c_1^2} - z_1^2\right)\exp\left(-x_3\sqrt{p^2 - z_1^2}\right) + B\left(\frac{\omega^2}{c_1^2} - z_2^2\right)\exp\left(-x_3\sqrt{p^2 - z_2^2}\right)\right],
\end{aligned}$$

where A, B and C are arbitrary constants, z_1^2 and z_2^2 are the roots of the equations

$$z^4 - \left[\frac{\omega^2}{c_1^2} + \frac{i\omega c}{k}(1+\varepsilon)\right]z^2 + \frac{i\omega^3 c}{k c_1^2} = 0\,, \tag{3.6.6}$$

and

$$z_3^2 = \frac{\omega^2}{c_2^2}\,.$$

From the constitutive equations we obtain

$$\begin{aligned}
t_{13} =& \varrho c_2^2(2\varphi_{,13} + \psi_{,11} - \psi_{,33})\,, \qquad t_{23} = 0\,,\\
t_{33} =& \varrho c_1^2 \Delta_2 \varphi + 2\varrho c_2^2(\psi_{,13} - \varphi_{,33}) - \beta\theta\,.
\end{aligned}$$

Thus, the boundary conditions (3.6.1) become

$$\begin{aligned}
(A+B)\left(2 - \frac{\omega^2}{p^2 c_2^2}\right) - 2ib_3 C =& 0\,,\\
2i(b_1 A + b_2 B) + \left(2 - \frac{\omega^2}{p^2 c_2^2}\right)C =& 0\,,\\
A\left(\frac{h}{p} - b_1\right)\left(b_1^2 - 1 + \frac{\omega^2}{p^2 c_2^2}\right) + B\left(\frac{h}{p} - b_2\right)\left(b_2^2 - 1 + \frac{\omega^2}{p^2 c_1^2}\right) =& 0\,,
\end{aligned}$$

where

$$b_i^2 = 1 - \frac{z_i^2}{p^2}\,. \tag{3.6.7}$$

Eliminating A, B and C from among these equations, we obtain the relation

$$\begin{aligned}&\left(2-\frac{\omega^2}{p^2c_2^2}\right)^2\left(b_1^2+b_1b_2+b_2^2-1+\frac{\omega^2}{p^2c_1^2}\right)-4b_1b_2b_3(b_1+b_2)=\\&=\frac{h}{p}\left[\left(2-\frac{\omega^2}{p^2c_2^2}\right)(b_1+b_2)-4b_3\left(b_1b_2+1-\frac{\omega^2}{p^2c_1^2}\right)\right],\end{aligned}\tag{3.6.8}$$

connecting the wave number p and the frequency of the wave ω. If we take ω to be the independent variable and write

$$v=\frac{\omega}{p},\tag{3.6.9}$$

then the equation (3.6.8) becomes

$$\begin{aligned}&\left(2-\frac{v^2}{c_2^2}\right)^2\left(b_1^2+b_1b_2+b_2^2-1+\frac{v^2}{c_1^2}\right)-4b_1b_2b_3(b_1+b_2)=\\&=\frac{hv}{\omega}\left[(b_1+b_2)\left(2-\frac{v^2}{c_2^2}\right)-4b_3\left(b_1b_2+1-\frac{v^2}{c_1^2}\right)\right].\end{aligned}\tag{3.6.10}$$

From (3.6.7) we get

$$b_1^2+b_2^2=2-\frac{1}{p^2}(z_1^2+z_2^2),\qquad b_1^2b_2^2=1-\frac{1}{p^2}(z_1^2+z_2^2)+\frac{1}{p^4}z_1^2z_2^2.$$

Clearly,

$$z_1^2+z_2^2=\frac{\omega^2}{c_1^2}+\frac{i\omega c}{k}(1+\varepsilon),\qquad z_1^2z_2^2=\frac{i\omega^3c}{kc_1^2}.$$

If we use the relations (3.2.15) and (3.6.9), then we obtain

$$\frac{\omega c}{kp^2}=\frac{v^2}{\chi c_1^2},\qquad \frac{\omega^3c}{kp^4c_1^2}=\frac{v^4}{\chi c_1^4},$$

so that we can write

$$\begin{aligned}b_1^2+b_2^2&=2-\left(\frac{v}{c_1}\right)^2-\frac{i}{\chi}\left(\frac{v}{c_1}\right)^2(1+\varepsilon),\\b_1^2b_2^2&=1-\left(\frac{v}{c_1}\right)^2-\frac{i}{\chi}\left(\frac{v}{c_1}\right)^2(1+\varepsilon)+\frac{i}{\chi}\left(\frac{v}{c_1}\right)^4.\end{aligned}$$

We note that

$$b_3^2=1-\left(\frac{v}{c_1}\right)^2.$$

Thus, we may rid equation (3.6.10) of b_1, b_2, b_3 and obtain an equation for v. It follows from (3.6.4) that $1/\operatorname{Re} v^{-1}$ is the phase velocity and $\omega \operatorname{Im} v^{-1}$ is a measure of the attenuation in the x_1-direction. Thus, the thermoelastic surface wave is subject to dispersion since these quantities depend on the frequency ω. If we expand in powers of χ and neglect terms of order $\chi^{1/2}$ we find that the velocity v does not depend upon the frequency and the constant h. In this case we obtain

$$\left(2-\frac{v^2}{c_2^2}\right)^2 = 4\left(1-\frac{v^2}{c_2^2}\right)\left[1-\frac{v^2}{(1+\varepsilon)c_1^2}\right]^{1/2}.$$

If we write $(1+\varepsilon)c_1^2 = v_p^2$, $\quad c_2^2 = v_s^2$, the above equation is the well-known relation connecting v, the velocity of Rayleigh waves, with the velocities of longitudinal and transverse elastic waves.

2. *A homogeneous and isotropic thermoelastic material occupies the region $\{\boldsymbol{x} : x_1^2 + x_2^2 < a^2,\ x_3 \in \mathbb{R}\}$, where a is a positive constant. The lateral surface is free of tractions and body forces and heat sources are absent. Investigate the vibrations of the cylinder when the body is free to exchange heat by radiation.*

Solution. We assume that there is radiation from the lateral surface of the cylinder into a medium maintained at the reference temperature. We shall use the cylindrical coordinate $(r,\ \varphi,\ z)$ referred to the axis of the cylinder. The conditions on the surface $r = a$ are given by

$$\sigma_{rr} = 0\,, \qquad \sigma_{rz} = 0\,, \qquad \frac{\partial \theta}{\partial r} + h\theta = 0 \qquad \text{for } r = a\,. \tag{3.6.11}$$

We note that

$$\operatorname{curl}\,\bar{g} = \left(\frac{1}{r}\frac{\partial g_z}{\partial \varphi} - \frac{\partial g_\varphi}{\partial z}\right)\bar{\varepsilon}_r + \left(\frac{\partial g_r}{\partial z} - \frac{\partial g_r}{\partial r}\right)\bar{\varepsilon}_\varphi + \left[\frac{1}{r}\frac{\partial}{\partial r}(r g_\varphi) - \frac{1}{r}\frac{\partial g_r}{\partial \varphi}\right]\bar{\varepsilon}_z\,,$$

where g_r, g_φ and g_z are the physical components of the vector $\bar{g}$. We use the reppresentation $(3.1.20)_1$ where we replace φ by f and take $\bar{g} = g\bar{\varepsilon}_\varphi$. If we assume that f and g are functions of t, r and z only, then we have

$$u_r = \frac{\partial f}{\partial r} - \frac{\partial g}{\partial z}\,, \qquad u_\varphi = 0\,, \qquad u_z = \frac{\partial f}{\partial z} + \frac{1}{r}\frac{\partial}{\partial r}(rg)\,. \tag{3.6.12}$$

The equations (3.1.13) become

$$\begin{aligned} &c_1^2 \Lambda f - \ddot{f} - \frac{\beta}{\varrho}\theta = 0\,, \\ &k\Lambda\theta - \beta T_0 \Lambda \dot{f} - c\dot{\theta} = 0\,, \\ &c_2^2\left(\Lambda - \frac{1}{r^2}\right)g - \ddot{g} = 0\,, \end{aligned} \tag{3.6.13}$$

where

$$\Lambda = \frac{\partial^2}{\partial r^2} - \frac{1}{r}\frac{\partial}{\partial r} + \frac{\partial^2}{\partial z^2}\,. \tag{3.6.14}$$

Assuming a solution of the equations (3.6.14) of the form

$$\{f,\ g,\ \theta\} = \mathrm{Re}\ \big\{\big[F(r),\ G(r),\ H(r)\big]\exp\big[i(pz-\omega t)\big]\big\}\,, \tag{3.6.15}$$

we find that the functions F, G and H satisfy the system of equations

$$F'' + \frac{1}{r}F' + \Big(\frac{\omega^2}{c_1^2} - p^2\Big)F = \frac{\beta}{\varrho c_1^2}H\,,$$

$$H'' + \frac{1}{r}H' + \Big(\frac{i\omega c}{k} - p^2\Big)H = -\frac{i\omega\beta T_0}{k}\Big(H'' + \frac{1}{r}H' - p^2H\Big)\,,$$

$$G'' + \frac{1}{r}G' + \Big[\Big(\frac{\omega^2}{x^2} - p^2\Big) - \frac{1}{r^2}\Big]G = 0\,.$$

We assume that u_r, u_z and θ remain finite along the z-axis. Then, we obtain

$$\begin{aligned} F &= AJ_\circ(\lambda_1 r) + BJ_\circ(\lambda_2 r)\,,\\ G &= CJ_1(\lambda_3 r)\,,\\ H &= \frac{\varrho c_1^2}{\beta}\Big[\Big(\frac{\omega_2}{c_1^2} - z_1^2\Big)AJ_\circ(\lambda_1 r) + \Big(\frac{\omega_2}{c_1^2} - z_2^2\Big)BJ_\circ(\lambda_2 r)\Big]\,, \end{aligned} \tag{3.6.16}$$

where A, B and C are arbitrary constants, J_s is the Bessel function of first kind and s order, z_α^2 are the roots of the equation (3.6.6), and

$$\lambda_j^2 = z_j^2 - p^2\,, \qquad z_3^2 = \frac{\omega_2}{c_2^2}\,. \tag{3.6.17}$$

The constitutive equations imply

$$\begin{aligned} \sigma_{rr} &= \varrho c_1^2\Lambda f - 2\varrho c_2^2\Big(\frac{1}{r}\frac{\partial f}{\partial r} + \frac{\partial^2 f}{\partial z^2} + \frac{\partial^2 g}{\partial r\partial z}\Big) - \beta\theta\,,\\ \sigma_{rz} &= \varrho c_1^2\Big(2\frac{\partial^2 f}{\partial r\partial z} + \frac{\partial^2 g}{\partial r^2} + \frac{1}{r}\frac{\partial g}{\partial r} - \frac{\partial^2 g}{\partial z^2} - \frac{1}{r^2}g\Big)\,. \end{aligned} \tag{3.6.18}$$

If we use (3.6.15)–(3.6.18), then the boundary conditions (3.6.11) become

$$\big[hJ_\circ(\lambda_1 a) - \lambda_1 J_1(\lambda_1 a)\big]\Big(\frac{\omega^2}{c_1^2} - p^2 - \lambda_1^2\Big)A +$$

$$+\ \big[hJ_\circ(\lambda_2 a) - \lambda_2 J_1(\lambda_2 a)\big]\Big(\frac{\omega^2}{c_1^2} - p^2 - \lambda_2^2\Big)B = 0\,,$$

$$\Big[\Big(2p^2 - \frac{\omega^2}{c_2^2}\Big)aJ_\circ(\lambda_1 a) + 2\lambda_1 J_1(\lambda_1 a)\Big]A + \Big[\Big(2p^2 - \frac{\omega^2}{c_2^2}\Big)aJ_\circ(\lambda_2 a) +$$

$$+\ 2\lambda_2 J_1(\lambda_2 a)\Big]B - 2ipa\big[\lambda_3 aJ_\circ(\lambda_3 a) - J_1(\lambda_3 a)\big]C = 0\,,$$

$$p\lambda_1 J_1(\lambda_1 a)A + p\lambda_2 J_1(\lambda_2 a)B - \Big(p^2 - \frac{\omega^2}{2c_2^2}\Big)J_1(\lambda_3 a)C = 0\,.$$

Eliminating A, B and C from these equations we obtain the equation

$$\begin{aligned}&\Big(2p^2-\frac{\omega^2}{c_2^2}\Big)M+\lambda_1^2\lambda_2^2(\lambda_1^2-\lambda_2^2)J_1(\lambda_1 a)J_1(\lambda_2 a)N=\\&=h\Big(2p^2-\frac{\omega^2}{c_2^2}\Big)(\lambda_1^2-\lambda_2^2)J_\circ(\lambda_1 a)J_1(\lambda_2 a)-hMN\,,\end{aligned}\tag{3.6.19}$$

where

$$\begin{aligned}M=&\lambda_1\Big(p^2+\lambda_1^2-\frac{\omega^2}{c_1^2}\Big)J_1(\lambda_1 a)J_\circ(\lambda_2 a)-\lambda_2\Big(p^2+\lambda_2^2-\frac{\omega^2}{c_1^2}\Big)J_\circ(\lambda_1 a)J_\circ(\lambda_2 a)\,,\\N=&4\frac{p^2}{a}\Big[\lambda_3 a\frac{J_\circ(\lambda_3 a)}{J_1(\lambda_3 a)}-\frac{\omega^2}{c_2^2}\Big]\,.\end{aligned}$$

Equations (3.6.6), (3.6.17) and (3.6.19), together determine the wave number p as a function of the frequency ω.

3. *A homogeneous and isotropic thermoelastic body is in a state of plane strain parallel to the x_1Ox_2-plane. Determine the fundamental solutions in the theory of steady vibrations.*

Solution. In the theory of plane strain, parallel to the x_1Ox_2-plane, we have

$$u_\alpha=u_\alpha(x_1,\,x_2,\,t)\,,\qquad u_3=0\,,\qquad \theta=\theta(x_1,\,x_2,\,t)\,.\tag{3.6.20}$$

We assume that the body force and the heat source are independent of x_3,and that $f_3=0$. The equations (3.1.13) reduce to

$$\begin{aligned}\mu\Delta u_\alpha+(\lambda+\mu)u_{\varrho,\varrho\alpha}-\beta\theta_{,\alpha}+f_\alpha=\varrho\ddot u_\alpha\,,\\k\Delta\theta-\beta T_0\dot u_{\varrho,\varrho}-aT_0\dot\theta=-S\,,\end{aligned}\tag{3.6.21}$$

where $\Delta g=g_{,\alpha\alpha}$. From (3.1.16) we obtain the representation

$$\begin{aligned}u_\alpha=&(\lambda+2\mu)\Big[\Box_1\,\Box_1-\varepsilon\Delta\frac{\partial}{\partial t}\Big]F_\alpha-(\lambda+\mu)\,\Box_2\,F_{\varrho,\varrho\alpha}+\frac{\beta}{\lambda+2\mu}G_{,\alpha}\,,\\\theta=&\frac{\beta T_0\mu}{c}\,\Box_2\,\dot F_{\varrho,\varrho}+\Box_1\,G\,,\end{aligned}\tag{3.6.22}$$

where the fields F_1, F_2 and G are functions on x_1, x_2 and t and satisfy the equations

$$\begin{aligned}\Box_2\Big(\Box_1\,\Box_1-\varepsilon\Delta\frac{\partial}{\partial t}\Big)F_\alpha=-\frac{1}{\mu(\lambda+2\mu)}f_\alpha\,,\\\Big(\Box_1\,\Box_1-\varepsilon\Delta\frac{\partial}{\partial t}\Big)G=-\frac{1}{c}S\,.\end{aligned}\tag{3.6.23}$$

We assume that

$$f_\alpha = \text{Re}\left[f^*_\alpha(x_1,x_2)\exp(-i\omega t)\right], \quad S = \text{Re}\left[S^*(x_1,x_2)\exp(-i\omega t)\right], \qquad (3.6.24)$$

where ω is a given frequency of vibration. We seek for solutions to (3.6.21) in the form

$$u_\alpha = \text{Re}\left[u^*_\alpha(x_1,x_2;\omega)\exp(-i\omega t)\right], \quad \theta = \text{Re}\left[\theta^*(x_1,x_2;\omega)\exp(-i\omega t)\right], \quad (3.6.25)$$

If we introduce the notations

$$F_\alpha = \text{Re}\left[F^*_\alpha(x_1,x_2;\omega)\exp(-i\omega t)\right], \quad G = \text{Re}\left[G^*(x_1,x_2;\omega)\exp(-i\omega t)\right], \quad (3.6.26)$$

then from (3.6.23) we find that the functions F^*_α and G^* satisfy the equations

$$\begin{aligned}(\Delta-\lambda_3^2)\left[(\Delta+\sigma^2)(\Delta+q)+\varepsilon q\Delta\right]F^*_\alpha &= -\frac{c}{k\mu(\lambda+2\mu)}f^*_\alpha,\\ \left[(\Delta+\sigma^2)(\Delta+q)+\varepsilon q\Delta\right]G^* &= -\frac{1}{k}S^*,\end{aligned} \qquad (3.6.27)$$

where we have used the notations (3.2.3) and $\lambda_3 = i\omega/c_2$. We can write

$$(\Delta+\sigma^2)(\Delta+q)+\varepsilon q\Delta = (\Delta-\lambda_1^2)(\Delta-\lambda_2^2),$$

where λ_1^2 and λ_2^2 are defined by

$$\lambda_{1,2}^2 = \frac{1}{2}\left\{-\sigma^2-q-\varepsilon q \pm \left[(\sigma^2+q+\varepsilon q)^2-4q\sigma^2\right]^{\frac{1}{2}}\right\}.$$

Thus the equations (3.6.27) become

$$\begin{aligned}(\Delta-\lambda_1^2)(\Delta-\lambda_2^2)(\Delta-\lambda_3^2)F^*_\alpha &= -\frac{c}{k\mu(\lambda+2\mu)}f^*_\alpha,\\ (\Delta-\lambda_1^2)(\Delta-\lambda_2^2)G^* &= -\frac{1}{k}S^*.\end{aligned} \qquad (3.6.28)$$

From (3.6.22) we obtain

$$\begin{aligned}u^*_\alpha =& \frac{1}{c}(\lambda+2\mu)k(\Delta-\lambda_1^2)(\Delta-\lambda_2^2)F^*_\alpha - \frac{1}{c}(\lambda+\mu)k\Big(\Delta+\\ &+q+\frac{\lambda+2\mu}{\lambda+\mu}q\varepsilon\Big)F^*_{\varrho,\varrho,\alpha} + \frac{\beta}{\lambda+2\mu}G^*_{,\alpha},\\ \theta^* =& -\frac{1}{c}\beta T_0\mu i\omega(\Delta-\lambda_3^2)F^*_{\varrho,\varrho}+(\Delta+\sigma^2)G^*.\end{aligned} \qquad (3.6.29)$$

If $f^*_\alpha = \delta_{\alpha\beta}\delta(x-y)$, $S^* = 0$, then we take $F^*_\alpha = g\delta_{\alpha\beta}$, $G^* = 0$. From (3.6.29) we obtain

$$\begin{aligned} u^*_\alpha(\beta) =& \frac{1}{c}(\lambda+2\mu)k(\Delta-\lambda_1^2)(\Delta-\lambda_2^2)g\delta_{\alpha\beta} - \frac{1}{c}(\lambda+\mu)k\Big(\Delta + \\ &+ q + \frac{\lambda+2\mu}{\lambda+\mu}q\varepsilon\Big)g_{,\alpha\beta}\,, \\ \theta^{*(\alpha)} =& -\frac{1}{c}\beta T_0\mu i\omega(\Delta-\lambda_3^2)g_{,\alpha}\,, \end{aligned} \tag{3.6.30}$$

where g satisfies the equation

$$(\Delta-\lambda_1^2)(\Delta-\lambda_2^2)(\Delta-\lambda_3^2)g = -a\delta(x-y)\,. \tag{3.6.31}$$

We have used the notation $a = c[k\mu(\lambda+2\mu)]^{-1}$.
In what follows we asume that $\lambda_1^2 \neq \lambda_2^2 \neq \lambda_3^2$. If the functions g_1, g_2, g_3 satisfy the equations

$$\begin{aligned} &(\Delta-\lambda_1^2)g_1 = -a\delta(x-y)\,, \quad (\Delta-\lambda_2^2)g_2 = -a\delta(x-y)\,, \\ &(\Delta-\lambda_3^2)g_3 = -a\delta(x-y)\,, \end{aligned} \tag{3.6.32}$$

then the solution of the equation (3.6.3) can be written in the form

$$g = \frac{g_1}{(\lambda_2^2-\lambda_1^2)(\lambda_3^2-\lambda_1^2)} + \frac{g_2}{(\lambda_1^2-\lambda_2^2)(\lambda_3^2-\lambda_2^2)} + \frac{g_3}{(\lambda_1^2-\lambda_3^2)(\lambda_2^2-\lambda_3^2)}\,. \tag{3.6.33}$$

From (3.6.32) we obtain

$$g_j = \frac{a}{2\pi}K_\circ(\lambda_j r)\,, \tag{3.6.34}$$

where $K_\circ$ denotes the modified Bessel function of zeroth order and $r = (x_1^2+x_2^2)^{1/2}$. Thus, from (3.6.33) and (3.6.34) we obtain the functions $u^{*(\beta)}_\alpha$. Next we suppose that $f^*_\alpha = 0$ and $S^* = \delta(x-y)$. In this case we take $F^*_\alpha = 0$ and from (3.6.29) we obtain

$$u^{*(3)}_\alpha = \frac{\beta}{\lambda+2\mu}G^*_{,\alpha}\,, \qquad \theta^{*(3)} = (\Delta+\sigma^2)G^*\,, \tag{3.6.35}$$

where the function G^* satisfies the equation

$$(\Delta-\lambda_1^2)(\Delta-\lambda_2^2)G^* = -\frac{1}{k}\delta(x-y)\,. \tag{3.6.36}$$

The solution of the equation (3.6.35) can be expressed in the form

$$G^* = \frac{1}{\lambda_1^2-\lambda_2^2}(G_1-G_2)\,,$$

where G_α satisfy the equations

$$(\Delta - \lambda_1^2)G_1 = -\frac{1}{k}\delta(x-y)\,, \qquad (\Delta - \lambda_2^2)G_2 = -\frac{1}{k}\delta(x-y)\,.$$

Thus we find

$$G^* = \frac{1}{2\pi k(\lambda_1^2 - \lambda_2^2)}\left[K_\circ(\lambda_1 r) - K_\circ(\lambda_2 r)\right]. \tag{3.6.37}$$

From (3.6.35) and (3.6.37) we obtain the functions $u_\alpha^{*(3)}$, $\theta^{*(3)}$. The functions $u_\alpha^{*(j)}$, $\theta^{*(j)}$ $(j = 1, 2, 3)$ are the fundamental solutions in the two-dimensional theory of steady vibrations.

4. *The surface of the half-space $x_1 > 0$ is free of tractions and is subjected to an impulsive uniform temperature. Investigate the dyamical problem assuming that the body forces, the heat source and the initial data are zero.*

Solution. We assume that the relations (3.4.2) hold. The boundary conditions are

$$t_{11}(0,t) = 0\,, \qquad \theta(0,t) = \theta_0\delta(t)\,, \tag{3.6.38}$$

where θ_0 is a prescribed constant and δ is the Dirac delta.
The initial conditions are

$$u_1(x_1,0) = 0\,, \qquad \dot{u}_1(x_1,0) = 0\,, \qquad \theta(x_1,0) = 0\,. \tag{3.6.39}$$

In the context of the uncoupled theory, the equation of energy becomes

$$\theta_{,11} - \frac{c}{k}\dot{\theta} = 0\,. \tag{3.6.40}$$

The equations of motion reduce to

$$u_{1,11} - \frac{1}{c_1^2}\ddot{u}_1 = m\theta_{,1}\,, \tag{3.6.41}$$

where $m = \beta/(\lambda+2\mu)$. Taking the Laplace transform of the equations (3.6.40) and (3.6.41), in view of (3.6.39), we obtain

$$\begin{aligned} &\bar{\theta}_{,11} - \frac{pc}{k}\bar{\theta} = 0\,, \\ &\bar{u}_{1,11} - \frac{p^2}{c_1^2}\bar{u}_1 = m\bar{\theta}_{,1}\,. \end{aligned} \tag{3.6.42}$$

From (3.6.42) we find

$$\bar{\theta} = C\exp\left(-x_1\sqrt{\frac{pc}{k}}\right),$$

where C is an arbitrary constant. The condition $(3.6.38)_2$ implies that $\bar{\theta}(0,\, p) = \theta_0$ so that

$$\bar{\theta} = \theta_0 \exp\left(-x_1\sqrt{\frac{pc}{k}}\right). \tag{3.6.43}$$

The function $\bar{u}_1$ satisfies the equation

$$\bar{u}_{1,11} - \frac{p^2}{c_1^2}\bar{u}_1 = -m\theta_0\sqrt{\frac{pc}{k}}\exp\left(-x_1\sqrt{\frac{pc}{k}}\right),$$

so that

$$\bar{u}_1 = B\exp\left(-\frac{p}{c_1}x_1\right) - \frac{m\theta_0 c_1^2\sqrt{ck}}{(cc_1^2 - kp)\sqrt{p}}\exp\left(-x_1\sqrt{\frac{pc}{k}}\right),$$

where B is an arbitrary constant. Thus, we have

$$\begin{aligned}\bar{t}_{11} =&(\lambda+2\mu)\bar{u}_{1,11} - \beta\bar{\theta} = \\ =& -p\frac{(\lambda+2\mu)}{c_1}B\exp\left(-\frac{p}{c_1}x_1\right) - \frac{\beta\theta_0 kp}{cc_1^2 - kp}\exp\left(-x_1\sqrt{\frac{pc}{k}}\right).\end{aligned} \tag{3.6.44}$$

The boundary condition $\bar{t}_{11}(0,\, p) = 0$ implies that

$$B = \frac{\beta\theta_0 k c_1}{(\lambda+2\mu)(cc_1^2 - kp)}. \tag{3.6.45}$$

From (3.6.44) and (3.6.45) we obtain

$$\bar{t}_{11}(x_1,\, p) = \frac{\beta\theta_0 kp}{cc_1^2 - kp}\left[\exp\left(-x_1\sqrt{\frac{pc}{k}}\right) - \exp\left(-\frac{p}{c_1}x_1\right)\right]. \tag{3.6.46}$$

We introduce the notations $y = c_1 c x_1/k, \quad s = cc_1^2 t/k.$
The original functions θ and t_{11} can be expressed in the form

$$\begin{aligned}\theta(y,\, s) =& \frac{\theta_0}{2\sqrt{\pi s}}\exp\left(-y^2/4s\right), \\ t_{11}(y,\, s) =& -\frac{c_1^2\beta c\theta_0}{k}\left[K_1(y,\, s) - K_2(y,\, s)\right],\end{aligned}$$

where

$$\begin{aligned}K_1(y,\, s) =& \frac{1}{2}\exp(s)\left[\exp(y)\,\mathrm{erfc}\left(\frac{y}{2\sqrt{s}} + \sqrt{s}\right) + \exp(-y)\,\mathrm{erfc}\left(\frac{y}{2\sqrt{s}} - \sqrt{s}\right)\right] + \\ &+ \frac{1}{2s\sqrt{\pi s}}\exp\left(-y^2/4s\right), \\ K_2(y,\, s) =& \exp(s-y)H(s-y) + \delta(s-y),\end{aligned}$$

H being the Headviside function.

In the framework of coupled theory of thermoelasticity the problem can be studied by using the procedure of Section 3.4.

5. *A homogeneous and isotropic thermoelastic material occupies the region* $\{\boldsymbol{x} : x_1^2+x_2^2+x_3^2 > a^2\}$, *where* $a > 0$. *The surface of the cavity is subjected to a spatially uniform pressure and to a spatially uniform temperature. Discuss the deformation of the body assuming that the body forces, the heat source and the initial data are zero.*

Solution. We illustrate the use of the spherical coordinates (R, φ, ψ) in the solution of the coupled equations of thermoelasticity. From hypotheses we can see that the thermoelastic disturbances are radially symmetrical. the conditions on the surface of cavity are

$$\sigma_{RR}(a, t) = -g(t), \qquad \theta(a, t) = f(t), \qquad t \geq 0, \tag{3.6.47}$$

where f and g are prescribed functions. The initial conditions are

$$u_R(R, 0) = 0, \qquad \dot{u}_R(R, 0) = 0, \qquad \theta(R, 0) = 0, \qquad R > a. \tag{3.6.48}$$

We seek the solution of equations (3.1.13) in the form

$$u_R = \frac{\partial}{\partial r}\left[R^{-1}F(R, t)\right], \qquad u_\varphi = u_\psi = 0, \qquad \theta = \frac{1}{R}G(R, t), \tag{3.6.49}$$

where F and G are unknown functions. The field equations reduce to

$$\begin{aligned} c_1^2\frac{\partial^2 F}{\partial R^2} - \frac{\partial^2 F}{\partial t^2} &= \frac{\beta}{\varrho}G, \\ k\frac{\partial^2 G}{\partial R^2} - c\frac{\partial G}{\partial t} &= \beta T_0\frac{\partial^3 F}{\partial R^2 \partial t}. \end{aligned} \tag{3.6.50}$$

The conditions (3.6.48) are easily shown to be equivalent to

$$F(R, 0) = 0, \qquad \dot{F}(R, 0) = 0, \qquad G(R, 0) = 0, \qquad R > a. \tag{3.6.51}$$

so that if we introduce the Laplace transforms

$$\bar{F}(R, p) = \int_0^\infty \exp(-ps)F(R, s)ds, \qquad \bar{G}(R, p) = \int_0^\infty \exp(-ps)G(R, s)ds,$$

we find that the pair of partial differential equations (3.6.51) imply that

$$\begin{aligned} \left(c_1^2\frac{d^2}{dR^2} - p^2\right)\bar{F} &= \frac{\beta}{\varrho}\bar{G}, \\ \left(k\frac{d}{dR} - cp\right)\bar{G} &= \beta T_0 p\frac{d^2\bar{F}}{dR^2}. \end{aligned} \tag{3.6.52}$$

Similarly the Laplace transforms of the conditions (3.6.47) can be shown to be equivalent to

$$\varrho c_1^2 R^2\frac{d^2\bar{F}}{dR^2} - 4\varrho c_2^2\left(R\frac{d}{dR} - 1\right)\bar{F} - \beta R^2\bar{G} = -R^3\bar{g}, \quad \bar{G} = a\bar{f} \quad \text{for } R = a. \tag{3.6.53}$$

If we impose the conditions at infinity

$$u_R(R,\,t) \to 0, \qquad \theta(R,\,t) \to 0 \qquad \text{as } R \to \infty,$$

we see that the appropriate solution of the equations (3.6.52) is

$$\begin{aligned} \bar{F} =& A\exp\left[-z_1(R-a)\right] + B\exp\left[-z_2(R-a)\right], \\ \bar{G} =& \frac{\varrho}{\beta}c_1^2\left\{\left(z_1^2 - \frac{p^2}{c_1^2}\right)A\exp\left[-z_1(R-a)\right] + \left(z_2^2 - \frac{p^2}{c_1^2}\right)B\exp\left[-z_2(R-a)\right]\right\}, \end{aligned}$$

where z_1 and z_2 are positive square roots of the square roots of the equation

$$z^4 - \left[\frac{p^2}{c_1^2} + \frac{cp}{k}(1+\varepsilon)\right]z^2 + \frac{cp^3}{kc_1^3} = 0,$$

and A and B are arbitrary constants. From (3.6.53) we find that A and B are given by

$$\begin{aligned} dA =& \frac{a^3}{\varrho}\left(z_2^2 - \frac{p^2}{c_1^2}\right)\bar{g} + \frac{a\beta}{\varrho c_1^2}\left[p^2a^2 + 4c_2^2(az_2+1)\right]\bar{f}, \\ dB =& -\frac{a^3}{\varrho}\left(z_1^2 - \frac{p^2}{c_1^2}\right)\bar{g} - \frac{a\beta}{\varrho c_1^2}\left[p^2a^2 + 4c_2^2(az_1+1)\right]\bar{f}, \end{aligned}$$

where

$$d = (z_1 - z_2)\left[(z_1+z_2)(p^2a^2 + 4c_2^2) + 4\left(z_1z_2 + \frac{p^2}{c_1^2}\right)ac_2^2\right].$$

The calculation of the inverse Laplace transforms for special forms of the functions f and g can be made by the approximate methods presented in Section 3.4.

4 Problems of Equilibrium

4.1 Thermoelastic states. Mean strain and mean stress

In this chapter, we consider the equilibrium theory of classical thermoelasticity. In the absence of time dependence, the basic equations from Section 3.1 reduce to the equations of equilibrium

$$t_{ji,j} + f_i = 0, \tag{4.1.1}$$

the equilibrium energy equation

$$q_{i,i} + S = 0, \tag{4.1.2}$$

the constitutive equations

$$t_{ij} = C_{ijrs}e_{rs} - \beta_{ij}\theta, \tag{4.1.3}$$

$$q_i = k_{ij}\theta_{,j}, \tag{4.1.4}$$

and the geometrical equations

$$e_{ij} = \frac{1}{2}(u_{i,j} + u_{j,i}), \tag{4.1.5}$$

on B. The constitutive coefficients satisfy the properties (3.1.4).

To the field equations (4.1.1)–(4.1.5) we adjoin the boundary conditions

$$u_i = \tilde{u}_i \quad \text{on } \bar{S}_1, \qquad t_{ji}n_j = \tilde{t}_i \quad \text{on } S_2, \tag{4.1.6}$$

$$\theta = \tilde{\theta} \quad \text{on } \bar{S}_3, \qquad q_i n_i = \tilde{q} \quad \text{on } S_4, \tag{4.1.7}$$

where $\tilde{u}_i$, $\tilde{t}_i$, $\tilde{\theta}$ and $\tilde{q}$ are prescribed functions and S_i $(i = 1, 2, 3, 4)$ are parts of ∂B such that $\bar{S}_1 \cup S_2 = \bar{S}_3 \cup S_4 = \partial B$, $S_1 \cap S_2 = S_3 \cap S_4 = 0$.

The temperature field θ is the solution of the boundary value problem (4.1.2), (4.1.4), (4.1.7). Generally, we shall treat the temperature field as having already been so determined.

By an admissible state we mean an ordered array $\{u_i, e_{ij}, t_{ij}\}$ with the properties : (i) $u_i \in C^2(B)$, $u_i \in C^1(\bar{B})$; (ii) e_{ij} is a continuous symmetric tensor field on $\bar{B}$; (iii) $t_{ij} \in C^1(B)$, $t_{ij}, t_{ij,j} \in C^0(\bar{B})$, $t_{ij} = t_{ji}$. If we define the addition of admissible states and the multiplication of an admissible state with a scalar by

$$\{u_i, e_{ij}, t_{ij}\} + \{u'_i, e'_{ij}, t'_{ij}\} = \{u_i + u'_i, e_{ij} + e'_{ij}, t_{ij} + t'_{ij}\},$$

$$\lambda\{u_i, e_{ij}, t_{ij}\} = \{\lambda u_i, \lambda e_{ij}, \lambda t_{ij}\},$$

then the set of all admissible states is a vector space.

We assume that: (*i*) C_{ijrs} and β_{ij} are continuously differentiable on $\bar{B}$ and satisfy (3.1.4); (*ii*) f_i are continuous on $\bar{B}$; (*iii*) $\tilde{u}_i$ are continuous on $\bar{S}_1$, and $\tilde{t}_i$ are piecewise regular on S_2.

By a *thermoelastic state* corresponding to the body force $\boldsymbol{f}$ and temperature field θ we mean an admissible state $\{u_i, e_{ij}, t_{ij}\}$ that satisfies the equations (4.1.1), (4.1.3), (4.1.5) on B.

Thus, the determination of the thermoelastic equilibrium is reduced to the solution of two boundary-value problems. The first boundary value problem is characterized by the equations (4.1.2), (4.1.4) on B and boundary conditions (4.1.7). The second boundary value problem consists in the finding of a thermoelastic state corresponding to the body force $\boldsymbol{f}$ and temperature field θ which satisfies the boundary conditions (4.1.6). If such a thermoelastic state exists, it is called a *solution of the mixed problem.*

The ordered array $\Lambda = (f_i, \tilde{u}_i, \tilde{t}_i, \theta)$ is called *external data system* on B for the mixed problem of thermoelastostatics. As in Section 2.9, we can see that $\boldsymbol{u}$ is a displacement field corresponding to a solution of thermoelastic mixed problem for B with the external data system $\Lambda = (f_i, \tilde{u}_i, \tilde{t}_i, \tilde{\theta})$ if and only if $\boldsymbol{u}$ is a displacement field corresponding to a solution of the mixed problem of elastostatics for B with the external data system $\Lambda^* = \left(f_i - (\beta_{ij}\theta)_{,j}, \tilde{u}_i, \tilde{t}_i + \beta_{ij}\theta n_j\right)$. This result is known as the body force analogy, and it allows us to bring all of the results of isothermal elastostatics to bear on the equilibrium theory of thermoelasticity.

When $S_2 = \partial B$ (i.e. S_1 is empty), the resulting problem is called the *traction problem.* Let us present a characterization of the traction problem in terms of stress. First, we record without proof the following well-known proposition.

Theorem 4.1.1. *The strain tensor e_{ij} corresponding to a class C^3 displacement field satisfies the equations of compatibility*

$$\varepsilon_{ijk}\varepsilon_{smn}e_{jm,kn} = 0\,. \tag{4.1.8}$$

Conversely, let B be simply-connected, and let e_{ij} be a class C^2 symmetric tensor field on B that satisfies the conditions (4.1.8). Then there exists a displacement field $\boldsymbol{u}$ of class C^3 on B such that e_{ij} and u_k satisfy the equations (4.1.5).

We assume that the tensor C_{ijrs} is invertible. Then, the constitutive equations (4.1.3) can be written in the form

$$e_{ij} = A_{ijrs}t_{rs} + \alpha_{ij}\theta\,, \tag{4.1.9}$$

where

$$\alpha_{ij} = A_{ijrs}\beta_{rs}\,. \tag{4.1.10}$$

We suppose that the domain B is simply-connected and that A_{ijrs}, α_{ij} and θ are of class C^2 on $\bar{B}$. In view of Theorem 4.1.1 we conclude that the symmetric tensor t_{ij} of class C^2 on $\bar{B}$ is a stress tensor field corresponding to a solution of the traction problem if and only if

$$t_{ji,j} + f_i = 0 \qquad \text{on } B\,, \tag{4.1.11}$$

$$\varepsilon_{ijk}\varepsilon_{smn}(A_{jmpq}t_{pq} + \alpha_{jm}\theta)_{,kn} = 0 \qquad \text{on } \bar{B}, \tag{4.1.12}$$

$$t_{ji}n_j = \tilde{t}_i \qquad \text{on } \partial B\,. \tag{4.1.13}$$

We denote by g^* the mean value of the continuous function g on $\bar{B}$, i.e.

$$g^*(B) = \frac{1}{V(B)}\int_B g\,dv\,,$$

where $V(B)$ is the volume of B.

Theorem 4.1.2. *If u_i and e_{rs} are continuous functions on $\bar{B}$ that satisfy the equations (4.1.5), then*

$$e^*_{ij}(B) = \frac{1}{2V(B)}\int_{\partial B}(u_i n_j + u_j n_i)\,da\,. \tag{4.1.14}$$

Moreover, if θ is continuous on B and B is occupied by a homogeneous material, then

$$t^*_{ij}(B) = -\beta_{ij}\theta^*(B) + \frac{1}{V(B)}C_{ijrs}\int_{\partial B} u_r n_s\,da\,. \tag{4.1.15}$$

Proof. The first assertion follows by using the divergence theorem. Then, since B is homogeneous,

$$t^*_{ij}(B) = C_{ijrs}e^*_{rs}(B) - \beta_{ij}\theta^*(B)\,. \tag{4.1.16}$$

It follows from (3.1.4), (4.1.14) and (4.1.16) that (4.1.15) holds. □

If $u_i = 0$ on ∂B, then (4.1.15) implies that

$$t^*_{ij}(B) = -\beta_{ij}\theta^*(B)\,. \tag{4.1.17}$$

Theorem 4.1.3. *If $\{u_i,\, e_{ij},\, t_{ij}\}$ is a thermoelastic state corresponding to the body force $\boldsymbol{f}$ and temperature field θ, then*

$$t^*_{ij}(B) = \frac{1}{V(B)}\Big(\int_B x_i f_j\,dv + \int_{\partial B} x_i t_j\,da\Big)\,. \tag{4.1.18}$$

Moreover, if the body is homogeneous and C_{ijrs} is invertible, then

$$e^*_{ij}(B) = \frac{1}{V(B)}A_{ijrs}\Big(\int_B x_r f_s\,dv + \int_{\partial B} x_r t_s\,da\Big) + \alpha_{ij}\theta^*(B)\,. \tag{4.1.19}$$

Proof. In view of the relation

$$\int_{\partial B} x_i t_{sj} n_s \, da + \int_B x_i f_j \, dv = \int_B \big[x_i(t_{sj,s} + f_j) + t_{ij}\big] dv \,,$$

we obtain (4.1.18). Since B is homogeneous,

$$e^*_{ij}(B) = A_{ijrs} t^*_{rs}(B) + \alpha_{ij}\theta^*(B) \,. \tag{4.1.20}$$

The relations (4.1.18) and (4.1.20) imply (4.1.19). □

If the body forces are zero and $t_{ji}n_j = 0$ on ∂B, then from (4.1.18) we obtain

$$t^*_{ij}(B) = 0 \,,$$

and the relation (4.1.19) implies that

$$e^*_{ij}(B) = \alpha_{ij}\theta^*(B) \,.$$

4.2 Reciprocity. Variational theorems

While the results presented in this section can be obtained by applying the body force analogy to their counterparts in the isothermal theory, we prefer to derive them directly in order to facilitate the reading.

Let us assume that the body B is subject to two external data systems

$$L^{(\alpha)} = \big(f_i^{(\alpha)},\, \tilde{u}_i^{(\alpha)},\, \tilde{t}_i{}^{(\alpha)},\, \theta^{(\alpha)}\big) \qquad (\alpha = 1,\, 2) \,. \tag{4.2.1}$$

Let $s^{(\alpha)} = \{u_i^{(\alpha)},\, e_{ij}^{(\alpha)},\, t_{ij}^{(\alpha)}\}$ be a solution corresponding to the external data system $L^{(\alpha)}$.

Theorem 4.2.1. *If $s^{(\alpha)}$ are elastic states corresponding to the external data systems $L^{(\alpha)}$, $(\alpha = 1,\, 2)$, then*

$$\begin{aligned} &\int_B \big[f_i^{(1)} - (\beta_{ij}\theta^{(1)})_{,j}\big] u_i^{(2)} \, dv + \int_{\partial B} \big(t_i^{(1)} + \beta_{ij}\theta^{(1)} n_j\big) u_i^{(2)} \, da = \\ &= \int_B \big[f_i^{(2)} - (\beta_{ij}\theta^{(2)})_{,j}\big] u_i^{(1)} \, dv + \int_{\partial B} \big(t_i^{(2)} + \beta_{ij}\theta^{(2)} n_j\big) u_i^{(1)} \, da \,, \end{aligned} \tag{4.2.2}$$

where

$$t_i^{(\alpha)} = t_{ji}^{(\alpha)} n_j^{(\alpha)} \,. \tag{4.2.3}$$

Proof. We introduce the notations

$$2W_{\alpha\beta} = (t_{ij}^{(\alpha)} + \beta_{ij}\theta^{(\alpha)})e_{ij}^{(\beta)} . \tag{4.2.4}$$

It follows from (4.1.3) and (4.2.3) that

$$2W_{\alpha\beta} = C_{ijrs}e_{ij}^{(\beta)}e_{rs}^{(\alpha)} . \tag{4.2.5}$$

In view of (3.1.4) and (4.2.5), we get

$$W_{12} = W_{21} . \tag{4.2.6}$$

On the other hand, by (4.1.1) and (4.1.5) we obtain

$$\begin{aligned} 2W_{\alpha\beta} &= (t_{ij}^{(\alpha)} + \beta_{ij}\theta^{(\alpha)})u_{i,j}^{(\beta)} = (t_{ij}^{(\alpha)}u_i^{(\beta)} + \beta_{ij}\theta^{(\alpha)}u_i^{(\beta)})_{,j} - \\ &- t_{ij,j}^{(\alpha)}u_i^{(\beta)} - (\beta_{ij}\theta^{(\alpha)})_{,j}u_i^{(\beta)} = (t_{ij}^{(\alpha)}u_i^{(\beta)} + \beta_{ij}\theta^{(\alpha)}u_i^{(\beta)})_{,j} + \\ &+ f_i^{(\alpha)}u_i^{(\beta)} - (\beta_{ij}\theta^{(\alpha)})_{,j}u_i^{(\beta)} . \end{aligned} \tag{4.2.7}$$

By the divergence theorem and (4.2.7),

$$\begin{aligned} 2\int_B W_{\alpha\beta}\, dv &= \int_{\partial B} (t_i^{(\alpha)} + \beta_{ij}\theta^{(\alpha)}n_j)u_i^{(\beta)}\, da + \\ &+ \int_B [f_i^{(\alpha)} - (\beta_{ij}\theta^{(\alpha)})_{,j}]u_i^{(\beta)}\, dv . \end{aligned} \tag{4.2.8}$$

From (4.2.6) and (4.2.8) we obtain the desired result. □

We note that the relation (4.2.1) can be written in the form

$$\begin{aligned} &\int_B f_i^{(1)}u_i^{(2)}\, dv + \int_{\partial B} t_i^{(1)}u_i^{(2)}\, da + \int_B \beta_{ij}\theta^{(1)}e_{ij}^{(2)}\, dv = \\ &= \int_B f_i^{(2)}u_i^{(1)}\, dv + \int_{\partial B} t_i^{(2)}u_i^{(1)}\, da + \int_B \beta_{ij}\theta^{(2)}e_{ij}^{(1)}\, dv . \end{aligned} \tag{4.2.9}$$

Let us assume that

$$f_i^{(2)} = \delta_{ik}\delta(x-y) , \quad \tilde{u}_i^{(2)} = 0 , \quad \tilde{t}_i{}^{(2)} = 0 , \quad \theta^{(2)} = 0 , \tag{4.2.10}$$

where $\delta(\cdot)$ is the Dirac measure. The displacement field corresponding to this external data system is denoted by $U_i^{(k)}$. Let u_i be the displacement field corresponding to the following external data system

$$f_i^{(1)} = 0\,, \quad \tilde{u}_i^{(1)} = 0\,, \quad \tilde{t}_i{}^{(1)} = 0\,, \quad \theta^{(1)} = \theta\,. \tag{4.2.11}$$

If we apply the reciprocity relation (4.2.9) for the external data systems (4.2.10) and (4.2.11), then we obtain

$$u_k(y) = \int_B \beta_{ij}(x)\theta(x)U_{i,j}^{(k)}(x,y)\,dv_x\,. \tag{4.2.12}$$

This result is due to Maysel [248].

Let $L^{(1)} = (f_i,\, \tilde{u}_i,\, \tilde{t}_i,\, \theta)$ and $s^{(1)} = \{u_i,\, e_{ij},\, t_{ij}\}$. Then, from (4.2.5) and (4.2.8) we obtain

$$\begin{aligned} 2\int_B C_{ijrs}e_{ij}e_{rs}\,dv = \int_{\partial B}&(t_i + \beta_{ij}\theta)u_i\,da\, + \\ &+ \int_B \big[f_i - (\beta_{ij}\theta)_{,j}\big]u_i\,dv\,. \end{aligned} \tag{4.2.13}$$

This relation leads to the following uniqueness result.

Theorem 4.2.2. *If the tensor C_{ijrs} is positive definite, then any two solutions of the mixed problem are equal modulo a rigid displacement. Moreover, if S_1 is non-empty, then the mixed problem has at most one solution.*

Proof. Let $s = \{u_i^0,\, e_{ij}^0,\, t_{ij}^0\}$ be the difference of two solutions of the mixed problem.

According to the linearity of the problem, s is a solution corresponding to null external data. From (4.2.13) we get

$$\int_B C_{ijrs}e_{ij}^0 e_{rs}^0\,dv = 0\,.$$

Since C_{ijrs} is positive definite we deduce that $e_{ij}^0 = 0$ and therefore

$$u_i^0 = a_i + \varepsilon_{ijr}b_j x_r\,, \tag{4.2.14}$$

where a_i and b_i are arbitrary constants. If S_1 is non-empty, then $a_i = b_i = 0$. □

In what follows we will establish several variational principles for thermoelastostatics.

By a kinematically admissible state we mean an admissible state $A = \{u_i,\, e_{ij},\, t_{ij}\}$ that satisfies the strain-displacement relation (4.1.5), the stress-strain-temperature relation (4.1.3) and the boundary condition

$$u_i = \tilde{u}_i \qquad \text{on} \quad \bar{S}_1\,. \tag{4.2.15}$$

Theorem 4.2.3. *Let M be the set of all kinematically admissible states, and let F be the functional defined on M by*

$$\begin{aligned} F(A) = \frac{1}{2}\int_B C_{ijrs}e_{ij}e_{rs}\,dv + \int_B (\beta_{ij}\theta)_{,j}u_i\,dv - \\ - \int_B f_iu_i\,dv - \int_{S_2}\beta_{ij}\theta n_j\,da - \int_{S_2}\tilde{t}_iu_i\,da\,. \end{aligned} \tag{4.2.16}$$

Then A is a solution of the mixed problem if and only if

$$F(A) \le F(A^*)\,, \tag{4.2.17}$$

for any $A^* \in M$.

Proof. Let $A = \{u_i,\, e_{ij},\, t_{ij}\}$ be a solution of the mixed problem, let $A^* = \{u_i^*,\, e_{ij}^*,\, t_{ij}^*\} \in M$, and define $A' = A^* - A$. We have $A' = \{u_i',\, e_{ij}',\, t_{ij}'\}$ where $u_i' = u_i^* - u_i$, $e_{ij}' = e_{ij}^* - e_{ij}$, $t_{ij}' = t_{ij}^* - t_{ij}$. Clearly, A' is an admissible state and

$$2e_{ij}' = u_{i,j}' + u_{j,i}'\,, \qquad t_{ij}' = C_{ijrs}e_{rs}' \quad \text{on } B\,, \qquad u_i' = 0 \quad \text{on } \bar{S}_1\,. \tag{4.2.18}$$

By (4.2.16),

$$\begin{aligned} F(A^*) =& F(A) + \frac{1}{2}\int_B C_{ijrs}e_{ij}'e_{rs}'\,dv + \int_B (\beta_{ij}\theta)_{,j}u_i'\,dv - \int_B f_iu_i'\,dv - \\ & - \int_{S_2}\beta_{ij}\theta n_ju_i'\,da - \int_{S_2}\tilde{t}_iu_i'\,da + \int_B C_{ijrs}e_{rs}e_{ij}'\,dv = F(A) + \\ & + \frac{1}{2}\int_B C_{ijrs}e_{ij}'e_{rs}'\,dv + \int_B (\beta_{ij}\theta)_{,j}u_i'\,dv - \int_B f_iu_i'\,dv - \\ & - \int_{S_2}\beta_{ij}\theta n_ju_i'\,da - \int_{S_2}\tilde{t}_iu_i'\,da + \int_B (t_{ij} + \beta_{ij}\theta)u_{i,j}'\,dv = \\ & = F(A) + \frac{1}{2}\int_B C_{ijrs}e_{ij}'e_{rs}'\,dv - \int_B (t_{ji,j} + f_i)u_i'\,dv + \int_{S_2}(t_{ji}n_j - \tilde{t}_i)u_i'\,da\,. \end{aligned} \tag{4.2.19}$$

Since A is a solution of the problem, (4.2.19) implies

$$F(A^*) - F(A) = \frac{1}{2}\int_B C_{ijrs} e'_{ij} e'_{rs}\, dv\,. \tag{4.2.20}$$

Thus, since C_{ijrs} is positive definite we conclude that (4.2.17) holds. □

Theorem 4.2.3 is a counterpart of principle of minimum potential energy in the isothermal theory.

The standard method of obtaining an approximate solution to the mixed problem is to minimize the functional F over a restricted class of functions (see, e.g. Zienkiewicz [370], Mikhlin [253]).

Theorem 4.2.4. *Let K denote the set of all admissible states, and let G be the functional on K defined by*

$$\begin{aligned} G(A) = \frac{1}{2}\int_B C_{ijrs} e_{ij} e_{rs}\, dv - \int_B (t_{ij} + \beta_{ij}\theta) e_{ij}\, dv - \\ - \int_B (t_{ji,j} + f_i) u_i\, dv + \int_{S_1} t_i \tilde{u}_i\, da + \int_{S_2} (t_i - \tilde{t}_i)\, da\,, \end{aligned} \tag{4.2.21}$$

for every $A = \{u_i,\, e_{ij},\, t_{ij}\} \in K$. Then

$$\delta G\{A\} = 0\,, \tag{4.2.22}$$

at an admissible state A if and only if A is a solution of the mixed problem.

Proof. Let $A = \{u_i,\, e_{ij},\, t_{ij}\}$ and $A' = \{u'_i,\, e'_{ij},\, t'_{ij}\}$ be admissible states. Then $A + \lambda A' \in K$ for every scalar λ, and in view of (2.6.1) and the symmetry of C_{ijrs},

$$\begin{aligned} \delta_{A'} G(A) = \int_B \left[(C_{ijrs} e_{rs} - t_{ij} - \beta_{ij}\theta) e'_{ij} - (t_{ji,j} + f_i) u'_i - t'_{ij} e_{ij} - t'_{ji,j} u_i \right] dv + \\ + \int_{S_1} t'_i \tilde{u}_i\, da + \int_{S_2} \left[t'_i u_i + (t_i - \tilde{t}_i) u'_i \right] da\,, \end{aligned}$$

where $t'_i = t'_{ji} n_j$. If we use the relation

$$\int_B t'_{ji,j} u_i\, da = \int_{\partial B} t'_i u_i\, da - \int_B u_{i,j} t'_{ij}\, dv\,,$$

then we get

$$\delta_{A'}G(A) = \int_B (C_{ijrs}e_{rs} - \beta_{ij}\theta - t_{ij})e'_{ij}dv -$$
$$- \int_B (t_{ji,j} + f_i)u'_i\, dv + \int_B \Big[\frac{1}{2}(u_{i,j} + u_{j,i}) - e_{ij}\Big] t'_{ij}\, dv + \qquad (4.2.23)$$
$$+ \int_{S_1} (\tilde{u}_i - u_i)t'_i\, da + \int_{S_2} (t_i - \tilde{t}_i)u'_i\, da\,.$$

If A is a solution to the mixed problem, then (4.2.23) yields

$$\delta_{A'}G(A) = 0 \qquad \text{for every } A' \in K\,, \qquad (4.2.24)$$

which implies (4.2.22). To prove the converse assertion assume that (4.2.22), and hence (4.2.24), holds. If we choose $A' = \{u'_i,\, 0,\, 0\}$ and let $\boldsymbol{u}'$ vanish near ∂B, then it follows from (4.2.23) and (4.2.24) that

$$\int_B (t_{ji,j} + f_i)u'_i\, dv = 0\,.$$

Since this relation must hold for every u'_i of class C^1 on $\bar{B}$, we conclude from the fundamental lemma that

$$t_{ji,j} + f_i = 0 \qquad \text{on } B\,.$$

Next, let $A' = \{u'_i,\, 0,\, 0\}$ where u'_i vanish near S_1. Then (4.2.23) and (4.2.24) imply that $t_i = \tilde{t}_i$ on S_2. Now, let $A' = \{0,\, e'_{ij},\, 0\}$ and suppose that e'_{ij} vanishes near ∂B. Then (4.2.23) and (4.2.24) imply (4.1.3). Similarly, we can show that the relations (4.1.5) and the condition (4.2.15) are satisfied. Thus A is a solution of the mixed problem. □

Theorem 4.2.4 represents the thermoelastic version of the Hu-Washizu theorem from linear elastostatics.

4.3 Homogeneous and isotropic bodies

a) **Basic equations**. In this section we assume that the body B is homogeneous and isotropic. In this case the constitutive equations are given by (3.1.9) where λ, μ, β and k are prescribed constants.

The boundary value problem (4.1.2), (4.1.4), (4.1.7) reduces to

$$k\Delta\theta = -S \qquad \text{on} \quad B\,,$$
$$\theta = \tilde{\theta} \qquad \text{on} \quad \bar{S}_3\,, \qquad k\frac{\partial\theta}{\partial n} = \tilde{q} \qquad \text{on} \quad S_4\,. \qquad (4.3.1)$$

After the determination of the temperature field we have to study the field equations

$$t_{ji,j} + f_i = 0,$$

$$t_{ij} = \lambda e_{rr}\delta_{ij} + 2\mu e_{ij} - \beta\theta\delta_{ij}, \tag{4.3.2}$$

$$e_{ij} = \frac{1}{2}(u_{i,j} + u_{j,i}),$$

on B, with the boundary conditions (4.1.6). From (4.3.2) we obtain the field equations expressed in terms of the displacement vector field

$$\mu\Delta u_i + (\lambda + \mu)u_{r,ri} + f_i - \beta\theta_{,i} = 0 \qquad \text{on} \quad B. \tag{4.3.3}$$

b) **Thermoelastic potential displacement.** We assume that the body forces are absent. Then the equations (4.3.3) reduce to

$$\mu\Delta u_i + (\lambda + \mu)u_{r,ri} - \beta\theta_{,i} = 0 \qquad \text{on} \quad B. \tag{4.3.4}$$

A method for the reduction of this problem to that for an unheated body with zero body forces has been given by Goodier (see, e.g. Booley and Weiner [28], p.78). A particular solution of the system (4.3.4) can be taken in the form

$$u_i^* = \Phi_{,i}\,. \tag{4.3.5}$$

where Φ is a function of class C^3 on B. A function Φ related in this manner to the displacements components is termed a displacement potential. Substitution of u_i^* into equations (4.3.4) yields the following equation for the function Φ

$$\Delta\Phi = \frac{\beta}{\lambda + 2\mu}\theta \qquad \text{on } B. \tag{4.3.6}$$

A solution of (4.3.6) is given by

$$\Phi(\boldsymbol{x}) = -\frac{\beta}{4\pi(\lambda + 2\mu)}\int_B \frac{1}{r(\boldsymbol{x}, \boldsymbol{y})}\theta(\boldsymbol{y})\, dv_y\,.$$

where $r(\boldsymbol{x}, \boldsymbol{y}) = |\boldsymbol{x} - \boldsymbol{y}|$.

c) **Boussinesq-Papkovitch-Neuber solution.** Here we study a general solution of the equations (4.3.3). The next result is a counterpart of the Boussinesq-Papkovitch-Neuber solution of the isothermal theory.

Theorem 4.3.1. *Let*

$$u_i = \psi_i - \frac{1}{4(1-\nu)}(x_r\psi_r + \varphi)_{,i}, \tag{4.3.7}$$

where φ and ψ_i are class C^3 fields on B that satisfy

$$\Delta\psi_i = -\frac{1}{\mu}(f_i - \beta\theta_{,i}), \qquad \Delta\varphi = \frac{1}{\mu}x_j(f_j - \beta\theta_{,j}). \tag{4.3.8}$$

Then u_i satisfy the equations (4.3.3).
Proof. By (3.1.10) and (4.3.7),

$$\begin{aligned}\mu\Delta u_i + &(\lambda+\mu)u_{r,ri} + f_i - \beta\theta_{,i} = \mu\Delta\psi_i + (\lambda+\mu)\psi_{j,ji} - \\ &- (\lambda+2\mu)\frac{\lambda+\mu}{2(\lambda+2\mu)}\left(x_j\Delta\psi_j + 2\psi_{j,j} + \Delta\varphi\right).\end{aligned} \tag{4.3.9}$$

Since

$$x_j\Delta\psi_j = -\Delta\varphi,$$

from (4.3.8) and (4.3.9) we obtain the desired result. □

We note that the Boussinesq-Papkovitch-Neuber solution is complete (see, e.g. Gurtin [158], p.141).

d) **Characterization of the traction problem in terms of stress.** We note that for homogeneous and isotropic bodies the constitutive equations (4.1.9) take the form (3.1.10). From (3.1.10) and (4.1.8) we can derive the stress equation of compatibility. Thus we obtain the following result (cf. Boley and Weiner [28], p.95).

Theorem 4.3.2. *Assume that $\nu \neq 1$, $k \neq 0$, and that B is simply-connected. Then a symmetric tensor field t_{ij} that is of class C^2 on $\bar{B}$ is a stress field corresponding to a solution of the traction problem if and only if*

$$t_{ji,j} + f_i = 0 \qquad \text{on } B,$$

$$\begin{aligned}t_{ij,ss} &+ \frac{1}{1+\nu}t_{ss,ij} + E_\alpha\left[\frac{1}{1+\nu}\theta_{,ij} - \frac{1}{k(1-\nu)}S\delta_{ij}\right] + \\ &+ \frac{\nu}{1-\nu}\delta_{ij}f_{m,m} + f_{i,j} + f_{j,i} = 0 \qquad \text{on } B,\end{aligned} \tag{4.3.10}$$

$$t_{ji}n_j = \tilde{t}_i \qquad \text{on } \partial B.$$

The proof follows at once from the body force analogy and the characterization of the traction problem in terms of stress within the isothermal theory.

If the domain B is (N+1)-tuply connected, then to (4.3.10) we must adjoin the conditions

$$\frac{1+\nu}{E}\int_{C_p}(t_{ik}-\varepsilon_{ijr}\varepsilon_{rsm}\xi_j t_{ks,m})\,d\xi_k-\frac{\nu}{E}\int_{C_p}(t_{rr}\delta_{ik}-$$

$$-\varepsilon_{ijr}\varepsilon_{rkm}\xi_j t_{ss,m})\,d\xi_k+\alpha\int_{C_p}(\theta\delta_{ik}-\varepsilon_{ijr}\varepsilon_{rkm}\xi_j\theta_{,m})\,d\xi_k=0\,, \qquad (4.3.11)$$

$$\frac{1+\nu}{E}\int_{C_p}\varepsilon_{ism}t_{ks,m}d\xi_k-\frac{\nu}{E}\int_{C_p}\varepsilon_{ikm}t_{rr,m}d\xi_k+\alpha\int_{C_p}\varepsilon_{ikm}\theta_{,m}d\xi_k=0\,,$$

where C_p ($p = 1, 2, \ldots, N$) are simply closed curves in $\bar{B}$ which each surround only a single cavity.

e) **Temperature fields that induce stress free states.** For the remainder of this section, we consider the traction problem and assume that $f_i = 0$ and $\tilde{t}_i = 0$. We seek a temperature field that induces a stress free state in B. If we take $t_{ij} = 0$ on B, then from (4.3.10) we obtain

$$\theta_{,ij} = 0 \qquad \text{on } B\,,$$

and therefore

$$\theta = a_s x_s + b\,, \qquad (4.3.12)$$

where a_i and b are arbitrary constants.
It follows from (3.1.10) and (4.3.12) that

$$e_{ij} = \alpha(a_s x_s + b)\delta_{ij}\,. \qquad (4.3.13)$$

The corresponding displacement field is given by

$$u_i = \alpha\Big[(a_s x_s + b)x_i - \frac{1}{2}a_i x_j x_j\Big] + w_i\,, \qquad (4.3.14)$$

where $\boldsymbol{w}$ is an arbitrary rigid displacement field. Conversely, if θ is given by (4.3.12), then the state $\{u_i,\, e_{ij},\, t_{ij}\}$ with u_i given by (4.3.14), e_{ij} given by (4.3.13) and $t_{ij} = 0$ is a solution of the traction problem corresponding to zero body force and zero surface traction.

4.4 Special problems

In this section we study a number of special problems of thermoelastic equilibrium. We consider homogeneous and isotropic bodies which are in equilibrium in the absence of body force and heat supply.

a) **Thermal stresses in a thick walled spherical shell.** We assume that B is a hollow sphere. Let the internal and external radii of the shell be a_1 and a_2, respectively. We choose the system Ox_i such that the origin O is at the center of the shell. We assume that the surface of the shell is free of tractions. The flow of heat in the shell is produced by maintaining the outer surface of the shell at the constant temperature T_2 and the inner surface at the constant temperature T_1. In this case

$$\theta = \Lambda_1 R^{-1} + \Lambda_2 , \tag{4.4.1}$$

where $R^2 = x_i x_i$, and

$$\Lambda_1 = \frac{a_1 a_2}{a_2 - a_1}(T_2 - T_1) , \qquad \Lambda_2 = \frac{T_2 a_2 - T_1 a_1}{a_2 - a_1} .$$

We seek the solution of the equations (4.3.4) in the form

$$u_i = x_i F(R) , \tag{4.4.2}$$

where F is an unknown function. The substitution from (4.4.2) into (4.3.4) yields the equation

$$F'' + \frac{4}{R} F' = \frac{\beta \theta'}{(\lambda + 2\mu) R} , \tag{4.4.3}$$

where prime denotes the derivative with respect to R. The general solution of this equation is

$$F(R) = C_1 + C_2 R^{-3} + \frac{\beta}{\lambda + 2\mu} F_0(R) , \tag{4.4.4}$$

where C_α are arbitrary constants and F_0 is defined by

$$F_0(R) = R^{-3} \int_{a_1}^{R} x^2 \theta(x) \, dx .$$

It follows from (4.3.2) and (4.4.2) that

$$t_{ij} = \lambda(3F + RF')\delta_{ij} + 2\mu(F\delta_{ij} + R^{-1} x_i x_j F') - \beta\theta\delta_{ij} . \tag{4.4.5}$$

The boundary conditions

$$t_{ji} n_j = 0 \qquad \text{on} \qquad R = a_1 \quad \text{and} \quad R = a_2 , \tag{4.4.6}$$

are satisfied if and only if

$$C_1 = \frac{4\mu\beta}{(\lambda+2\mu)(3\lambda+2\mu)(a_2^3-a_1^3)}\left[a_2^3F_0(a_2)-a_1F_0(a_1)\right],$$

$$C_2 = \frac{\beta a_1^3a_2^3}{(\lambda+2\mu)(a_2^3-a_1^3)}\left[F_0(a_2)-F_0(a_1)\right].$$

We now present a different method to solve this problem. We use spherical coordinates (R, φ, ψ) and the relations (3.1.36)–(3.1.39). We seek the solution in the form

$$u_R = u(R)\,, \qquad u_\varphi = u_\psi = 0\,, \tag{4.4.7}$$

where u_R, u_φ and u_ψ are the physical components of the displacement field. From (3.1.36) and (4.4.7) we obtain

$$\begin{aligned} &\varepsilon_{RR} = u'\,, \qquad \varepsilon_{\varphi\varphi} = \varepsilon_{\psi\psi} = \frac{1}{R}u\,,\\ &\varepsilon_{\varphi\psi} = \varepsilon_{\psi R} = \varepsilon_{\varphi R} = 0\,. \end{aligned} \tag{4.4.8}$$

By (3.1.38) and (4.4.8),

$$\begin{aligned} &\sigma_{RR} = \lambda\left(\frac{2}{R}u+u'\right)+2\mu u'-\beta\theta\,,\\ &\sigma_{\varphi\varphi} = \sigma_{\psi\psi} = \lambda\left(\frac{2}{R}u+u'\right)+\frac{2}{R}\mu u-\beta\theta\\ &\sigma_{\varphi R} = \sigma_{\psi R} = \sigma_{\varphi\psi} = 0\,. \end{aligned} \tag{4.4.9}$$

The equilibrium equations reduce to

$$\sigma'_{RR} + \frac{2}{R}(\sigma_{RR}-\sigma_{\varphi\varphi}) = 0\,. \tag{4.4.10}$$

It follows from (4.4.9) and (4.4.10) that u satisfies the equation

$$\frac{d}{dR}\left[R^{-2}\frac{d}{dR}(R^2u)\right] = \frac{\beta}{\lambda+2\mu}\theta'\,. \tag{4.4.11}$$

The general solution of the equation (4.4.11) is

$$u = A_1R + A_2R^{-2} + \frac{\beta}{\lambda+2\mu}R^{-2}\int_{a_1}^{R}x^2\theta(x)\,dx\,, \tag{4.4.12}$$

where A_α are arbitrary constants. From (4.4.9) and (4.4.12) we obtain

$$\sigma_{RR} = (3\lambda + 2\mu)A_1 - 4\mu A_2 R^{-3} - \frac{4\mu\beta}{\lambda + 2\mu} R^{-3} \int_{a_1}^{R} x^2\theta(x)\,dx\,,$$

$$\sigma_{\varphi\varphi} = \sigma_{\psi\psi} = (3\lambda + 2\mu)A_1 + 2\mu A_2 R^{-3} - \frac{2\mu\beta\theta}{\lambda + 2\mu} + \tag{4.4.13}$$

$$+ \frac{2\mu\beta}{(\lambda + 2\mu)R^3} \int_{a_1}^{R} x^2\theta(x)\,dx\,,$$

$$\sigma_{\varphi R} = \sigma_{\psi R} = \sigma_{\varphi\psi} = 0\,.$$

The boundary conditions (4.4.6) reduce to

$$\sigma_{RR} = 0 \qquad \text{for} \qquad R = a_1 \quad \text{and} \quad R = a_2\,. \tag{4.4.14}$$

From (4.4.13) and (4.4.14) we determine the constants A_α. The final result is

$$u = \frac{\beta}{(\lambda + 2\mu)(a_2^3 - a_1^3)} \Big[a_1^3 R^{-2} \int_R^{a_2} x^2\theta(x)\,dx +$$

$$+ a_2 R^{-2} \int_{a_1}^{R} x^2\theta(x)\,dx + \frac{4\mu R}{3\lambda + 2\mu} \int_{a_1}^{a_2} x^2\theta(x)\,dx \Big]\,,$$

$$\sigma_{RR} = \frac{4\mu\beta}{(a_1^3 - a_2^3)(\lambda + 2\mu)} \Big[a_1^3 R^{-3} \int_R^{a_2} x^2\theta(x)\,dx +$$

$$+ a_2^3 R^{-3} \int_{a_1}^{R} x^2\theta(x)\,dx - \int_{a_1}^{a_2} x^2\theta(x)\,dx \Big]\,,$$

$$\sigma_{\varphi\varphi} = \sigma_{\psi\psi} = \frac{2\mu\beta}{(a_2^3 - a_1^3)(\lambda + 2\mu)} \Big[a_1^3 R^{-3} \int_R^{a_2} x^2\theta(x)\,dx +$$

$$+ a_2^3 R^{-3} \int_{a_1}^{R} x^2\theta(x)\,dx + 2\int_{a_1}^{a_2} x^2\theta(x)\,dx - (a_2^3 - a_1^3)\theta \Big]\,.$$

The results established here can be used to obtain the solution in the case of the solid sphere and in the case of a thermoelastic space with a spherical cavity.

b) **Thermal stresses in a half-space.** We now consider the problem of determining the stress induced in the half-space $x_3 \geq 0$ when the temperature of the boundary $x_3 = 0$ is subjected to a prescribed variation, that boundary being free from applied stress. The boundary conditions therefore assume the form

$$\theta(x_1,\, x_2,\, 0) = \tilde{\theta}(x_1,\, x_2)\,, \tag{4.4.15}$$

and

$$t_{i3}(x_1,\, x_2,\, 0) = 0\,, \tag{4.4.16}$$

for all $(x_1,\, x_2,) \in \mathbb{R}^2$. We impose the conditions $u_i = 0$, $\theta = 0$ at infinity.

The problem of thermal stresses in a half-space has been studied in various papers (see, e.g. Sternberg and McDowell [336], Sneddon and Lockett [330]). Here we present the results of Sneddon and Lockett [330].

We denote

$$a^2 = 1 + \frac{\lambda + \mu}{\mu}\,, \qquad b = \frac{\beta}{\mu}\,. \tag{4.4.17}$$

Then, the constitutive equations become

$$\frac{1}{\mu} t_{ij} = u_{i,j} + u_{j,i} + \big[(a^2 - 2)e_{rr} - b\theta\big]\delta_{ij}\,. \tag{4.4.18}$$

The equations (4.3.4) can be expressed in the form

$$\Delta u_i + (a^2 - 1)u_{r,ri} = b\theta_{,i}\,. \tag{4.4.19}$$

To integrate these equations we use the Fourier transform. If we denote

$$\begin{aligned} u_r^*(\alpha_1,\, \alpha_2,\, x_3) &= \frac{1}{2\pi}\int_{-\infty}^{\infty}\int_{-\infty}^{\infty} u_r(x_1,\, x_2,\, x_3)\exp\left(i\alpha_\rho x_\rho\right) dx_1 dx_2\,, \\ \theta^*(\alpha_1,\, \alpha_2,\, x_3) &= \frac{1}{2\pi}\int_{-\infty}^{\infty}\int_{-\infty}^{\infty} \theta(x_1,\, x_2,\, x_3)\exp\left(i\alpha_\rho x_\rho\right) dx_1 dx_2\,, \end{aligned} \tag{4.4.20}$$

g then from (4.3.1) and (4.4.19) we obtain the ordinary differential equations

$$\left(\frac{d^2}{dx_3^2} - \alpha^2\right)\theta^* = 0\,, \tag{4.4.21}$$

$$\left(\frac{d^2}{dx_3^2} - \alpha^2\right)u_r^* - (a^2 - 1)\alpha_r\alpha_j u_j^* = -ib\alpha_r\theta^*\,, \tag{4.4.22}$$

where

$$\alpha^2 = \alpha_1^2 + \alpha_2^2\,, \qquad \alpha_3 = i\frac{d}{dx_3}\,.$$

The solution of the equation (4.4.21) is

$$\theta^* = A\exp(-\alpha x_3) + B\exp(\alpha x_3)\,, \tag{4.4.23}$$

where A and B are arbitrary constants. It follows from (4.4.22) and (4.4.23) that

$$\begin{aligned} u_\beta^* &= (C_\beta + D\alpha_\beta x_3)\exp(-\alpha x_3) + (C'_\beta + D'\alpha_\beta x_3)\exp(\alpha x_3)\,,\\ u_3^* &= (C_3 - i\alpha D x_3)\exp(-\alpha x_3) + (C'_3 + i\alpha D' x_3)\exp(\alpha x_3)\,, \end{aligned} \tag{4.4.24}$$

where C_r and C'_r are arbitrary constants, and

$$\begin{aligned} D &= -\frac{a^2-1}{(a^2+1)\alpha}\left(\alpha_\beta C_\beta - i\alpha C_3 - \frac{ibA}{a^2-1}\right),\\ D' &= -\frac{a^2-1}{(a^2+1)\alpha}\left(\alpha_\beta C'_\beta + i\alpha C'_3 - \frac{ibB}{a^2-1}\right). \end{aligned} \tag{4.4.25}$$

If we impose the conditions at infinity, then we get

$$\begin{aligned} \theta^* &= A\exp(-\alpha x_3)\,,\\ u_\beta^* &= (C_\beta + D\alpha_\beta x_3)\exp(-\alpha x_3)\,,\\ u_3^* &= (C_3 - i\alpha D x_3)\exp(-\alpha x_3)\,. \end{aligned} \tag{4.4.26}$$

From (4.4.18) and (4.4.26) we obtain

$$\begin{aligned} t_{\beta 3}^* &= -\mu(\alpha C_\beta + \alpha\alpha_\beta D x_3 - \alpha_\beta D + i\alpha_\beta C_3 + \alpha_\beta \alpha D x_3)\exp(-\alpha x_3)\,,\\ t_{33}^* &= \mu\big[i(2-a^2)(\alpha_\beta C_\beta + \alpha^2 D x_3) + a^2(i\alpha^2 D x_3 - \alpha C_3 -\\ &\quad - i\alpha D) - bA\big]\exp(-\alpha x_3)\,. \end{aligned} \tag{4.4.27}$$

By (4.4.15) and (4.4.26),

$$A = \tilde{\theta}^*(\alpha_1, \alpha_2)\,. \tag{4.4.28}$$

The boundary conditions (4.4.16) are satisfied if

$$C_\beta = \frac{ib\tilde{\theta}^*\alpha_\beta}{2(a^2-1)\alpha^2}\,, \qquad C_3 = -\frac{b\tilde{\theta}^*}{2(a^2-1)\alpha}\,. \tag{4.4.29}$$

With the aid of (4.4.28) and (4.4.29), we find from (4.4.25) that $D=0$. Thus, the relations (4.4.26) become

$$\theta^* = \tilde{\theta}^*(\alpha_1, \alpha_2)\exp(-\alpha x_3),$$

$$u_\beta^* = \frac{ib\tilde{\theta}^*\alpha_\beta}{2(a^2-1)\alpha^2}\exp(-\alpha x_3), \tag{4.4.30}$$

$$u_3^* = -\frac{b\tilde{\theta}^*}{2(a^2-1)\alpha}\exp(-\alpha x_3).$$

It follows from (4.4.27)–(4.4.29) that

$$t_{i3}^* = 0, \qquad x_3 \geq 0. \tag{4.4.31}$$

This result shows that the stress field due to arbitrary surface temperatures is plane and parallel to the boundary.

The calculation of the inverse Fourier transforms has been made by Sternberg and McDowell [336] for the case in which the temperature field is created by the surface condition

$$\theta(x_1, x_2, 0) = \theta_0 H(c-r),$$

where θ_0 and c are prescribed constants, $r^2 = x_\alpha x_\alpha$ and H is the Heaviside function. If we use the cylindrical coordinates (r, φ, z), then from (4.4.29) and the formulae of Section 3.1 we obtain

$$u_r = \frac{bc\theta_0}{2(a^2-1)}J(1, 1; -1),$$

$$u_z = -\frac{bc\theta_0}{2(a^2-1)}J(1, 0; -1),$$

where

$$J(\mu, \nu; \lambda) = \int_0^\infty J_\mu(\alpha c)J_\nu(\alpha r)\exp(-\alpha x_3)\alpha^\lambda\, d\alpha.$$

Here J_μ is the Bessel function of first kind and μ order. For the non-zero components of the stress tensor we get

$$\tau_{rr} + \tau_{\varphi\varphi} = -\frac{cb\theta_0\mu}{(a^2-1)}J(1, 0; 0),$$

$$\tau_{rr} - \tau_{\varphi\varphi} = \frac{cb\theta_0\mu}{(a^2-1)}\left[J(1, 0; 0) - \frac{2}{r}J(1, 1; -1)\right].$$

The integrals $J(1,\,1;\,-1)$, $J(1,\,0;\,-1)$, $J(1,\,0;\,0)$ have been tabulated for various values of r/c and z/c by Eason, Noble and Sneddon [118].

c) **Elastic layer resting on a rigid foundation**. We consider an elastic layer which occupies the region $0 \leq x_3 \leq d$, and which rests on a rigid, frictionless foundation. We assume that the surface $x_3 = 0$ is free of tractions, and the surface $x_3 = d$ is in contact with the foundation. The surface $x_3 = 0$ is kept at prescribed temperature, while on the surface $x_3 = d$ is prescribed the heat flux. Thus, the boundary conditions are

$$\begin{aligned} &\theta = \tilde{\theta}\,, \qquad t_{i3} = 0 \qquad \text{on} \quad x_3 = 0\,, \\ &\frac{\partial \theta}{\partial x_3} = Q\,, \qquad u_3 = 0\,, \qquad t_{\alpha 3} = 0 \qquad \text{on} \quad x_3 = d\,. \end{aligned} \tag{4.4.32}$$

This problem has been studied by Sneddon and Lockett [331].

We have seen that the solutions of the equations (4.4.21) and (4.4.22) are given by (4.4.23) and (4.4.24). These solutions can be written in the form

$$\theta^* = E\cosh \alpha x_3 + E'\sinh \alpha x_3\,, \tag{4.4.33}$$

and

$$\begin{aligned} u_\beta^* &= (A_\beta + P'\alpha_\beta x_3)\cosh \alpha x_3 + (A'_\beta + P\alpha_\beta x_3)\sinh \alpha x_3\,, \\ u_3^* &= (A_3 + i\alpha P' x_3)\sinh \alpha x_3 + (A'_3 + i\alpha P x_3)\cosh \alpha x_3\,, \end{aligned} \tag{4.4.34}$$

respectively. In the above relations A_i, A'_i, E and E' are arbitrary constants and P and P' are given by

$$\begin{aligned} P &= \frac{(a^2-1)}{(a^2+1)\alpha}\Big(A_\beta\alpha_\beta + i\alpha A_3 - \frac{ibE}{a^2-1}\Big)\,, \\ P' &= \frac{(a^2-1)}{(a^2+1)\alpha}\Big(A'_\beta\alpha_\beta + i\alpha A'_3 - \frac{ibE'}{a^2-1}\Big)\,, \end{aligned} \tag{4.4.35}$$

From (4.4.18) we obtain

$$\begin{aligned} \frac{1}{\mu}t_{\beta 3}^* &= (\alpha A_\beta - i\alpha_\beta A_3 + P\alpha_\beta)\sinh \alpha x_3 + 2P\alpha\alpha_\beta x_3\cosh \alpha x_3 + \\ &\quad + (\alpha A'_\beta - i\alpha_\beta A'_3 + P'\alpha_\beta)\cosh \alpha x_3 + 2P'\alpha_\beta \alpha x_3 \sinh \alpha x_3 \\ \frac{1}{\mu}t_{33}^* &= -i(a^2-2)(A_\beta x_\beta \cosh \alpha x_3 + P\alpha^2 x_3\sinh \alpha x_3) - bE\cosh \alpha x_3 + \\ &\quad + a^2\alpha(A_3\cosh \alpha x_3 + iP\cosh \alpha x_3 + i\alpha x_3 P\sinh \alpha x_3) - \\ &\quad - i(a^2-2)(A'_\beta \alpha_\beta \sinh \alpha x_3 + P'\alpha^2 x_3\cosh \alpha x_3) - bE'\sinh \alpha x_3 + \\ &\quad + a^2\alpha(A'_3\sinh \alpha x_3 + iP'\sinh \alpha x_3 + i\alpha x_3 P'\cosh \alpha x_3)\,. \end{aligned} \tag{4.4.36}$$

The thermal boundary conditions imply that

$$E = \tilde{\theta}^* , \qquad E' = \frac{Q^*}{\alpha C} - \frac{\tilde{\theta}^* S}{C} , \tag{4.4.37}$$

where $S = \sinh \alpha d$, $C = \cosh \alpha d$. The mechanical boundary conditions reduce to

$$\begin{aligned} & - i(a^2 - 2)A_\beta \alpha_\beta - bE + a^2 A_3 \alpha + a^2 i\alpha P = 0 , \\ & \alpha A'_\beta - i\alpha_\beta A'_3 + P' \alpha_\beta = 0 , \\ & (\alpha A_\beta - i\alpha_\beta A_3 + P\alpha_\beta) S + 2P\alpha_\beta \alpha d C + 2P' \alpha_\beta \alpha d S = 0 , \\ & (A_3 + iP' ad) S + (A'_3 + iP\alpha d) C = 0 . \end{aligned} \tag{4.4.38}$$

If we denote

$$D = -\frac{b\tilde{\theta}^*}{2i\alpha^2(a^2-1)} - \frac{bdSC^{-1}Q^*}{2i\alpha^2(a^2-1)(S+\alpha dC^{-1})} ,$$

then we obtain

$$\begin{aligned} & A_\beta = \alpha_\beta D , \qquad A_3 = \frac{1}{2a^2\alpha}(bE - 2i\alpha^2 D) , \\ & A'_\beta = \frac{1}{2a^2\alpha^2}(ibE' - 2\alpha^2 DSC^{-1})\alpha_\beta , \qquad SA_3 = i\alpha DSC , \\ & P = \frac{1}{2a^2\alpha}\left[2\alpha^2 D(a^2-1) - ibE\right] , \\ & P' = -\frac{1}{2a^2\alpha}\left[2\alpha^2 ASC^{-1}(a^2-1) + ibE'\right] . \end{aligned}$$

If the foundation is made by a non-conductor of heat, then $Q = 0$. In this case we obtain

$$D = \frac{ib\tilde{\theta}^*}{2\alpha^2(a^2-1)} , \qquad P = P' = 0 ,$$

$$A'_\beta = -\frac{ib\tilde{\theta}^* SC^{-1}}{2\alpha^2(a^2-1)}\alpha_\beta , \qquad A_3 = \frac{b\tilde{\theta}^*}{2\alpha(a^2-1)} , \qquad A'_3 = -\frac{b\tilde{\theta}^* SC}{2\alpha(a^2-1)} .$$

In view of (4.4.33), (4.4.34) we obtain

$$\theta^* = \tilde{\theta}^*(\cosh \alpha x_3 - SC^{-1} \sinh \alpha x_3),$$

$$u_\beta^* = \frac{ib\tilde{\theta}^* \alpha_\beta}{2\alpha^2(a^2-1)}(\cosh \alpha x_3 - SC^{-1} \sinh \alpha x_3),$$

$$u_3^* = \frac{b\tilde{\theta}^*}{2\alpha(a^2-1)}(\sinh \alpha x_3 - SC^{-1} \cosh \alpha x_3).$$

The calculation of the inverse Fourier transforms has been made for special cases by Sneddon and Lockett [330], [331].

d) **Boussinesq's problem for a heated punch**. We now consider the problem of determining the distribution of stress in a homogeneous and isotropic elastic half-space when a rigid punch in the form of a surface of revolution is pressed against it. The isothermal problem was first discussed in detail by J. Boussinesq.

We assume that the rigid punch is heated and so produces a non-uniform distribution in the elastic body. We suppose that the axis of the punch is normal to the boundary plane of the elastic body. If we take this boundary to be the plane $x_3 = 0$ and the point at which the tip of the punch begins to indent the solid to be the origin of coordinates. Since the punch is a solid of revolution and it is pressed normally against the boundary $x_3 = 0$, it will fit the elastic half-space over the area between the apex of the punch and a certain circle of radius a. For convenience, we shall take the length a as the unit of length. In what follows we shall use the cylindrical coordinates (r, φ, z).

We assume that the surfaces in contact are smooth and that displacement u_z is specified over the disk $0 \leq r \leq 1$, $z = 0$, while outside that region we impose the condition of zero normal stress. These assumptions imply the following boundary conditions

$$\begin{aligned} &\sigma_{rz}(r, 0) = 0, \quad r \geq 0, \qquad \sigma_{zz}(r, 0) = 0, \quad r > 1, \\ &u_z(r, 0) = g(r), \qquad 0 \leq r \leq 1, \end{aligned} \tag{4.4.39}$$

We consider the following thermal boundary condition

$$\theta(r, 0) = \theta_0(r) H(1-r). \tag{4.4.40}$$

Let

$$\begin{aligned} u_\alpha &= \Phi_{,\alpha} + (a^2-1) x_3 \Phi_{,\alpha 3} + x_3 \psi_{,\alpha}, \\ u_3 &= -a^2 \Phi_{,3} + (a^2-1) x_3 \Phi_{,33} + x_3 \psi_{,3} - \psi, \\ \theta &= \frac{2}{b} \psi_{,3}, \end{aligned} \tag{4.4.41}$$

where ψ and Φ are class C^3 fields on B that satisfy the equations

$$\Delta\psi = 0\,, \qquad \Delta\Phi = 0\,. \tag{4.4.42}$$

Then u_i and θ satisfy the equations (4.3.4) and $\Delta\theta = 0$ on B. The solution (4.4.41) has been established by Sneddon [332]. From (4.4.18) and (4.4.41) we obtain

$$\begin{aligned} t_{\alpha 3} &= 2\mu\big[(a^2-1)x_3\Phi_{,\alpha 33} + x_3\psi_{,\alpha 3}\big]\,, \\ t_{33} &= 2\mu\big[(a^2-1)x_3\Phi_{,333} - (a^2-1)\Phi_{,33} + x_3\psi_{,33} - \psi_{,3}\big]\,. \end{aligned} \tag{4.4.43}$$

The equations (4.4.42) can be written in the form

$$\Big(\frac{\partial^2}{\partial r^2} + \frac{1}{r}\frac{\partial}{\partial r} + \frac{\partial^2}{\partial z^2}\Big)\psi = 0\,, \qquad \Big(\frac{\partial^2}{\partial r^2} + \frac{1}{r}\frac{\partial}{\partial r} + \frac{\partial^2}{\partial z^2}\Big)\Phi = 0\,. \tag{4.4.44}$$

We denote by $\hat{f}$ the Hankel transform of order zero of the function f,

$$\hat{f}(\xi,\, z) = \int_0^\infty rf(r,\, z)J_0(\xi r)\, dr\,, \qquad f(r,\, z) = \int_0^\infty \xi\hat{f}(\xi,\, z)J_0(\xi r)\, d\xi\,. \tag{4.4.45}$$

From (4.4.44) we obtain

$$\Big(\frac{d^2}{dz^2} - \xi^2\Big)\hat{\psi} = 0\,. \tag{4.4.46}$$

The solution of the equation (4.4.46) which vanishes at infinity is

$$\hat{\psi} = A(\xi)\exp\left(-\xi z\right), \tag{4.4.47}$$

where A is an unknown function. Similarly,

$$\hat{\Phi} = B(\xi)\exp\left(-\xi z\right). \tag{4.4.48}$$

We introduce the functions q and f by

$$q(\xi) = -\frac{1}{a^2-1}\xi^2 A(\xi)\,, \qquad f(\xi) = a^2\big[\xi^2 B(\xi) + q(\xi)\big]\,. \tag{4.4.49}$$

The functions Φ and ψ have the form

$$\begin{aligned} \Phi &= \int_0^\infty \xi^{-2}\big[a^{-2}f(\xi) - q(\xi)\big]\exp\left(-\xi z\right)J_0(\xi r)\, d\xi\,, \\ \psi &= -(a^2-1)\int_0^\infty \xi^{-1}q(\xi)\exp\left(-\xi z\right)J_0(\xi r)\, d\xi\,. \end{aligned} \tag{4.4.50}$$

It follows from (4.4.41), (4.4.43) and (4.4.50) that

$$u_z(r,\,0) = \int_0^\infty \xi^{-1}\big[f(\xi) - q(\xi)\big]\,J_0(\xi r)\,d\xi\,,$$

$$\sigma_{zz}(r,\,0) = -2\mu(1-a^{-2})\int_0^\infty f(\xi)J_0(\xi r)\,d\xi\,, \qquad \sigma_{rz}(r,\,0) = 0\,, \tag{4.4.51}$$

$$\theta(r,\,0) = 2b^{-1}(a^2-1)\int_0^\infty q(\xi)J_0(\xi r)\,d\xi\,.$$

From (4.4.41), (4.4.47) and (4.4.49) we obtain

$$q(\xi) = \frac{b\xi}{2(a^2-1)}\,, \qquad \hat{\theta}(\xi,\,0) = \frac{b\xi}{2(a^2-1)}\int_0^\infty r\theta(r,\,0)J_0(\xi r)\,dr\,. \tag{4.4.52}$$

If we denote

$$D(r) = \int_0^\infty \xi^{-1}q(\xi)J_0(\xi r)\,d\xi\,, \tag{4.4.53}$$

then from (4.4.39) and (4.4.51) we obtain for the function f the following pair of dual integral equations

$$\int_0^\infty \xi^{-1}f(\xi)J_0(\xi r)\,d\xi = D(r) + g(r)\,, \qquad 0 \le r \le 1\,,$$

$$\int_0^\infty f(\xi)J_0(\xi r)\,d\xi = 0\,, \qquad r > 1\,.$$

We write the solution of this pair of dual integral equations in the form

$$f = f_1 + f_2\,,$$

where f_1 is the solution of the pair

$$\begin{aligned} &\int_0^\infty \xi^{-1}f_1(\xi)J_0(\xi r)\,d\xi = D(r)\,, \qquad 0 \le r \le 1\,,\\ &\int_0^\infty f_1(\xi)J_0(\xi r)\,d\xi = 0\,, \qquad r > 1\,, \end{aligned} \tag{4.4.54}$$

and f_2 that of the pair

$$\begin{aligned} &\int_0^\infty \xi^{-1}f_2(\xi)J_0(\xi r)\,d\xi = g(r)\,, \qquad 0 \le r \le 1\,,\\ &\int_0^\infty f_2(\xi)J_0(\xi r)\,d\xi = 0\,, \qquad r > 1 \end{aligned} \tag{4.4.55}$$

The solution of (4.4.55) does not depend on the thermal field.

From (4.4.40) we obtain

$$\hat{\theta}(\xi,0)=\int_0^1 r\theta_0(r)J_0(\xi r)\,dr\,, \tag{4.4.56}$$

so that by (4.4.52)–(4.4.54) we find that

$$f_1(\xi)=\frac{b\xi}{2(a^2-1)}\hat{\theta}(\xi,0)\,.$$

The solution of the equations (4.4.55) is (cf. Sneddon [329])

$$f_2(\xi)=\frac{2}{\pi}\int_0^1 \sin\xi t\,dt\int_0^t \frac{rg(r)}{\sqrt{t^2-r^2}}\,dr\,.$$

The solutions for other special types of temperature fields have been obtained by Sneddon [329]. Thermoelastic contact problems have been intensively studied (see, e. g. Barber [14], Comninou and Dundurs [87], Dundurs [114], Duvaut [117]).

4.5 The plane strain problem

We assume that the region B refers to the interior of a right cylinder of length h with open cross-section Σ and the smooth lateral boundary Π. The rectangular Cartesian coordinate frame is supposed to be chosen in such a way that the x_3-axis is parallel to the generators of B. We denote by L the boundary of Σ.

a) **Statement of the problem.** We assume that the domain B is occupied by a homogeneous and isotropic material. The state $\{u_i,\,e_{ij},\,t_{ij}\}$ is a state of plane strain, parallel to the x_1, x_2-plane, provided

$$u_\alpha=u_\alpha(x_1,x_2)\,,\qquad u_3=0\,,\qquad \theta=\theta(x_1,x_2)\,. \tag{4.5.1}$$

From (4.1.5) and (4.5.1) we obtain

$$e_{\alpha\beta}=\frac{1}{2}(u_{\alpha,\beta}+u_{\beta,\alpha})\,, \tag{4.5.2}$$

and $e_{i3}=0$. By (4.3.2),

$$t_{\alpha\nu}=\lambda e_{\rho\rho}\delta_{\alpha\nu}+2\mu e_{\alpha\nu}-\beta\theta\delta_{\alpha\nu}\,, \tag{4.5.3}$$

and

$$t_{33}=\lambda e_{\rho\rho}-\beta\theta\,,\qquad t_{\alpha3}=0\,. \tag{4.5.4}$$

We note that e_{ij} and t_{ij} are independent of x_3.

We assume that the body force $\boldsymbol{f}$ is independent of x_3 and that $f_3 = 0$. In view of (4.5.3) and (4.5.4), the equations of equilibrium reduce to

$$t_{\beta\alpha,\beta} + \varrho_0 f_\alpha = 0 \qquad \text{on} \quad \Sigma. \tag{4.5.5}$$

We assume that $k \neq 0$ and that the heat-supply S is independent of x_3. Then, the equation $(4.3.1)_1$ reduces to

$$\theta_{,\alpha\alpha} = -\frac{1}{k} S \qquad \text{on } \Sigma. \tag{4.5.6}$$

We consider the traction problem. Given $\tilde{\boldsymbol{t}}$, $\tilde{\theta}$ and $\tilde{q}$ on Π, with $\tilde{\boldsymbol{t}}$, $\tilde{\theta}$, $\tilde{q}$ independent of x_3 and $\tilde{t}_3 = 0$, the boundary conditions on the lateral surface become

$$t_{\beta\alpha} n_\beta = \tilde{t}_\alpha \qquad \text{on } L, \tag{4.5.7}$$

$$\theta = \tilde{\theta} \qquad \text{on } L_1, \qquad k\theta_{,\alpha} n_\alpha = \tilde{q} \qquad \text{on } L_2, \tag{4.5.8}$$

where $\bar{L}_1 \cup L_2 = L$, $L_1 \cap L_2 = \emptyset$.

Thus, the temperature field θ is the solution of the boundary value problem (4.5.6), (4.5.8). Generally, we shall treat the temperature field as having already been so determined. The determination of the thermoelastic equilibrium consists in finding an admissible state $\{u_\alpha, e_{\alpha\beta}, t_{\alpha\beta}\}$ which satisfies the equations (4.5.2), (4.5.3), (4.5.5) in Σ and the boundary conditions (4.5.7) on Γ. The function t_{33} can be calculated after the functions u_α and θ have been determined. Thus we can determine the tractions over the bases which maintain B in equilibrium.

We note that if the domain Σ is smooth, f_α are continuous on $\bar{\Sigma}$ and $\tilde{t}_\alpha$ are piecewise regular on L, then the necessary and sufficient conditions for the existence of a solution to the boundary value problem (4.5.2), (4.5.3), (4.5.5), (4.5.7) are (cf. Muskhelishvili [264], p.283, Fichera [128], p.385)

$$\int_\Sigma f_\alpha \, da + \int_L \tilde{t}_\alpha \, ds = 0, \qquad \int_\Sigma \varepsilon_{\alpha\beta 3} x_\alpha f_\beta \, da + \int_L \varepsilon_{\alpha\beta 3} x_\alpha \tilde{t}_\beta \, ds = 0. \tag{4.5.9}$$

From (4.5.2), (4.5.3) and (4.5.5) we obtain the equations of equilibrium expressed in terms of the displacement field

$$\mu u_{\alpha,\rho\rho} + (\lambda + \mu) u_{\rho,\rho\alpha} + f_\alpha - \beta\theta_{,\alpha} = 0 \qquad \text{on } \Sigma. \tag{4.5.10}$$

Generally, in the problems of thermoelastostatics it is assumed that the mechanical loads are zero. Extensive reviews of the subject are given by Parkus [292], Boley and Weiner [28] and Nowacki [277].

b) **Reduction of the problem to the isothermal theory**. We assume that $S=0$. Then, the function θ satisfies the equation

$$\theta_{,\alpha\alpha}=0 \qquad \text{on } \Sigma. \tag{4.5.11}$$

We denote by $F(z)$ the following analytic function of the complex variable $z = = x_1 + ix_2$,

$$F(z)=\theta+i\psi. \tag{4.5.12}$$

We introduce the functions u^*_α by

$$u^*_1+iu^*_2=\int F(z)\,dz. \tag{4.5.13}$$

Clearly, we have

$$u^*_{1,1}=u^*_{2,2}=0, \qquad u^*_{1,2}=-u^*_{2,1}. \tag{4.5.14}$$

We introduce the functions u'_α by

$$u_\alpha=u'_\alpha+\frac{\beta}{2(\lambda+\mu)}u^*_\alpha, \tag{4.5.15}$$

From (4.5.2), (4.5.14) and (4.5.15) we obtain

$$\varepsilon_{\alpha\rho}=\varepsilon^*_{\alpha\varrho}+\frac{\beta\theta}{2(\lambda+\mu)}\delta_{\alpha\varrho}, \tag{4.5.16}$$

where

$$2\varepsilon'_{\alpha\beta}=u'_{\alpha,\beta}+u'_{\beta,\alpha}. \tag{4.5.17}$$

We denote

$$t'_{\alpha\rho}=\lambda\varepsilon'_{\nu\nu}\delta_{\alpha\rho}+2\mu\varepsilon'_{\alpha\rho}. \tag{4.5.18}$$

From (4.5.3) and (4.5.16) we find that

$$t_{\alpha\beta}=t'_{\alpha\beta} \qquad \text{on } \Sigma. \tag{4.5.19}$$

Thus, from (4.5.5) we obtain

$$t'_{\beta\alpha,\beta}+\varrho_0 f_\alpha=0 \qquad \text{on } \Sigma \tag{4.5.20}$$

and (4.5.7) leads to

$$t'_{\beta\alpha}n_\beta=\tilde{t}_\alpha \qquad \text{on } L. \tag{4.5.21}$$

In this way the thermoelastic problem is reduced to the problem of isothermal elastostatics characterized by the field equations (4.5.17), (4.5.18) and (4.5.20) on Σ, and the boundary conditions (4.5.21) on L.

Let us assume that the domain Σ is simply-connected and that $f_\alpha = 0$, $S = 0$, $\tilde{t}_\alpha = 0$. Then we conclude that $t'_{\alpha\beta} = 0$ and that $u'_\alpha = 0$ *modulo* a plane rigid displacement. Thus, if a simply-connected body is subjected to a state of plane strain in the absence of mechanical loads and heat supply, then $t_{\alpha\beta} = 0$ and

$$u_\alpha = \frac{\beta}{2(\lambda+\mu)} u^*_\alpha , \tag{4.5.22}$$

modulo a plane rigid displacement. The function t_{33} is given by

$$t_{33} = -\frac{\beta\mu\theta}{\lambda+\mu} . \tag{4.5.23}$$

Consider now a finite multiply-connected domain Σ. Inside Σ there are n contours Γ_k $(k = 1, 2, \cdots, n)$ which have no common points; the outer contour is denoted by Γ_0. Since the temperature must be single-valued, the function $F(z)$ has the form

$$F(z) = \sum_{k=1}^{n} B_k \log(z - z_k) + F_0(z) , \tag{4.5.24}$$

where z_k are points inside the contours Γ_k, B_k are real constants and $F_0(z)$ is holomorphic in the domain Σ. In this case we have

$$u^*_1 + iu^*_2 = z\sum_{k=1}^{n} B_k \log(z - z_k) + \sum_{k=1}^{n} (a_k + ib_k) \log(z - z_k) + f_0(z) , \tag{4.5.25}$$

where a_k, b_k are real constants and $f_0(z)$ is a holomorphic function. From (4.5.25) we get

$$[u^*_1 + iu^*_2]_{L_k} = 2\pi i(zB_k + a_k + ib_k) , \tag{4.5.26}$$

where $[\,]$ denotes the change in value of the function inside on passing once round a contour in the conventional positive sense which keeps the area enclosed on the left, the contour lying entirely in the body. Since the the functions u_α are assumed to be single-valued, it follows that the functions u'_α are multi-valued

$$\begin{aligned} [u'_1]_{L_k} &= \frac{\pi\beta}{\lambda+\mu}(B_k x_2 + b_k) , \\ [u'_2]_{L_k} &= -\frac{\pi\beta}{\lambda+\mu}(B_k x_1 + a_k) . \end{aligned} \tag{4.5.27}$$

Thus, in this case the isothermal problem is a problem with dislocations with the characteristics

$$\varepsilon_k = -\frac{\pi\beta}{\lambda+\mu} B_k\,, \qquad a_k = \frac{\pi\beta b_k}{\lambda+\mu}\,, \qquad \beta_k = \frac{\pi\beta a_k}{\lambda+\mu}\,. \tag{4.5.28}$$

These results have been established by Muskhelishvili [264], p.159.

We consider now the plane strain problem with null mechanical data, i.e. $f_\alpha = 0$, $\tilde{t}_\alpha = 0$. In order to reduce the problem to an isothermal problem we look for the functions u_α in the form

$$u_\alpha = u'_\alpha + \Phi_{,\alpha}\,, \tag{4.5.29}$$

where Φ is a function of class C^2 which satisfies in Σ the equation

$$\Phi_{,\alpha\alpha} = \frac{\beta}{\lambda+2\mu}\theta\,. \tag{4.5.30}$$

It follows from (4.5.10), (4.5.29) and (4.5.30) that the functions u'_α satisfy the equations

$$\mu u'_{\alpha,\beta\beta} + (\lambda+\mu)u'_{\rho,\rho\alpha} = 0\,. \tag{4.5.31}$$

By (4.5.3) and (4.5.29),

$$\begin{aligned} t_{\alpha\beta} &= t'_{\alpha\beta} + 2\mu(\Phi_{,\alpha\beta} - \delta_{\alpha\beta}\Phi_{,rr})\,,\\ t_{33} &= \lambda\varepsilon'_{\beta\beta} - 2\mu\Phi_{,\alpha\alpha}\,, \end{aligned} \tag{4.5.32}$$

where

$$\begin{aligned} t'_{\alpha\beta} &= \lambda\varepsilon'_{\rho\rho}\delta_{\alpha\beta} + 2\mu\varepsilon'_{\alpha\beta}\,,\\ 2\varepsilon'_{\alpha\beta} &= u'_{\alpha,\beta} + u'_{\beta,\alpha}\,. \end{aligned} \tag{4.5.33}$$

The equations of equilibrium (4.5.31) are equivalent to the equations

$$t'_{\beta\alpha,\beta} = 0\,. \tag{4.5.34}$$

The boundary conditions reduce to

$$t'_{\beta\alpha}n_\beta = -2\mu(\Phi_{,\alpha\beta}n_\beta - \Phi_{,\gamma\gamma}n_\alpha) \qquad \text{on } L\,. \tag{4.5.35}$$

Thus, the problem of thermoelasticity is reduced to the determination of the function Φ which satisfies the equation (4.5.30) and to solving the isothermal plane strain problem (4.5.33)–(4.5.35). It is well-known (see, e.g. Gurtin [158], p.156) that the problem of plane strain of elastostatics can be solved with the aid of the Airy function. Thus, the equations (4.5.34) and the compatibility equations are satisfied if

$$t'_{\alpha\beta} = \delta_{\alpha\beta} A_{,\rho\rho} - A_{,\alpha\rho}\,, \tag{4.5.36}$$

where A is a scalar field of class C^4 on Σ which satisfies the equation

$$\Delta\Delta A = 0 \qquad \text{on } \Sigma\,. \tag{4.5.37}$$

It follows from (4.5.36) that

$$t'_1 \equiv t'_{\beta 1} n_\beta = 2\mu \frac{d}{ds}(\Phi_{,2})\,, \qquad t'_2 \equiv t'_{\beta 2} n_\beta = 2\mu \frac{d}{ds}(\Phi_{,1}) \qquad \text{on } L\,. \tag{4.5.38}$$

In view of Goursat's theorem (see, e.g. Gurtin [158], p.160), there exist analytic complex functions ψ and χ on Σ such that

$$2A = \bar{z}\psi(z) + z\overline{\psi(z)} + \chi(z) + \overline{\chi(z)}\,. \tag{4.5.39}$$

The displacements u'_α and the stresses $t'_{\alpha\beta}$ can be expressed in the form

$$\begin{aligned} 2\mu(u'_1 + iu'_2) &= \frac{\lambda + 3\mu}{\lambda + \mu}\psi(z) - z\overline{\psi'(z)} - \overline{\chi'(z)}\,,\\ t'_{11} + t'_{22} &= 4\mathrm{Re}\left[\psi'(z)\right],\\ t'_{22} - t'_{11} + 2it'_{12} &= 2\left[\bar{z}\psi''(z) + \chi'(z)\right]. \end{aligned} \tag{4.5.40}$$

From (4.5.38) we obtain

$$t'_1 + it'_2 = -i\frac{d}{ds}(A_{,1} + iA_{,2}) = -i\frac{d}{ds}\left[\psi(z) + z\overline{\psi'(z)} + \overline{\chi'(z)}\right]. \tag{4.5.41}$$

By (4.5.29), (4.5.32) and (4.5.40),

$$\begin{aligned} 2\mu(u_1 + iu_2) &= \frac{\lambda + 3\mu}{\lambda + \mu}\psi(z) - z\overline{\psi'(z)} - \overline{\chi'(z)} + 2\mu(\Phi_{,1} + i\Phi_{,2})\,,\\ t_{11} + t_{22} &= 4\mathrm{Re}\left[\psi'(z)\right] - \frac{2\mu\beta}{\lambda + 2\mu}\theta\,,\\ t_{22} - t_{11} + 2it_{12} &= 2\left[\bar{z}\psi''(z) + \chi''(z)\right] + 2\mu(\Phi_{,22} - \Phi_{,11} + 2i\Phi_{,12})\,. \end{aligned} \tag{4.5.42}$$

A detailed study of the isothermal plane strain problem in elastostatics may be found in the monographs of Muskhelishvili [264], Green and Zerna [152] and Solomon [333].

c) **Thermal stresses in a hollow cylinder.** Let B be a right hollow cylinder with the generic cross-section Σ. We assume that the domain Σ is bounded by two concentric circles of radius R_1 and R_2, where $R_1 < R_2$. The rectangular Cartesian coordinate frame is chosen such that the x_3 axis coincides with the center line of the cylinder. We assume that the body forces are absent, and that the cylinder is free of external stresses. We shall use the cylindrical coordinates (r, φ, z) and the relations of Section 3.1. We assume that the temperature θ is a known function which depends only on the variable r. We assume that

$$u_r = u(r), \qquad u_\varphi = u_\psi = 0, \tag{4.5.43}$$

where u is an unknown function.
It follows from (4.5.43) and (3.1.29) that

$$\varepsilon_{rr} = \frac{du}{dr}, \qquad \varepsilon_{\varphi\varphi} = \frac{1}{r}u, \qquad \varepsilon_{r\varphi} = 0. \tag{4.5.44}$$

By (3.1.35) we get

$$\begin{aligned} \sigma_{rr} &= \lambda(\varepsilon_{rr} + \varepsilon_{\varphi\varphi}) + 2\mu\varepsilon_{rr} - \beta\theta \\ \sigma_{\varphi\varphi} &= \lambda(\varepsilon_{rr} + \varepsilon_{\varphi\varphi}) + 2\mu\varepsilon_{\varphi\varphi} - \beta\theta, \\ \sigma_{r\varphi} &= 2\mu\varepsilon_{r\varphi}, \end{aligned} \tag{4.5.45}$$

so that

$$\begin{aligned} \sigma_{rr} &= \frac{1}{r}\lambda\frac{d}{dr}(ru) + 2\mu\frac{du}{dr} - \beta\theta, \\ \sigma_{\varphi\varphi} &= \frac{1}{r}\lambda\frac{d}{dr}(ru) + \frac{2\mu}{r}u - \beta\theta, \\ \sigma_{r\varphi} &= 0. \end{aligned} \tag{4.5.46}$$

The equilibrium equations reduce to

$$\frac{d\sigma_{rr}}{dr} + \frac{1}{r}(\sigma_{rr} - \sigma_{\varphi\varphi}) = 0. \tag{4.5.47}$$

The boundary conditions become

$$\sigma_{rr}(R_1) = 0, \qquad \sigma_{rr}(R_2) = 0. \tag{4.5.48}$$

From (4.5.46) and (4.5.47) we obtain the following equation for the function u

$$\frac{d}{dr}\left[\frac{1}{r}\frac{d}{dr}(ru)\right] = \frac{\beta}{\lambda + 2\mu}\frac{d\theta}{dr}. \tag{4.5.49}$$

The general solution of the equation (4.5.49) is

$$u = A_1 r + \frac{1}{r} A_2 + \frac{\beta}{(\lambda + 2\mu) r} \int_{R_1}^{r} r\theta \, dr \,, \tag{4.5.50}$$

where A_α are arbitrary constants. From (4.5.46) and (4.5.50) we get

$$\begin{aligned} \sigma_{rr} &= 2(\lambda + \mu) A_1 - \frac{2\mu}{r^2} A_2 - \frac{2\mu\beta}{(\lambda + 2\mu) r^2} \int_{R_1}^{r} r\theta \, dr \,, \\ \sigma_{\varphi\varphi} &= 2(\lambda + \mu) A_1 + \frac{2\mu}{r^2} A_2 - \frac{2\mu\beta\theta}{\lambda + 2\mu} + \frac{2\mu\beta}{(\lambda + 2\mu) r^2} \int_{R_1}^{r} r\theta \, dr \,. \end{aligned} \tag{4.5.51}$$

Clearly,

$$t_{33} = \sigma_{zz} = \lambda(\varepsilon_{rr} + \varepsilon_{\varphi\varphi}) - \beta\theta = 2\lambda A_1 - \frac{2\mu\beta\theta}{\lambda + 2\mu} \,. \tag{4.5.52}$$

The boundary conditions (4.5.42) reduce to

$$\begin{aligned} (\lambda + \mu) A_1 &= \frac{\mu\beta}{(\lambda + 2\mu)(R_2^2 - R_1^2)} \int_{R_1}^{R_2} r\theta \, dr \,, \\ A_2 &= \frac{\beta R_1^2}{(\lambda + 2\mu)(R_2^2 - R_1^2)} \int_{R_1}^{R_2} r\theta \, dr \,. \end{aligned} \tag{4.5.53}$$

From (4.5.51) and (4.5.53) we obtain

$$\begin{aligned} \sigma_{rr} &= \frac{2\mu\beta}{(\lambda + 2\mu) r^2} \left[\frac{r^2 - R_1^2}{R_2^2 - R_1^2} \int_{R_1}^{R_2} r\theta \, dr - \int_{R_1}^{r} r\theta \, dr \right] , \\ \sigma_{\varphi\varphi} &= \frac{2\mu\beta}{(\lambda + 2\mu) r^2} \left[\frac{r^2 + R_1^2}{R_2^2 - R_1^2} \int_{R_1}^{R_2} r\theta \, dr + \int_{R_1}^{r} r\theta \, dr - r^2\theta \right] . \end{aligned} \tag{4.5.54}$$

The function u is given by

$$u = \frac{\beta}{(\lambda + 2\mu)(R_2^2 - R_1^2)} \left(\frac{\mu r}{\lambda + \mu} + \frac{R_1^2}{r} \right) \int_{R_1}^{R_2} r\theta \, dr + \frac{\beta}{(\lambda + 2\mu) r} \int_{R_1}^{r} r\theta \, dr \,. \tag{4.5.55}$$

We now assume that $S = 0$, and consider the following thermal boundary conditions

$$\theta(R_1) = T_1 \,, \qquad \theta(R_2) = T_2 \,, \tag{4.5.56}$$

where T_1 and T_2 are prescribed constants. In this case we obtain

$$\theta = C_1 \ln r + C_2 \,, \tag{4.5.57}$$

where

$$C_1 = \frac{T_2 - T_1}{\ln R_2 - \ln R_1}, \qquad C_2 = \frac{T_1 \ln R_2 - T_2 \ln R_1}{\ln R_2 - \ln R_1}. \tag{4.5.58}$$

If $T_2 = 0$, then from (4.5.54) we obtain

$$\begin{aligned} \sigma_{rr} &= \frac{\mu\beta T_1}{(\lambda+2\mu)\ln R_2/R_1}\left[\ln\frac{r}{R_2} - \frac{R_1^2}{R_2^2 - R_1^2}\left(1 - \frac{R_2^2}{r^2}\right)\ln\frac{R_2}{R_1}\right], \\ \sigma_{\varphi\varphi} &= \frac{\mu\beta T_1}{(\lambda+2\mu)\ln R_2/R_1}\left[1 - \ln\frac{R_2}{r} - \frac{R_1^2}{R_2^2 - R_1^2}\left(1 + \frac{R_2^2}{r^2}\right)\ln\frac{R_2}{R_1}\right]. \end{aligned} \tag{4.5.59}$$

The problem may be studied with the aid of functions of a complex variable. Thus, if the function θ is given by (4.5.57) then the function F defined by (4.5.12) is

$$F(z) = C_1 \log z + C_2. \tag{4.5.60}$$

From (4.5.13) and (4.5.60) we obtain

$$u_1^* + iu_2^* = z\left(C_2 - C_1 + C_1 \log z\right), \tag{4.5.61}$$

so that the constants B_k, a_k and b_k which appear in (4.5.25) are given by

$$B_1 = C_1, \qquad a_1 = b_1 = 0.$$

The solution of the isothermal plane strain problem corresponding to the characteristics of dislocation

$$\varepsilon_1 = -\frac{\pi\beta}{\lambda+\mu}C_1, \qquad \alpha_1 = \beta_1 = 0,$$

was established by Muskhelishvili [264]. This solution can be expressed in the form

$$2\mu\left(u_r' + iu_\varphi'\right) = \left[\frac{\lambda+3\mu}{\lambda+\mu}\psi(z) - z\overline{\psi'(z)} - \overline{\chi_1(z)}\right]\exp(-i\varphi),$$

$$\psi(z) = \int \Psi(z)\,dz, \qquad \chi_1(z) = \int \Lambda(z)\,dz,$$

$$\Psi(z) = \frac{\varepsilon_1\mu(\lambda+\mu)}{2\pi(\lambda+2\mu)}\left(\frac{1}{2} - \frac{R_2^2\ln R_2 - R_1^2\ln R_1}{R_2^2 - R_1^2}\right) + \frac{\varepsilon_1\mu(\lambda+\mu)}{2\pi(\lambda+2\mu)}\log z,$$

$$\Lambda(z) = -\frac{\varepsilon_1\mu(\lambda+\mu)R_1^2R_2^2}{\pi z^2(\lambda+2\mu)(R_2^2-R_1^2)}\ln\frac{R_2}{R_1}, \tag{4.5.62}$$

$$\sigma'_{rr} = \frac{\varepsilon_1\mu(\lambda+\mu)}{\pi(\lambda+2\mu)}\left(\ln r + \frac{R_1^2R_2^2}{r^2(R_2^2-R_1^2)}\ln\frac{R_2}{R_1} - \frac{R_2^2\ln R_2 - R_1^2\ln R_1}{R_2^2-R_1^2}\right),$$

$$\sigma'_{\varphi\varphi} = \frac{\varepsilon_1\mu(\lambda+\mu)}{\pi(\lambda+2\mu)}\left(\ln r - \frac{R_1^2R_2^2}{r^2(R_2^2-R_1^2)}\ln\frac{R_2}{R_1} - \frac{R_2^2\ln R_2 - R_1^2\ln R_1}{R_2^2-R_1^2} + 1\right),$$

$$\sigma'_{r\varphi} = 0.$$

The functions u_α can be obtained from (4.5.15), (4.5.61) and (4.5.62).

d) **Generalized plane strain problem**. We assume that the cylinder B is occupied by a nonhomogeneous and anisotropic elastic material. We suppose that the constitutive coefficients C_{ijrs}, β_{ij} and k_{ij} are independent of the axial coordinate x_3,

$$C_{ijrs} = C_{ijrs}(x_1, x_2), \qquad \beta_{ij} = \beta_{ij}(x_1, x_2),$$
$$k_{ij} = k_{ij}(x_1, x_2), \qquad (x_1, x_2) \in \Sigma. \tag{4.5.63}$$

The state of generalized plane strain of the cylinder B, parallel to the plane x_1Ox_2, is characterized by

$$u_i = u_i(x_1, x_2), \qquad \theta = \theta(x_1, x_2), \quad (x_1, x_2) \in \Sigma. \tag{4.5.64}$$

This restriction, in conjunction with the relations (4.1.5), (4.1.3) implies that $e_{ij} = e_{ij}(x_1, x_2)$, $t_{ij} = t_{ij}(x_1, x_2)$, $(x_1, x_2) \in \Sigma$. Further,

$$t_{i\alpha} = C_{i\alpha k\rho}u_{k,\rho} - \beta_{i\alpha}\theta, \tag{4.5.65}$$

$$q_i = k_{i\alpha}\theta_{,\alpha}, \tag{4.5.66}$$

and

$$t_{33} = C_{33r\alpha}u_{r,\alpha} - \beta_{33}\theta.$$

We consider the traction problem and assume that the body forces f_i, the heat supply S and the boundary data $\tilde{t}_i$, $\tilde{\theta}$ and $\tilde{q}$ are independent of x_3. The equations of equilibrium reduce to

$$t_{\alpha i,\alpha} + f_i = 0 \qquad \text{on } \Sigma. \tag{4.5.67}$$

The energy equation (4.1.2) becomes

$$q_{\alpha,\alpha} + S = 0 \qquad \text{on } \Sigma . \tag{4.5.68}$$

The boundary conditions take the form

$$t_{\beta i} n_\beta = \tilde{t}_i \qquad \text{on } L , \tag{4.5.69}$$

$$\theta = \tilde{\theta} \qquad \text{on } L_1 , \qquad q_\alpha n_\alpha = \tilde{q} \qquad \text{on } L_2 . \tag{4.5.70}$$

The temperature field is the solution of the boundary value problem characterized by the equations (4.5.66) and (4.5.68) on Σ, and the boundary conditions (4.5.70) on $\partial\Sigma$.

The determination of the thermoelastic equilibrium consists in solving the equations (4.5.65) and (4.5.67) on Σ with the boundary conditions (4.5.69).

The generalized plane strain problem has been studied in various papers (see, e.g. Lekhnitskii [235], Iesan [180], [186]). We note that the conditions of equilibrium for the cylinder B are equivalent to

$$\int_\Sigma f_i \, da + \int_L \tilde{t}_i \, ds = 0 , \qquad \int_\Sigma \varepsilon_{\alpha\beta 3} x_\alpha f_\beta \, da + \int_L \varepsilon_{\alpha\beta 3} x_\alpha f_\beta \, ds = 0 , \tag{4.5.71}$$

$$\int_\Sigma x_\alpha f_3 \, da + \int_L x_\alpha \tilde{t}_3 \, ds = \int_\Sigma t_{\alpha 3} \, da . \tag{4.5.72}$$

It follows from (4.5.67) and (4.5.69) that

$$\begin{aligned} \int_\Sigma t_{\alpha 3} \, da &= \int_\Sigma \left[t_{\alpha 3} + x_\alpha (t_{3\beta,\beta} + f_3) \right] da = \\ &= \int_\Sigma \left[(x_\alpha t_{3\beta})_{,\beta} + x_\alpha f_3 \right] da = \int_L x_\alpha \tilde{t}_3 \, ds + \int_\Sigma x_\alpha f_3 \, da . \end{aligned} \tag{4.5.73}$$

Thus, the conditions (4.5.72) are identically satisfied. If the domain Σ is smooth, f_i are continuous on $\bar{\Sigma}$, and $\tilde{t}_i$ are piecewise regular on L, then (4.5.71) are the necessary and sufficient conditions for the existence of a solution to the boundary value problem (4.5.65), (4.5.67) and (4.5.69) (cf. Fichera [128], p.385, Iesan [191]).

From (4.5.65) and (4.5.67) we obtain the equations of equilibrium expressed in terms of the displacement,

$$\left(C_{i\alpha k\beta} u_{k,\beta} \right)_{,\alpha} - \left(\beta_{i\alpha} \theta \right)_{,\alpha} + f_i = 0 \qquad \text{on } \Sigma . \tag{4.5.74}$$

The boundary condition (4.5.69) can be expressed in the form

$$C_{i\alpha k\beta}u_{k,\beta}n_\alpha = \tilde{t}_i + \beta_{i\alpha}\theta n_\alpha \qquad \text{on } L. \tag{4.5.75}$$

For the remainder of this section, we assume that B is homogeneous and that

$$f_i = 0, \qquad S = 0, \qquad \tilde{t}_i = 0. \tag{4.5.76}$$

We present the method given in [179] to reduce the thermoelastic boundary value problem to an isothermal one, with zero body forces and with certain known boundary data. It follows from (4.5.66), (4.5.68) and (4.5.76) that θ satisfies the equation

$$k_{\alpha\beta}\theta_{,\alpha\beta} = 0 \qquad \text{on } \Sigma. \tag{4.5.77}$$

We denote λ_α the roots of the equation

$$k_{\alpha\beta}x^{4-(\alpha+\beta)} = 0. \tag{4.5.78}$$

We have

$$\lambda_\alpha = a - (-1)^\alpha ib, \tag{4.5.79}$$

where a and b are real constants. If we introduce the variables y_α by

$$y_1 = bx_1, \qquad y_2 = x_2 + ax_1, \tag{4.5.80}$$

then the equation (4.5.77) becomes

$$\frac{\partial^2\theta}{\partial y_1^2} + \frac{\partial^2\theta}{y_2^2} = 0. \tag{4.5.81}$$

We denote by $F(\zeta)$ the following analytic function of the complex variable $\zeta = y_1 + iy_2$,

$$F(\zeta) = \theta + iP. \tag{4.5.82}$$

Clearly,

$$\frac{\partial\theta}{\partial y_1} = \frac{\partial P}{\partial y_2}, \qquad \frac{\partial\theta}{\partial y_2} = -\frac{\partial P}{\partial y_1}. \tag{4.5.83}$$

The above relations may be written in the form

$$P_{,1} = \frac{a}{b}\theta_{,1} - \frac{a^2+b^2}{b}\theta_{,2}, \qquad P_{,2} = \frac{1}{b}\theta_{,1} - \frac{a}{b}\theta_{,2}. \tag{4.5.84}$$

Thus, we have

$$P_{,\alpha} = c_{\alpha\rho}\theta_{,\rho}, \tag{4.5.85}$$

where

$$c_{11} = \frac{a}{b}, \qquad c_{12} = -\frac{a^2+b^2}{b}, \qquad c_{21} = \frac{1}{b}, \qquad c_{22} = -\frac{a}{b}.$$

We introduce the functions u^*_α by

$$u^*_1 + iu^*_2 = \int F(\zeta)\,d\zeta\,, \tag{4.5.86}$$

so that

$$\frac{\partial u^*_1}{\partial y_1} = \frac{\partial u^*_2}{\partial y_2} = \theta\,, \qquad \frac{\partial u^*_1}{\partial y_2} = -\frac{\partial u^*_2}{\partial y_1} = -P\,. \tag{4.5.87}$$

The relations (4.5.87) can be written in the form

$$u^*_{\alpha,\beta} = a_{\alpha\beta}\theta + b_{\alpha\beta}P\,, \tag{4.5.88}$$

where

$$a_{11} = b\,, \quad a_{12} = 0\,, \quad a_{21} = a\,, \quad a_{22} = 1\,,$$
$$b_{11} = -a\,, \quad b_{12} = -1\,, \quad b_{21} = b\,, \quad b_{22} = 0\,.$$

We introduce the functions u'_i by

$$u_i = u'_i + p_{i\beta}u^*_\beta\,, \tag{4.5.89}$$

where $p_{i\beta}$ are constants which will be specified in what follows. If we denote

$$t'_{i\alpha} = C_{i\alpha k\rho}u'_{k,\rho}\,, \tag{4.5.90}$$

then from (4.5.65) and (4.5.89) we obtain

$$t_{i\alpha} = t'_{i\alpha} + C_{i\alpha k\rho}p_{k\nu}u^*_{\nu,\rho} - \beta_{i\alpha}\theta\,. \tag{4.5.91}$$

In view of (4.5.88) we can write

$$t_{i\alpha} = t'_{i\alpha} + (C_{i\alpha k\rho}p_{k\nu}a_{\nu\rho} - \beta_{i\alpha})\,\theta + C_{i\alpha k\rho}p_{k\nu}b_{\nu\rho}P\,. \tag{4.5.92}$$

We assume that

$$t'_{i\alpha,\alpha} = 0 \qquad \text{on } \Sigma\,. \tag{4.5.93}$$

From the equations of equilibrium, (4.5.92) and (4.5.93) we obtain

$$(C_{i\alpha k\rho}p_{k\nu}a_{\nu\rho} - \beta_{i\alpha})\,\theta_{,\alpha} + C_{i\alpha k\rho}p_{k\nu}b_{\nu\rho}P_{,\alpha} = 0\,. \tag{4.5.94}$$

It follows from (4.5.85) and (4.5.94) that

$$[C_{i\alpha k\rho}p_{k\nu}(a_{\nu\rho}\delta_{\alpha\mu}+b_{\nu\rho}c_{\alpha\mu})-\beta_{i\mu}]\,\theta_{,\mu}=0\,.$$

The constants $p_{i\nu}$ will be determined such that

$$C_{i\alpha k\rho}(a_{\nu\rho}\delta_{\alpha\mu}+b_{\nu\rho}c_{\alpha\mu})\,p_{k\nu}=\beta_{i\mu}\,. \tag{4.5.95}$$

The boundary conditions (4.5.69) reduce to

$$t'_{i\alpha}n_\alpha=-\left[(C_{i\alpha k\rho}p_{k\nu}a_{\nu\rho}-\beta_{i\alpha})\,\theta+C_{i\alpha k\rho}p_{k\nu}b_{\nu\rho}P\right]n_\alpha \qquad \text{on } L\,. \tag{4.5.96}$$

Thus, the thermoelastic problem has been reduced to the isothermal plane strain problem (4.5.90), (4.5.93), (4.5.96). If the domain Σ is multiply-connected, as in (4.5.24), we have

$$F(\zeta)=\sum_{k=1}^{n}B_k\log(\zeta-\zeta_k)+F_0(\zeta)\,,$$

where ζ_k are points inside the contours Γ_k, B_k are real constants and $F_0(\zeta)$ is a holomorphic function. Thus, we have

$$u_1^*+iu_2^*=\zeta\sum_{k=1}^{n}B_k\log(\zeta-\zeta_k)+\sum_{k=1}^{n}(a_k+ib_k)\log(\zeta-\zeta_k)+f_0(\zeta)\,.$$

By (4.5.89),

$$\left[u'_j\right]_{L_k}=2\pi\left[(p_{j1}a-p_{j2}b)\,B_kx_1+p_{j1}B_kx_2+p_{j1}b_k-p_{j2}a_k\right]\,.$$

Thus, for multiply-connected regions the displacements u'_i are multi-valued.

4.6 Thermal stresses in cylinders

Throughout this section the region B refers to the interior of a right cylinder of length h with the generic cross-section Σ and the lateral boundary Π. The rectangular Cartesian coordinate frame is chosen such that the x_3-axis is parallel to the generators of B and the x_1Ox_2 plane contains one of the terminal cross-sections. We denote by Σ_1 and Σ_2 the cross-sections located at $x_3=0$ and $x_3=h$, respectively. We assume that the cross-section Σ is a simply-connected regular region, bounded by the closed curve L.

We suppose that B is in equilibrium in the absence of body forces, under the action of a prescribed temperature field.

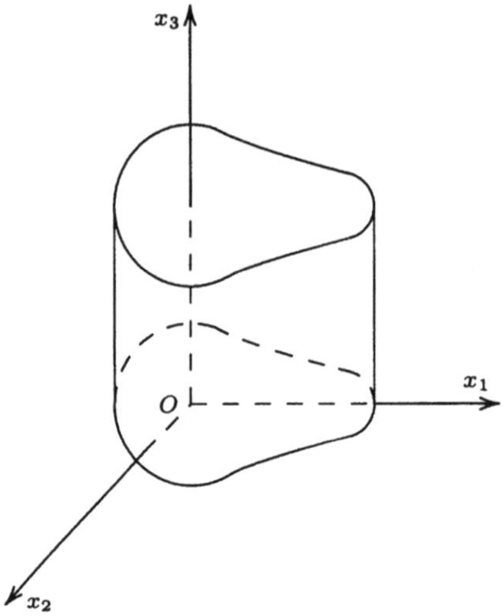

Fig. 4

The cylinder is supposed to be free of lateral tractions. We assume that the tractions applied over the ends are self-equilibrated, and that the temperature θ is a polynomial of degree r in the axial coordinate, namely

$$\theta = \sum_{k=0}^{r} T_k(x_1, x_2)x_3^k, \tag{4.6.1}$$

where T_k are prescribed.

The equations of equilibrium reduce to

$$t_{ji,j} = 0 \qquad \text{on } B. \tag{4.6.2}$$

Since the lateral boundary of the cylinder is free of loads we have the conditions

$$t_{\alpha i}n_\alpha = 0 \qquad \text{on } \Pi. \tag{4.6.3}$$

On the plane $x_3 = 0$ we have the conditions

$$\int_{\Sigma_1} t_{3\alpha}\, da = 0, \tag{4.6.4}$$

$$\int_{\Sigma_1} t_{33}\, da = 0, \tag{4.6.5}$$

$$\int_{\Sigma_1} x_\alpha t_{33}\, da = 0\,, \tag{4.6.6}$$

$$\int_{\Sigma_1} \varepsilon_{3\alpha\beta} x_\alpha t_{3\beta}\, da = 0\,, \tag{4.6.7}$$

which show that over the base Σ_1 the loads are specified to be self-equilibrated. In view of (4.6.2) and (4.6.3), the conditions (4.6.4)–(4.6.7) hold over each cross section Σ.

We denote by $P^{(n)}$ the problem of thermoelastic equilibrium corresponding to the temperature field

$$\theta = T_n(x_1,\, x_2)x_3^n\,, \tag{4.6.8}$$

where n is a positive integer or zero, and the function T_n is known. Obviously, if we know the solution of the problem $P^{(n)}$ for any n then, according to the linearity of the theory, we can determine the solution when the temperature has the form (4.6.1). In what follows we use the method of induction. First we establish the solution of the problem $P^{(0)}$. Then, we establish a solution of the problem $P^{(n+1)}$ once a solution of the problem $P^{(n)}$ is known.

If we denote by $\boldsymbol{u}^{(k)}$ the displacement field corresponding to a solution of the problem $P^{(k)}$, then the displacement field corresponding to a solution of the initial problem is

$$\boldsymbol{u} = \sum_{k=0}^{r} \boldsymbol{u}^{(k)}\,.$$

a) **Homogeneous and isotropic bodies.** In this case the constitutive equations can be written in the form

$$e_{ij} = \frac{1+\nu}{E} t_{ij} - \frac{\nu}{E} t_{rr}\delta_{ij} + \alpha\theta\delta_{ij}\,, \tag{4.6.9}$$

where

$$E = \frac{\mu(3\lambda+2\mu)}{\lambda+\mu}\,, \qquad \nu = \frac{\lambda}{2(\lambda+\mu)}\,, \qquad \alpha = \frac{\beta}{3\lambda+2\mu}\,. \tag{4.6.10}$$

First, we consider the problem $P^{(0)}$. We assume that

$$\theta = f(x_1,\, x_2) \qquad \text{on } B\,, \tag{4.6.11}$$

where f is a prescribed function.

We seek the solution of the problem $P^{(o)}$ in the form

$$u_1 = -\frac{1}{2}a_1\left[x_3^2 + \nu(x_1^2 - x_2^2)\right] - a_2\nu x_1 x_2 - a_3\nu x_1 + v_1(x_1, x_2).$$

$$u_2 = -a_1\nu x_1 x_2 - \frac{1}{2}a_2\left[x_3^2 - \nu(x_1^2 - x_2^2)\right] - a_3\nu x_2 + v_2(x_1, x_2), \tag{4.6.12}$$

$$u_3 = (a_1 x_1 + a_2 x_2 + a_3)\, x_3\,,$$

where ν is given by (4.6.10), $v_\alpha(x_1, x_2)$, $(x_1, x_2) \in \Sigma$ are unknown functions, and a_i are unknown constants.

If we introduce the notations

$$\gamma_{\alpha\beta} = \frac{1}{2}\left(v_{\alpha,\beta} + v_{\beta,\alpha}\right),$$

$$\sigma_{\alpha\beta} = \lambda\gamma_{\rho\rho}\delta_{\alpha\beta} + 2\mu\gamma_{\alpha\beta} - \beta f\delta_{\alpha\beta}\,, \tag{4.6.13}$$

then from (4.3.2), (4.6.12) and (4.6.10) we obtain

$$t_{\alpha\beta} = \sigma_{\alpha\beta}\,, \qquad t_{\alpha 3} = 0\,,$$

$$t_{33} = E(a_1 x_1 + a_2 x_2 + a_3) + \nu\sigma_{\beta\beta} - E\alpha f\,. \tag{4.6.14}$$

The equilibrium equations (4.6.2) become

$$\sigma_{\beta\alpha,\beta} = 0 \qquad \text{on } \Sigma\,. \tag{4.6.15}$$

The boundary conditions (4.6.3) reduce to

$$\sigma_{\beta\alpha} n_\beta = 0 \qquad \text{on } L\,. \tag{4.6.16}$$

Thus, we conclude that the functions v_α are the components of the displacement vector field in the thermoelastic plane strain problem for temperature distribution $\theta = f$, characterized by the equations (4.6.13), (4.6.15) on Σ and the boundary conditions (4.6.16) on L. In what follows we assume that v_α and $\sigma_{\alpha\beta}$ are known.

The conditions (4.6.4) and (4.6.7) are satisfied on the basis of the relations (4.6.14). The conditions (4.6.5) and (4.6.6) reduce to

$$a_1 = \frac{1}{Ed}\left[I_{22}M_1 - I_{12}M_2 + \left(x_2^0 I_{12} - x_1^0 I_{22}\right)P\right],$$

$$a_2 = \frac{1}{Ed}\left[I_{11}M_2 - I_{12}M_1 + \left(x_1^0 I_{12} - x_2^0 I_{11}\right)P\right],$$

$$a_3 = \frac{1}{EA}P - a_1 x_1^0 - a_2 x_2^0\,,$$

where

$$P = \int_\Sigma \Lambda\, da\,, \quad M_\rho = \int_\Sigma x_\rho \Lambda\, da\,, \quad \Lambda = E\alpha f - \nu\sigma_{\beta\beta}\,, \quad A = \int_\Sigma da\,,$$
$$I_{\alpha\beta} = \int_\Sigma (x_\alpha - x_\alpha^0)(x_\beta - x_\beta^0)\, da\,, \quad d = I_{11}I_{22} - I_{12}^2\,, \quad Ax_\alpha^0 = \int_\Sigma x_\alpha\, da\,. \tag{4.6.17}$$

Thus, the problem $P^{(0)}$ is reduced to solving a thermoelastic plane strain problem.

Let us now study the recurrence process. We denote by u_i^*, e_{ij}^* and t_{ij}^* the components of the displacement vector, the components of the strain tensor and the components of the stress tensor from the problem $P^{(n)}$, respectively, and by u_i, e_{ij}, t_{ij} the analogous functions from the problem $P^{(n+1)}$. Let us establish the solution of the problem $P^{(n+1)}$ when the solution of the problem $P^{(n)}$ is known. As the solution of the problem $P^{(n)}$ is supposed to be known for any T_n we know the solution of the problem $P^{(n)}$ when $\theta = T_{n+1}(x_1, x_2)x_3^n$. Thus, the problem can be presented as follows: find the functions u_i, e_{ij}, t_{ij} which satisfy the equations (4.1.5), (4.6.2), (4.6.9) on B and the conditions (4.6.3)–(4.6.7) when the temperature field has the form

$$\theta = g(x_1, x_2)x_3^{n+1}\,, \tag{4.6.18}$$

assuming that the functions u_i^*, e_{ij}^*, t_{ij}^* which satisfy the equations (4.1.5), (4.6.2), (4.6.9) on B and the conditions (4.6.3)–(4.6.7) for the temperature variation

$$\theta = g(x_1, x_2)x_3^n\,, \tag{4.6.19}$$

are known.

We seek the solution of this problem in the form

$$\begin{aligned} u_1 =& (n+1)\Big\{ \int_0^{x_3} u_1^*\, dx_3 - \frac{1}{2} b_1 \left[x_3^2 + \nu(x_1^2 - x_2^2)\right] - \\ & - b_2\nu x_1 x_2 - b_3 \nu x_1 - \tau x_2 x_3 + \omega_1(x_1, x_2)\Big\}\,, \\ u_2 =& (n+1)\Big\{ \int_0^{x_3} u_2^*\, dx_3 - \frac{1}{2} b_2 \left[x_3^2 - \nu(x_1^2 - x_2^2)\right] - \\ & - b_1\nu x_1 x_2 - b_3 \nu x_2 + \tau x_1 x_3 + \omega_2(x_1, x_2)\Big\}\,, \\ u_3 =& (n+1)\Big\{ \int_0^{x_3} u_3^*\, dx_3 + (b_1 x_1 + b_2 x_2 + b_3)x_3 + F(x_1, x_2)\Big\}\,, \end{aligned} \tag{4.6.20}$$

where $\omega_\alpha(x_1, x_2)$, $F(x_1, x_2)$, $(x_1, x_2) \in \Sigma$ are unknown functions, and b_i, τ are unknown constants. From (4.1.5) and (4.6.20) we obtain

$$e_{\alpha\beta} = (n+1)\left[\int_0^{x_3} e^*_{\alpha\beta}\, dx_3 - \nu(b_1x_1 + b_2x_2 + b_3)\delta_{\alpha\beta} + \gamma_{\alpha\beta}\right],$$

$$e_{\alpha 3} = (n+1)\left\{\int_0^{x_3} e^*_{\alpha 3}\, dx_3 + \frac{1}{2}\left[F_{,\alpha} - \tau\varepsilon_{3\alpha\beta}x_\beta + u^*_\alpha(x_1, x_2, 0)\right]\right\},$$

$$e_{33} = (n+1)\left[\int_0^{x_3} e^*_{33}\, dx_3 + b_1x_1 + b_2x_2 + b_3 + u^*_3(x_1, x_2, 0)\right],$$

where

$$2\gamma_{\alpha\beta} = \omega_{\alpha,\beta} + \omega_{\beta,\alpha}\,. \tag{4.6.21}$$

In view of (4.6.18)–(4.6.20) and $(4.3.2)_2$ we obtain

$$\begin{aligned} t_{\alpha\beta} =&(n+1)\left[\int_0^{x_3} t^*_{\alpha\beta}\, dx_3 + \pi_{\alpha\beta} + \lambda\delta_{\alpha\beta}u^*_3(x_1, x_2, 0)\right],\\ t_{\alpha 3} =&(n+1)\left\{\int_0^{x_3} t^*_{\alpha 3}\, dx_3 + \mu\left[F_{,\alpha} - \tau\varepsilon_{3\alpha\beta}x_\beta + u^*_\alpha(x_1, x_2, 0)\right]\right\},\\ t_{33} =&(n+1)\left[\int_0^{x_3} t^*_{33}\, dx_3 + E(b_1x_1 + b_2x_2 + b_3) + \nu\pi_{\alpha\alpha} + \right.\\ &\left. + (\lambda + 2\mu)u^*_3(x_1, x_2, 0)\right], \end{aligned} \tag{4.6.22}$$

where

$$\pi_{\alpha\beta} = \lambda\gamma_{\rho\rho}\delta_{\alpha\beta} + 2\mu\gamma_{\alpha\beta}\,. \tag{4.6.23}$$

The equations of equilibrium (4.6.2) reduce to

$$\pi_{\beta\alpha,\beta} + g_\alpha = 0 \qquad \text{on } \Sigma, \tag{4.6.24}$$

$$F_{,\alpha\alpha} = q_1 \qquad \text{on } \Sigma, \tag{4.6.25}$$

where

$$\begin{aligned} g_\alpha(x_1, x_2) &= t^*_{\alpha 3}(x_1, x_2, 0) + \lambda u^*_{3,\alpha}(x_1, x_2, 0)\,,\\ q_1(x_1, x_2) &= -u^*_{\alpha,\alpha}(x_1, x_2, 0) - \frac{1}{\mu}t^*_{33}(x_1, x_2, 0)\,. \end{aligned} \tag{4.6.26}$$

The boundary conditions (4.6.3) become

$$\pi_{\beta\alpha}n_\beta = p_\alpha \qquad \text{on } L, \tag{4.6.27}$$

$$\frac{\partial F}{\partial n} = q_2 \qquad \text{on } L, \tag{4.6.28}$$

where

$$p_\alpha = -\lambda u_3^*(x_1, x_2, 0)n_\alpha, \qquad q_2 = \tau\varepsilon_{3\alpha\beta}x_\beta n_\alpha - u_\alpha^*(x_1, x_2, 0)n_\alpha. \tag{4.6.29}$$

Thus, the functions w_α are the components of the displacement field in the isothermal plane strain problem characterized by the equations (4.6.22), (4.6.23), (4.6.24) on Σ and the boundary conditions (4.6.27) on L. It follows from (4.6.26), (4.6.29) and the divergence theorem that

$$\begin{aligned}
&\int_\Sigma g_\alpha\, da + \int_L p_\alpha\, ds = \int_\Sigma t_{\alpha 3}^*(x_1, x_2, 0)\, da,\\
&\int_\Sigma \varepsilon_{3\alpha\beta}x_\alpha g_\beta\, da + \int_L \varepsilon_{3\alpha\beta}x_\alpha p_\beta\, ds = \int_\Sigma \varepsilon_{3\alpha\beta}x_\alpha t_{\beta 3}^*(x_1, x_2, 0)\, da.
\end{aligned} \tag{4.6.30}$$

Since t_{ij}^* satisfy the conditions (4.6.4)–(4.6.7), from (4.6.30) we conclude that the necessary and sufficient conditions (4.5.9) for the existence of a solution to the boundary value problem (4.6.22), (4.6.23), (4.6.24), (4.6.27) are satisfied. We will assume that the functions w_α are known.

The necessary and sufficient condition for the existence of a solution to the boundary value problem (4.6.25), (4.6.28) is

$$\int_\Sigma q_1\, da = \int_L q_2\, ds. \tag{4.6.31}$$

By (4.6.26) and (4.6.29),

$$\int_\Sigma q_1\, da - \int_L q_2\, ds = -\frac{1}{\mu}\int_\Sigma t_{33}^*(x_1, x_2, 0)\, da.$$

Since t_{33}^* satisfies (4.6.5), we conclude that (4.6.31) holds. To determine the constant τ we consider the torsion function φ which is the solution of the boundary value problem

$$\begin{aligned}
&\varphi_{,\alpha\alpha} = 0 \qquad \text{on } \Sigma,\\
&\frac{\partial\varphi}{\partial n} = \varepsilon_{3\alpha\beta}x_\alpha n_\beta \qquad \text{on } L.
\end{aligned} \tag{4.6.32}$$

We introduce the function ψ by

$$F = \tau\varphi + \psi\,. \tag{4.6.33}$$

It follows from (4.6.25)–(4.6.29) and (4.6.33) that the function ψ is the solution of the following boundary value problem

$$\begin{aligned} &\psi_{,\alpha\alpha} = q_1 \qquad \text{on } \Sigma\,,\\ &\frac{\partial\psi}{\partial n} = -u_\alpha^*(x_1,\, x_2,\, 0)n_\alpha \qquad \text{on } L\,. \end{aligned} \tag{4.6.34}$$

The function ψ is independent of τ, and in what follows we will assume that ψ is known. From (4.6.22) and (4.6.33) we get

$$\begin{aligned} t_{\alpha 3} =&(n+1)\left[\int_0^{x_3} t_{\alpha 3}^*\,dx_3 + \mu\tau(\varphi_{,\alpha} - \varepsilon_{3\alpha\beta}x_\beta) + \right.\\ &\left. + \mu\psi_{,\alpha} + \mu u_\alpha^*(x_1,\, x_2,\, 0)\right]. \end{aligned} \tag{4.6.35}$$

By (4.6.7) and (4.6.35), we find that

$$\tau D = -\mu\int_\Sigma \varepsilon_{3\alpha\beta}x_\alpha\left[\psi_{,\beta} + u_\beta^*(x_1,\, x_2,\, 0)\right] da\,, \tag{4.6.36}$$

where D is the torsional rigidity,

$$D = \mu\int_\Sigma (\varepsilon_{3\alpha\beta}x_\alpha\varphi_{,\beta} + x_1^2 + x_2^2)\, da\,.$$

It is known that $D > 0$ (cf. Solomon [333], p.199) so that the equation (4.6.36) determines the constant τ.

From (4.6.5), (4.6.6) and (4.6.22) we obtain

$$\begin{aligned} b_1 &= \frac{1}{Ed}\left[I_{22}N_1 - I_{12}N_2 + (x_2^0 I_{12} - x_1^0 I_{22})Q\right],\\ b_2 &= \frac{1}{Ed}\left[I_{11}N_2 - I_{12}N_1 + (x_1^0 I_{12} - x_2^0 I_{11})Q\right],\\ b_3 &= \frac{1}{EA}Q - b_1 x_1^0 - b_2 x_2^0\,, \end{aligned} \tag{4.6.37}$$

where

$$Q = \int_\Sigma \Omega \, da\,, \quad N_\alpha = \int_\Sigma x_\alpha \Omega \, da\,, \quad \Omega = -\nu \pi_{\alpha\alpha} - (\lambda + 2\mu) u_3^*(x_1,\, x_2,\, 0)\,,$$

and we have used the notations (4.6.17).

We note that

$$\begin{gathered} \int_\Sigma t_{\alpha 3}\, da = \int_\Sigma (t_{\alpha 3} + x_\alpha t_{i3,i})\, da = \int_\Sigma [(x_\alpha t_{\beta 3})_{,\beta} + \\ + x_\alpha t_{33,3}]\, da = \int_L x_\alpha t_{\beta 3} n_\beta \, ds + (n+1) \int_\Sigma x_\alpha t_{33}^*\, da\,. \end{gathered} \tag{4.6.38}$$

In view of the conditions (4.6.3) and (4.6.6), the relations (4.6.38) imply that the conditions (4.6.4) are indentically satisfied.

The above results have been established by Boley and Weiner [28], p.328 and Iesan [177].

b) **Cylinders composed of different materials.** We now assume that the cross-section Σ consists of the assembly of the regions Σ_0 and Σ_j $(j = 1, 2, \ldots, m)$, Σ_0 being a multiply-connected region bounded by the closed curves L_j $(j = 1, 2, \ldots, m{+}1)$ possessing no common points. We assume that the domains Σ_j $(j = 1, 2, \ldots, m)$ are finite and simply-connected, bounded by the corresponding curves L_j $(j = 1, 2, \ldots, m)$, L_{m+1} is the boundary of the region Σ. We assume that the matter filling each of the regions Σ_0 and Σ_j $(j = 1, 2, \ldots, m)$ is homogeneous and isotropic, while passing from one medium to another the thermoelastic properties are different.

The problem of deformation of composite cylinders within the isothermal theory, has been studied by Muskhelishvili [264], p.546.

The displacement vector field and the stress vector field must be continuous in passing from one medium to another, so that we have the conditions

$$[u_k]_i = [u_k]_0\,, \tag{4.6.39}$$

$$[t_{\alpha k}]_i n_\alpha = [t_{\alpha k}]_0 n_\alpha \qquad \text{on } L_i\, (i = 1,\, 2, \ldots, m)\,, \tag{4.6.40}$$

$$[t_{\alpha k}]_0 n_\alpha = 0 \qquad \text{on } L_{m+1}\,, \tag{4.6.41}$$

for any $x_3 \in [0,\, h]$, where n_α are the direction cosines of the vector normal to the boundary of Σ_0, outward to Σ_0, and we have indicated that the expressions in brackets are calculated for the material corresponding to the regions Σ_i $(i = 1, 2, \ldots, m)$ and Σ_0.

We denote by λ_i, μ_i, β_i $(i = 0, 1, 2, \ldots, m)$ the constitutive coefficients for the material corresponding to the domain Σ_i $(i = 0, 1, 2, \ldots, m)$. We introduce also the constants E_i, ν_i and α_i corresponding to the constitutive equations (3.1.10).

We consider three problems of isothermal plane strain $A^{(k)}$, $(k = 1, 2, 3)$, characterized by the equations

$$\tau_{\alpha\beta}^{(k)} = \lambda_i \varepsilon_{\rho\rho}^{(k)} \delta_{\alpha\beta} + 2\mu_i \varepsilon_{\alpha\beta}^{(k)} , \tag{4.6.42}$$

$$2\varepsilon_{\alpha\beta}^{(k)} = v_{\alpha,\beta}^{(k)} + v_{\beta,\alpha}^{(k)} , \tag{4.6.43}$$

$$\tau_{\alpha\beta,\beta}^{(k)} = 0 , \tag{4.6.44}$$

in Σ_i $(i = 0, 1, 2, \ldots, m)$, and the boundary conditions

$$\begin{aligned} &\left[\tau_{\alpha\beta}^{(k)}\right]_i n_\beta = \left[\tau_{\alpha\beta}^{(k)}\right]_0 n_\beta , \\ &\left[v_\alpha^{(k)}\right]_i - \left[v_\alpha^{(k)}\right]_0 = g_{i\alpha}^{(k)} \qquad \text{on } L_i \, (i = 1, 2, \ldots, m) , \\ &\left[\tau_{\alpha\beta}^{(k)}\right]_0 n_\beta = 0 \qquad \text{on } L_{m+1} , \end{aligned} \tag{4.6.45}$$

where

$$\begin{aligned} &g_{i1}^{(1)} = \frac{1}{2}(\nu_i - \nu_0)(x_1^2 - x_2^2) , \qquad g_{i2}^{(1)} = (\nu_i - \nu_0) x_1 x_2 , \\ &g_{i1}^{(2)} = (\nu_i - \nu_0) x_1 x_2 , \qquad g_{i2}^{(2)} = -\frac{1}{2}(\nu_i - \nu_0)(x_1^2 - x_2^2) , \\ &g_{i\alpha}^{(3)} = (\nu_i - \nu_0) x_\alpha . \end{aligned} \tag{4.6.46}$$

The problems $A^{(k)}$ have been introduced by Muskhelishvili [264], p.549 to study Saint-Venant's problem in isothermal elasticity.

We seek the solution of the problem $P^{(0)}$ in the form

$$\begin{aligned} u_1 =& -\frac{1}{2} a_1 \left[x_3^2 + \nu_i (x_1^2 - x_2^2)\right] - a_2 \nu_i x_1 x_2 - a_3 \nu_i x_1 + \\ &+ \sum_{k=1}^{3} a_k v_1^{(k)} + v_1(x_1, x_2) , \\ u_2 =& - a_1 \nu_i x_1 x_2 - \frac{1}{2} a_2 \left[x_3^2 - \nu_i (x_1^2 - x_2^2)\right] - a_3 \nu_i x_2 + \\ &+ \sum_{k=1}^{3} a_k v_2^{(k)} + v_2(x_1, x_2) , \\ u_3 =& (a_1 x_1 + a_2 x_2 + a_3) x_3 \qquad \text{on } \Sigma_i \, (i = 0, 1, 2, \ldots, m) , \end{aligned} \tag{4.6.47}$$

where v_α are unknown functions, and a_i are unknown constants.

In view of (4.6.45), the functions (4.6.47) are continuous in Σ if

$$[v_\alpha]_i = [v_\alpha]_0 \qquad \text{on } L_i \, (i = 1, 2, \ldots, m). \tag{4.6.48}$$

From $(3.1.9)_1$ and (4.6.47) we obtain

$$\begin{aligned} t_{\alpha\beta} &= \sigma_{\alpha\beta} + \sum_{k=1}^{3} a_k \tau_{\alpha\beta}^{(k)}, \qquad\qquad t_{\alpha 3} = 0, \\ t_{33} &= E_i(a_1 x_1 + a_2 x_2 + a_3) + \nu_i \sum_{k=1}^{3} a_k \tau_{\beta\beta}^{(k)} + \nu_i \sigma_{\beta\beta} - E_i \alpha_i f, \end{aligned} \tag{4.6.49}$$

on $\Sigma_i \, (i = 0, 1, 2, \ldots, m)$, where

$$\begin{aligned} \sigma_{\alpha\beta} &= \lambda_i \gamma_{\rho\rho} \delta_{\alpha\beta} + 2\mu_i \gamma_{\alpha\beta} - \beta_i f \delta_{\alpha\beta}, \\ \gamma_{\alpha\beta} &= \frac{1}{2}(v_{\alpha,\beta} + v_{\beta,\alpha}), \end{aligned} \tag{4.6.50}$$

on $\Sigma_i \, (i = 0, 1, 2, \ldots, m)$.
If we take into account (4.6.44) and (4.6.49), we find that the equations of equilibrium (4.6.2) reduce to

$$\sigma_{\beta\alpha,\beta} = 0 \qquad \text{on } \Sigma_i \, (i = 0, 1, 2, \ldots, m). \tag{4.6.51}$$

The conditions (4.6.40) and (4.6.41) become

$$\begin{aligned} &[\sigma_{\beta\alpha}]_i n_\beta = [\sigma_{\beta\alpha}]_0 n_\beta \qquad \text{on } L_i \, (i = 1, 2, \ldots, m), \\ &[\sigma_{\alpha\beta}]_0 n_\alpha = 0 \qquad \text{on } L_{m+1}. \end{aligned} \tag{4.6.52}$$

Clearly, the functions v_α are the components of the displacement vector field in the thermoelastic plane strain problem (4.6.48), (4.6.50), (4.6.51), (4.6.52) for temperature field f.

The conditions (4.6.4) and (4.6.7) are satisfied on the basis of the relations (4.6.49). The conditions (4.6.5) and (4.6.6) reduce to

$$\begin{aligned} &a_1 = K_{22} m_1 - K_{21} m_2, \qquad a_2 = K_{11} m_2 - K_{12} m_1, \\ &a_3 = p - a_1 d_1 - a_2 d_2, \end{aligned} \tag{4.6.53}$$

where

$$
\begin{aligned}
&K_{\alpha\beta} = \frac{1}{d}\sum_{i=0}^{m}\int_{\Sigma_i} h_\alpha^{(i)} x_\beta da\,, \quad h_\alpha^{(i)} = E_i x_\alpha + \nu_i \tau_{\beta\beta}^{(\alpha)} - d_\alpha\big[E_i + \nu_i \tau_{\beta\beta}^{(3)}\big]\,,\\
&d_\alpha = \frac{1}{d_3}\sum_{i=0}^{m}\int_{\Sigma_i}\big[E_i x_\beta + \nu_i \tau_{\rho\rho}^{(\beta)}\big]da\,, \quad d_3 = \sum_{i=0}^{m}\int_{\Sigma_i}\big[E_i + \nu_i\tau_{\rho\rho}^{(3)}\big]da\,,\\
&d = K_{11}K_{22} - K_{12}^2\,, \quad p = \frac{1}{d_3}\sum_{i=0}^{m}\int_{\Sigma_i}\Lambda^{(i)}\,da\,, \quad \Lambda^{(i)} = E_i\alpha_i f - \nu_i\sigma_{\beta\beta}\,,\\
&m_\beta = \sum_{i=0}^{m}\int_{\Sigma_i}\big[\Lambda^{(i)} - p\big(E_i + \nu_i\tau_{\alpha\alpha}^{(3)}\big)\big]x_\beta\, da\,.
\end{aligned}
\tag{4.6.54}
$$

Muskhelishvili [264], p.548 has proved that $d_3 \neq 0$ and $d \neq 0$.

Let us now study the recurrence process. We seek the solution of the problem $P^{(n+1)}$ in the form

$$
\begin{aligned}
u_1 =&(n+1)\Big\{\int_0^{x_3} u_1^*\,dx_3 - \frac{1}{2}b_1\big[x_3^2 + \nu_i(x_1^2 - x_3^2)\big] -\\
&- b_2\nu_i x_1 x_2 - b_3\nu_i x_1 - \tau x_2 x_3 + \sum_{k=1}^{3} b_k v_1^{(k)} + w_1(x_1, x_2)\Big\}\,,\\
u_2 =&(n+1)\Big\{\int_0^{x_3} u_2^*\,dx_3 - b_1\nu_i x_1 x_2 - \frac{1}{2}b_2\big[x_3^2 - \nu_i(x_1^2 - x_2^2)\big] -\\
&- b_3\nu_i x_2 + \tau x_1 x_2 + \sum_{k=1}^{3} b_k v_2^{(k)} + w_2(x_1, x_2)\Big\}\,,\\
u_3 =&(n+1)\Big[\int_0^{x_3} u_3^*\,dx_3 + (b_1x_1 + b_2x_2 + b_3)x_3 + F(x_1, x_2)\Big]\,,
\end{aligned}
\tag{4.6.55}
$$

where the functions w_α, F and the constants b_i, τ are unknown.

The stress tensor is given by

$$
t_{\alpha\beta} = (n+1)\Big[\int_0^{x_3} t_{\alpha\beta}^*\,dx_3 + \pi_{\alpha\beta} + \sum_{k=1}^{3} b_k\tau_{\alpha\beta}^{(k)} + \lambda_i\delta_{\alpha\beta}u_3^*(x_1, x_2, 0)\Big]\,,
$$

$$t_{\alpha 3} =(n+1)\left\{ \int_0^{x_3} t^*_{\alpha 3}\, dx_3 + \mu_i \left[F_{,\alpha} - \tau \varepsilon_{3\alpha\beta} x_\beta + u^*_\alpha(x_1,\, x_2,\, 0)\right] \right\},$$

$$t_{33} =(n+1)\left\{ \int_0^{x_3} t^*_{33}\, dx_3 + E_i(b_1 x_1 + b_2 x_2 + b_3) + \nu_i \pi_{\alpha\alpha} + \right.$$

$$\left. + \nu_i \sum_{k=1}^{3} b_k \tau^{(k)}_{\alpha\alpha} + (\lambda_i + 2\mu_i) u^*_3(x_1,\, x_2,\, 0) \right\}, \quad \text{on } \Sigma_i \,(i = 0, 1, 2, \ldots, m), \tag{4.6.56}$$

where

$$\pi_{\alpha\beta} = \lambda \gamma_{\rho\rho} \delta_{\alpha\beta} + 2\mu \gamma_{\alpha\beta}, \qquad 2\gamma_{\alpha\beta} = w_{\alpha,\beta} + w_{\beta,\alpha}. \tag{4.6.57}$$

In view of (4.6.45), the conditions (4.6.39)–(4.6.41) reduce to

$$[w_\alpha]_i = [w_\alpha]_0, \qquad [\pi_{\alpha\beta}]_i n_\beta = [\pi_{\alpha\beta}]_0 n_\beta + \kappa^{(i)}_\alpha \quad \text{on } L_i, \tag{4.6.58}$$

$$[F]_i = [F]_0, \qquad \left[\mu \frac{\partial F}{\partial n}\right]_i - \left[\mu \frac{\partial F}{\partial n}\right]_0 = q_i \quad \text{on } L_i \,(i = 1, 2, \ldots, m), \tag{4.6.59}$$

$$[\pi_{\alpha\beta}]_0 n_\beta = p_\alpha \qquad \text{on } L_{m+1}, \tag{4.6.60}$$

$$\left[\mu \frac{\partial F}{\partial n}\right]_0 n_\beta = q \qquad \text{on } L_{m+1}, \tag{4.6.61}$$

where

$$\kappa^{(i)}_\alpha = (\lambda_0 - \lambda_i) u^*_3(x_1,\, x_2,\, 0) n_\alpha, \qquad p_\alpha = -\lambda_0 n_\alpha u^*_3(x_1,\, x_2,\, 0),$$

$$q_i = (\mu_0 - \mu_i)\left[u^*_\alpha(x_1,\, x_2,\, 0) - \tau \varepsilon_{3\alpha\beta} x_\beta\right] n_\alpha, \tag{4.6.62}$$

$$q = \mu_0 \left[\tau \varepsilon_{3\alpha\beta} x_\beta - u^*_\alpha(x_1,\, x_2,\, 0)\right] n_\alpha.$$

If we take into account (4.6.44) and (4.6.56), then the equations (4.6.2) reduce to

$$\pi_{\beta\alpha,\beta} + g^{(i)}_\alpha = 0 \qquad \text{on } \Sigma_i \,(i = 0, 1, \ldots, m), \tag{4.6.63}$$

$$F_{,\alpha\alpha} = h^{(i)} \qquad \text{on } \Sigma_i \,(i = 0, 1, \ldots, m), \tag{4.6.64}$$

where

$$g^{(i)}_\alpha = \lambda_i u^*_{3,\alpha}(x_1,\, x_2,\, 0) + t^*_{\alpha 3}(x_1,\, x_2,\, 0),$$

$$h^{(i)} = -u^*_{\alpha,\alpha}(x_1,\, x_2,\, 0) - \frac{1}{\mu_i} t^*_{33}(x_1,\, x_2,\, 0). \tag{4.6.65}$$

Thus, the functions w_α are the components of the displacement vector field in the isothermal plane strain problem (4.6.57), (4.6.58), (4.6.60), (4.6.63). The necessary and sufficient conditions for the existence of a solution to this problem are (cf. Sherman [323], Muskhelishvili [264], p.547)

$$R_\alpha \equiv \sum_{i=0}^{m} \int_{\Sigma_i} g_\alpha^{(i)}\, da + \int_{L_{m+1}} p_\alpha\, ds - \sum_{i=1}^{m} \int_{L_i} \kappa_\alpha^{(i)}\, ds = 0\,,$$

$$M \equiv \sum_{i=0}^{m} \int_{\Sigma_i} \varepsilon_{3\alpha\beta} x_\alpha g_\beta^{(i)}\, da + \int_{L_{m+1}} \varepsilon_{3\alpha\beta} x_\alpha p_\beta\, ds - \qquad (4.6.66)$$

$$- \sum_{i=1}^{m} \int_{L_i} \varepsilon_{3\alpha\beta} x_\alpha \kappa_\beta^{(i)}\, ds = 0\,.$$

By (4.6.62) and (4.6.65),

$$R_\alpha = \sum_{i=0}^{m} \int_{\Sigma_i} t_{3\alpha}^*(x_1,\, x_2,\, 0)\, da\,, \qquad M = \sum_{i=0}^{m} \int_{\Sigma_i} \varepsilon_{3\alpha\beta} x_\alpha t_{3\beta}^*(x_1,\, x_2,\, 0)\, da\,.$$

Since t_{ij}^* satisfy the conditions (4.6.4)–(4.6.7), we conclude that the conditions (4.6.66) are satisfied. In what follows we will assume that w_α are known functions.

The function F is determined by the boundary value problem (4.6.64), (4.6.59), (4.6.61). The necessary and sufficient condition for the existence of the solution to this problem is (cf. Muskhelishvili [264], p.527)

$$N \equiv \sum_{i=0}^{m} \int_{\Sigma_i} \mu_i h^{(i)}\, da - \int_{L_{m+1}} q\, ds + \sum_{i=0}^{m} \int_{L_i} q_i\, ds = 0\,. \qquad (4.6.67)$$

From (4.6.62), (4.6.65) and (4.6.5) we obtain

$$N = -\sum_{i=0}^{m} \int_{\Sigma_i} t_{33}^*(x_1,\, x_2,\, 0)\, da = 0\,,$$

so that we can consider that the function F is known. As in the case of homogeneous cylinders, the conditions (4.6.4) are satisfied on the basis of equations of equilibrium and the boundary conditions.

To determine the constant τ we introduce the torsion function φ which satisfies the equation

$$\Delta\varphi = 0 \qquad \text{on } \Sigma_i\, (i = 0, 1, \ldots, m)\,, \qquad (4.6.68)$$

and the boundary conditions

$$
\begin{aligned}
&[\varphi]_i = [\varphi]_0 \qquad \text{on } L_i \, (i = 1, 2, \ldots, m) \, , \\
&\left[\mu \frac{\partial \varphi}{\partial n}\right]_i - \left[\mu \frac{\partial \varphi}{\partial n}\right]_0 = -(\mu_0 - \mu_i)\varepsilon_{3\alpha\beta} x_\beta n_\alpha \quad \text{on } L_i \, (i = 1, 2, \ldots, m) \, , \\
&\frac{\partial \varphi}{\partial n} = \varepsilon_{3\alpha\beta} x_\beta n_\alpha \qquad \text{on } L_{m+1} \, .
\end{aligned} \tag{4.6.69}
$$

If we introduce the function ψ by (4.6.33), then we conclude that ψ is independent of τ. From (4.6.7), (4.6.33) and (4.6.56) we obtain

$$
\tau D = -\sum_{i=0}^{m} \int_{\Sigma_i} \varepsilon_{3\alpha\beta} x_\alpha \mu_i \left[\psi_{,\beta} + u_\beta^*(x_1, x_2, 0)\right] da \, , \tag{4.6.70}
$$

where D is the torsional rigidity

$$
D = \sum_{i=0}^{m} \int_{\Sigma_i} \mu_i (\varepsilon_{3\alpha\beta} x_\alpha \varphi_{,\beta} + x_1^2 + x_2^2) \, da \, . \tag{4.6.71}
$$

The relation (4.6.70) determines the constant τ. From (4.6.5), (4.6.6) and (4.6.56) we obtain for the constants b_i expressions similar to (4.6.53).

c) **Nonhomogeneous and anisotropic cylinders**. We now assume that the cylinder B is occupied by a nonhomogeneous and anisotropic material. We suppose that the constitutive coefficients are independent of x_3,

$$
C_{ijrs} = C_{ijrs}(x_1, x_2,) \, , \qquad \beta_{ij} = \beta_{ij}(x_1, x_2,) \qquad (x_1, x_2,) \in \Sigma \, . \tag{4.6.72}
$$

We assume that C_{ijrs}, $\beta_{ij} \in C^\infty(\overline{\Sigma})$ and that the domain Σ is C^∞-smooth.

The problem consists in the finding of an admissible state which satisfies the equations (4.1.3), (4.1.5) and (4.6.2) on B and the boundary conditions (4.6.3)–(4.6.7), when the temperature field has the form (4.6.1).

The state of generalized isothermal plane strain of the cylinder B is characterized by

$$
u_i = u_i(x_1, x_2,) \, , \qquad \theta = 0 \, .
$$

This restriction implies that $t_{ij} = t_{ij}(x_1, x_2,)$, and that

$$
t_{ij} = C_{ijk\alpha} u_{k,\alpha} \, . \tag{4.6.73}
$$

Given the body forces $R_i = R_i(x_1, x_2,)$, $(x_1, x_2,) \in \Sigma$, the equations of equilibrium reduce to

$$
t_{\alpha i,\alpha} + R_i = 0 \qquad \text{on } \Sigma \, . \tag{4.6.74}
$$

We consider the traction problem

$$t_{\alpha i}n_\alpha = z_i \qquad \text{on } L, \tag{4.6.75}$$

where z_i are prescribed functions independent of x_3. We assume that R_i and z_i are functions of class C^∞.

The necessary and sufficient conditions for the existence of a solution to the boundary value problem (4.6.73)–(4.6.75) are (cf. Lekhnitskii [235], Iesan [180], Fichera [128], p.385)

$$\int_\Sigma R_i\,da + \int_L z_i\,ds = 0, \qquad \int_\Sigma \varepsilon_{3\alpha\beta}x_\alpha R_\beta\,da + \int_L \varepsilon_{3\alpha\beta}x_\alpha z_\beta\,ds = 0. \tag{4.6.76}$$

We note that the displacements u_i which correspond to a solution of the boundary value problem (4.6.73)–(4.6.75) are functions of class C^∞ on $\overline{\Sigma}$ (cf. Fichera [128]).

In what follows we will use four special problems $D^{(s)}(s=1,2,3,4)$ of generalized isothermal plane strain. We denote by $v_i^{(s)}$ and $t_{ij}^{(s)}$ the components of the displacement field and the components of the stress tensor field, respectively, from the problems $D^{(s)}$. The problems $D^{(s)}$ are defined by the equations

$$\sigma_{ij}^{(s)} = C_{ijk\beta}v_{k,\beta}^{(s)} \qquad (s=1,2,3,4), \tag{4.6.77}$$

$$\sigma_{i\alpha,\alpha}^{(\beta)} + (C_{i\alpha 33}x_\beta),\alpha = 0 \qquad (\beta=1,2), \tag{4.6.78}$$

$$\sigma_{i\alpha,\alpha}^{(3)} + C_{i\alpha 33,\alpha} = 0, \tag{4.6.79}$$

$$\sigma_{i\alpha,\alpha}^{(4)} - \varepsilon_{\rho\beta 3}(C_{i\alpha\rho 3}x_\beta)_{,\alpha} = 0 \qquad \text{on } \Sigma, \tag{4.6.80}$$

and the boundary conditions

$$\begin{gathered}\sigma_{i\alpha}^{(\beta)}n_\alpha = -C_{i\alpha 33}x_\beta n_\alpha, \qquad \sigma_{i\alpha}^{(3)}n_\alpha = -C_{i\alpha 33}n_\alpha, \\ \sigma_{i\alpha}^{(4)}n_\alpha = \varepsilon_{\rho\beta 3}C_{i\alpha\rho 3}x_\beta n_\alpha \qquad \text{on } L.\end{gathered} \tag{4.6.81}$$

It is easy to show that the necessary and sufficient conditions (4.6.76) for the existence of the solution are satisfied for each boundary value problem $D^{(s)}$. In what follows we assume that the functions $v_i^{(s)}$ are known. For a homogeneous and isotropic body, the solutions of the problems $D^{(s)}$ are given by

$$v_1^{(1)} = -\frac{\lambda}{4(\lambda+\mu)}(x_1^2 - x_2^2), \quad v_2^{(1)} = -\frac{\lambda}{2(\lambda+\mu)}x_1x_2, \quad v_3^{(1)} = 0,$$

$$v_1^{(2)} = -\frac{\lambda}{2(\lambda+\mu)}x_1x_2, \quad v_2^{(2)} = \frac{\lambda}{4(\lambda+\mu)}(x_1^2 - x_2^2), \quad v_3^{(2)} = 0, \tag{4.6.82}$$

$$v_\alpha^{(3)} = -\frac{\lambda}{2(\lambda+\mu)}x_\alpha, \quad v_3^{(3)} = 0, \quad v_\alpha^{(4)} = 0, \quad v_3^{(4)} = \varphi(x_1, x_2),$$

where φ is the solution of the boundary value problem (4.6.32).

Let us study the problem $P^{(0)}$. We seek the solution of this problem in the form

$$u_\alpha = -\frac{1}{2}a_\alpha x_3^2 - a_4\varepsilon_{3\alpha\beta}x_\beta x_3 + \sum_{s=1}^{4} a_s v_\alpha^{(s)} + v_\alpha(x_1, x_2),$$

$$u_3 = (a_1x_1 + a_2x_2 + a_3)x_3 + \sum_{s=1}^{4} a_s v_3^{(s)} + v_3(x_1, x_2), \tag{4.6.83}$$

where $v_i^{(s)}$ are the components of the displacement field from the problem $D^{(s)}$, v_i are unknown functions, and $a_s\ (s = 1, 2, 3, 4)$ are unknown constants.
By (4.6.83),

$$u_{k,\alpha} = a_\alpha x_3\delta_{k3} - a_4\varepsilon_{\beta\alpha 3}x_3\delta_{k\beta} + \sum_{s=1}^{4} a_s v_{k,\alpha}^{(s)} + v_{k,\alpha},$$

$$u_{k,3} = (a_1x_1 + a_2x_2 + a_3)\delta_{k3} - \delta_{k\alpha}a_\alpha x_3 - \delta_{k\alpha}\varepsilon_{\alpha\beta 3}x_\beta,$$

so that (4.1.3) implies

$$t_{ij} = C_{ij33}(a_1x_1 + a_2x_2 + a_3) - C_{ij\alpha 3}\varepsilon_{\alpha\beta 3}a_4x_\beta + \sum_{s=1}^{4} a_s\sigma_{ij}^{(s)} + \pi_{ij}, \tag{4.6.84}$$

where

$$\pi_{ij} = C_{ijk\beta}v_{k,\beta} - \beta_{ij}f. \tag{4.6.85}$$

In view of (4.6.78)–(4.6.80) and (4.6.84), the equations of equilibrium reduce to

$$\pi_{\alpha i,\alpha} = 0 \qquad \text{on } \Sigma. \tag{4.6.86}$$

With the aid of (4.6.81) and (4.6.84), the boundary conditions (4.6.3) become

$$\pi_{\alpha i} n_\alpha = 0 \qquad \text{on } L. \tag{4.6.87}$$

We conclude that the functions v_i are the components of displacement field in the generalized thermoelastic plane strain problem (4.6.85)–(4.6.87). We will assume that the functions v_i are known.

From (4.6.5)–(4.6.7) and (4.6.84) we obtain the following system for the constants a_s $(s = 1, 2, 3, 4)$,

$$\sum_{s=1}^{4} L_{rs} a_s = H_r, \qquad (r = 1, 2, 3, 4), \tag{4.6.88}$$

where

$$\begin{aligned}
&L_{\alpha\beta} = \int_\Sigma x_\alpha \big[C_{3333} x_\beta + \sigma_{33}^{(\beta)}\big]\, da, \quad L_{\alpha 3} = \int_\Sigma x_\alpha \big[C_{3333} + \sigma_{33}^{(3)}\big]\, da,\\
&L_{\alpha 4} = \int_\Sigma x_\alpha \big[C_{33\rho 3}\varepsilon_{\beta\rho 3} x_\beta + \sigma_{33}^{(4)}\big]\, da, \quad L_{3\alpha} = \int_\Sigma \big[C_{3333} x_\alpha + \sigma_{33}^{(\alpha)}\big]\, da,\\
&L_{33} = \int_\Sigma \big[C_{3333} + \sigma_{33}^{(3)}\big]\, da, \qquad L_{34} = \int_\Sigma \big[C_{33\alpha 3}\varepsilon_{\beta\alpha 3} x_\beta + \sigma_{33}^{(4)}\big]\, da,\\
&L_{4\alpha} = \int_\Sigma \varepsilon_{3\rho\beta} x_\rho \big[C_{\beta 333} x_\alpha + \sigma_{\beta 3}^{(\alpha)}\big]\, da, \quad L_{43} \int_\Sigma \varepsilon_{3\rho\beta} x_\rho \big[C_{\beta 333} + \sigma_{\beta 3}^{(3)}\big]\, da,\\
&L_{44} = \int_\Sigma \varepsilon_{3\rho\beta} x_\rho \big[C_{\beta 3\nu 3}\varepsilon_{\lambda\nu 3} x_\lambda + \sigma_{\beta 3}^{(4)}\big]\, da,\\
&H_\alpha = -\int_\Sigma x_\alpha \pi_{33}\, da, \quad H_3 = -\int_\Sigma \pi_{33}\, da, \quad H_4 = -\int_\Sigma \varepsilon_{3\alpha\beta} x_\alpha \pi_{\beta 3}\, da.
\end{aligned} \tag{4.6.89}$$

It is known that $\det(L_{rs}) \neq 0$ (cf. Iesan [180]), so that the system (4.6.88) determines the constants a_s. The conditions (4.6.4) are satisfied on the basis of equilibrium equations and boundary conditions. Thus, the problem $P^{(0)}$ is solved.

In the case of homogeneous and isotropic bodies $v_i^{(s)}$ are given by (4.6.82) and the constants L_{rs} reduce to

$$L_{\alpha\beta} = E\int_\Sigma x_\alpha x_\beta\, da, \qquad L_{\alpha 3} = EAx_\alpha^{(0)}, \qquad L_{i4} = 0, \qquad L_{44} = D,$$

where E is Young's modulus, A and $x_\alpha^{(0)}$ are given by (4.6.17), and D is the torsional rigidity.

We now study the recurrence process. We seek the solution of the problem $P^{(n+1)}$ corresponding to the temperature field (4.6.18) in the form

$$
\begin{aligned}
u_\alpha =&(n+1)\Bigg[\int_0^{x_3} u_\alpha^*\, dx_3 - \frac{1}{2} b_\alpha x_3^2 - b_4 \varepsilon_{3\alpha\beta} x_\beta x_3 + \\
&+ \sum_{s=1}^{4} b_s v_\alpha^{(s)} + w_\alpha(x_1,\, x_2)\Bigg]\,, \\
u_3 =&(n+1)\Bigg[\int_0^{x_3} u_3^*\, dx + (b_1 x_1 + b_2 x_2 + b_3) x_3 + \\
&+ \sum_{s=1}^{4} b_s v_3^{(s)} + w_3(x_1,\, x_2)\Bigg]\,,
\end{aligned}
\tag{4.6.90}
$$

where u_i^* are the components of displacement field from the problem $P^{(n)}$ corresponding to the temperature field (4.6.19), w_i are unknown functions, and b_r $(r = 1,\, 2,\, 3,\, 4)$ are unknown constants.

It follows from (4.1.3), (4.6.77), (4.6.18), (4.6.19) and (4.6.90) that

$$
\begin{aligned}
t_{ij} =&(n+1)\Bigg[\int_0^{x_3} t_{ij}^*\, dx_3 + C_{ij33}(b_1 x_1 + b_2 x_2 + b_3) - C_{ij\alpha 3} \varepsilon_{\alpha\beta 3} b_4 x_\beta + \\
&+ \sum_{s=1}^{4} b_s \sigma_{ij}^{(s)} + \sigma_{ij} + k_{ij}\Bigg]\,,
\end{aligned}
\tag{4.6.91}
$$

where

$$
\sigma_{ij} = C_{ijk\beta} w_{k,\beta}\,, \tag{4.6.92}
$$

and

$$
k_{ij} = C_{ijk3} u_k^*(x_1,\, x_2,\, 0)\,. \tag{4.6.93}
$$

In view of (4.6.78)–(4.6.80) and (4.6.91), the equations of equilibrium reduce to

$$
\sigma_{\alpha i,\alpha} + G_i = 0 \qquad \text{on } \Sigma\,, \tag{4.6.94}
$$

where

$$
G_i = t_{i3}^*(x_1,\, x_2,\, 0) + k_{\alpha i,\alpha}\,. \tag{4.6.95}
$$

With the aid of (4.6.81) and (4.6.91), the conditions (4.6.3) become

$$
\sigma_{\alpha i} n_\alpha = p_i \qquad \text{on } L\,, \tag{4.6.96}
$$

where

$$p_i = -k_{i\alpha} n_\alpha \, . \tag{4.6.97}$$

Thus, the functions w_i are are the components of displacement field in the generalized isothermal plane strain problem (4.6.92), (4.6.94). By (4.6.95) and (4.6.97),

$$\begin{aligned} &\int_\Sigma G_i \, da + \int_L p_i \, ds = \int_\Sigma t^*_{i3}(x_1, x_2, 0) \, da \, , \\ &\int_\Sigma \varepsilon_{3\alpha\beta} x_\alpha G_\beta \, da + \int_L \varepsilon_{3\alpha\beta} x_\alpha p_\beta \, ds = \int_\Sigma \varepsilon_{3\alpha\beta} x_\alpha t^*_{\beta 3}(x_1, x_2, 0) \, da \, . \end{aligned} \tag{4.6.98}$$

Since t^*_{ij} , satisfy the conditions (4.6.4)–(4.6.7), we conclude from (4.6.98) that the necessary and sufficient conditions (4.6.76) for the existence of the solution to the boundary value problem (4.6.92), (4.6.94) are satisfied.

The conditions (4.6.5)–(4.6.7) reduce to

$$\sum_{s=1}^{4} L_{rs} b_s = K_r \qquad (r = 1, 2, 3, 4) \, , \tag{4.6.99}$$

where L_{rs} are given by (4.6.89), and

$$K_\alpha = -\int_\Sigma x_\alpha(\sigma_{33} + k_{33}) \, da \, , \qquad K_3 = -\int_\Sigma (\sigma_{33} + k_{33}) \, da \, ,$$

$$K_4 = -\int_\Sigma \varepsilon_{3\alpha\beta} x_\alpha (\sigma_{\beta 3} + k_{\beta 3}) \, da \, .$$

The system (4.6.99) determines the constants b_s .

The conditions (4.6.4) are satisfied on the basis of the equations of equilibrium and boundary conditions.

d) **Thermal stresses in cylinders composed of nonhomogeneous and anisotropic materials.** Let Σ be a C^1-smooth domain (Fichera [128], p.369), and let L_1 and L_2 be complementary subsets of L. Let Γ be a curve contained in Σ with the property that $\bar{\Gamma} \cup \bar{L}_\rho$ $(\rho = 1, 2)$ is the boundary of a regular domain Σ_ρ contained in Σ such that $\Sigma_1 \cap \Sigma_2 = \emptyset$. We denote by B_ρ the cylinder defined by $B_\rho = \{\boldsymbol{x} : (x_1, x_2) \in \Sigma_\rho, 0 < x_3 < h\}$, $(\rho = 1, 2)$. We assume that B_ρ is occupied by a thermoelastic material with the constitutive coefficients $C^{(\rho)}_{ijrs}$, $\beta^{(\rho)}_{ij}$ of class C^∞. Let $\Pi_0 = \{\boldsymbol{x} : (x_1, x_2) \in \Gamma, 0 \le x_3 \le h\}$. We suppose that $C^{(\rho)}_{ijrs}$ is positive definite and that

$$C^{(\rho)}_{ijkl} = C^{(\rho)}_{ijkl}(x_1, x_2) \, , \qquad \beta^{(\rho)}_{ij} = \beta^{(\rho)}_{ij}(x_1, x_2) \, , \qquad \boldsymbol{x} \in B_\rho \, . \tag{4.6.100}$$

We can consider that the cylinder B is composed of two materials which are welded together along Π_0.

The displacement vector field and the stress vector field must be continuous in passing from one medium to another, so that we have the conditions

$$[u_i]_1 = [u_i]_2\,, \qquad [t_{\beta i}]_1\nu_\beta = [t_{\beta i}]_2\nu_\beta \qquad \text{on } \Pi_0\,, \tag{4.6.101}$$

where ν_α are the direction cosines of the vector normal to Γ, outward to Σ_1.

Let us consider the problem of the isothermal generalized plane strain problem for the domain Σ, with the body forces $f_i^{(\rho)} \in C^\infty\left(\bar{\Sigma}_\rho\right)$. The equilibrium equations take the form

$$t_{\alpha i,\alpha} + f_i^{(\rho)} = 0 \qquad \text{on } \Sigma_\rho\,, \tag{4.6.102}$$

and the constitutive equations become

$$t_{\alpha i} = C^{(\rho)}_{\alpha ij\beta}u_{j,\beta} \qquad \text{on } \Sigma_\rho\,. \tag{4.6.103}$$

If the displacement vector and the stress vector are continuous in Σ, then we have the conditions

$$[u_i]_1 = [u_i]_2\,, \qquad [t_{\alpha i}]_1\nu_\alpha = [t_{\alpha i}]_2\nu_\alpha \qquad \text{on } \Gamma\,. \tag{4.6.104}$$

Let us consider the following boundary conditions

$$[t_{\alpha i}n_\alpha]_\rho = h_i^{(\rho)} \qquad \text{on } L_\rho\,, \tag{4.6.105}$$

where $h_i^{(\rho)}$ are C^∞-functions.

If the domains Σ_ρ satisfy some conditions of regularity (see Fichera [128], p.369), then the boundary value problem (4.6.102)–(4.6.105) has a solution $u_i \in C^\infty\left(\Sigma_1 \cup L_1\right) \cap C^\infty\left(\Sigma_2 \cup L_2\right) \cap C^\infty\left(\Sigma\right)$ if and only if

$$\begin{gathered}\sum_{\rho=1}^{2}\left[\int_{\Sigma_\rho} f_i^{(\rho)}\,da + \int_{L_\rho} h_i^{(\rho)}\,ds\right] = 0\,,\\ \sum_{\rho=1}^{2}\left[\int_{\Sigma_\rho} \varepsilon_{3\alpha\beta}x_\alpha f_\beta^{(\rho)}\,da + \int_{L_\rho} \varepsilon_{3\alpha\beta}x_\alpha h_\beta^{(\rho)}\,ds\right] = 0\,.\end{gathered} \tag{4.6.106}$$

If the conditions (4.6.104) are replaced by

$$[u_i]_1 = [u_i]_2\,, \qquad [t_{i\alpha}]_1\nu_\alpha = [t_{i\alpha}]_2\nu_\alpha + g_i \qquad \text{on } \Gamma\,. \tag{4.6.107}$$

where g_i are C^∞-functions, then the conditions (4.6.106) are replaced by

$$\sum_{\rho=1}^{2}\left[\int_{\Sigma_\rho} f_i^{(\rho)}\,da + \int_{L_\rho} h_i^{(\rho)}\,ds\right] + \int_\Gamma g_i\,ds = 0\,,$$
$$\sum_{\rho=1}^{2}\left[\int_{\Sigma_\rho} \varepsilon_{3\alpha\beta}x_\alpha f_\beta^{(\rho)}\,da + \int_{L_\rho} \varepsilon_{3\alpha\beta}x_\alpha h_\beta^{(\rho)}\,ds\right] + \int_\Gamma \varepsilon_{3\alpha\beta}x_\alpha g_\beta\,ds = 0\,. \tag{4.6.108}$$

We will use four special problems $Z^{(s)}$ $(s = 1, 2, 3, 4)$ of generalized plane strain. We denote by $u_i^{(s)}$ and $t_{ij}^{(s)}$ the components of the displacement vector and the components of the stress tensor, respectively, from the problems $Z^{(s)}$. The problems $Z^{(s)}$ are characterized by the equations

$$\sigma_{i\alpha,\alpha}^{(\beta)} + \left(C_{i\alpha 33}^{(\rho)}x_\beta\right)_{,\alpha} = 0 \quad (\beta = 1, 2)\,, \qquad \sigma_{i\alpha,\alpha}^{(3)} + C_{i\alpha 33,\alpha}^{(\rho)} = 0\,,$$
$$\sigma_{i\alpha,\alpha}^{(4)} - \varepsilon_{\eta\beta 3}\left(C_{i\alpha\eta 3}^{(\rho)}x_\beta\right)_{,\alpha} = 0\,, \qquad \sigma_{i\alpha}^{(s)} = C_{i\alpha k\beta}^{(\rho)}v_{k,\beta}^{(s)} \quad \text{on } \Sigma_\rho\,, \tag{4.6.109}$$

and the conditions

$$[v_i^{(\beta)}]_1 = [v_i^{(s)}]_2\,, \qquad [\sigma_{i\alpha}^{(s)}]_1\nu_\alpha = [\sigma_{i\alpha}^{(s)}]_1\nu_\alpha + g_i^{(s)} \quad \text{on } \Gamma\,, \tag{4.6.110}$$

$$[\sigma_{i\alpha}^{(\beta)}n_\alpha]_\rho = -C_{i\alpha 33}^{(\rho)}x_\beta n_\alpha \qquad (\beta = 1, 2)\,,$$
$$[\sigma_{i\alpha}^{(3)}n_\alpha]_\rho = -C_{i\alpha 33}^{(\rho)}n_\alpha\,, \qquad [\sigma_{i\alpha}^{(4)}n_\alpha]_\rho = \varepsilon_{\eta\beta 3}C_{i\alpha\eta 3}^{(\rho)}x_\beta n_\alpha \quad \text{on } L_\rho\,, \tag{4.6.111}$$

where

$$g_i^{(\beta)} = \left[C_{i\alpha 33}^{(2)} - C_{i\alpha 33}^{(1)}\right]x_\beta\nu_\alpha\,,$$
$$g_i^{(3)} = \left[C_{i\alpha 33}^{(2)} - C_{i\alpha 33}^{(1)}\right]\nu_\alpha\,, \qquad g_i^{(4)} = \varepsilon_{\eta\beta 3}\left[C_{i\alpha\eta 3}^{(1)} - C_{i\alpha\eta 3}^{(2)}\right]x_\beta\nu_\alpha\,. \tag{4.6.112}$$

It is easy to prove that the necessary and sufficient conditions (4.6.108) for the existence of the solution are satisfied for each boundary value problem $Z^{(s)}$. In what follows we assume that the functions $u_i^{(s)}$ are known.

We seek the solution of the problem $P^{(0)}$ in the form (4.6.83), where $v_i^{(s)}$ satisfy the problem $Z^{(s)}$, v_i are unknown functions, and a_s $(s = 1, 2, 3, 4)$ are unknown constants. In this case the stress tensor is given by

$$t_{ij} = C_{ij33}^{(\rho)}(a_1x_1 + a_2x_2 + a_3) - C_{ij\alpha 3}^{(\rho)}\varepsilon_{\alpha\beta 3}a_4x_\beta + \sum_{s=1}^{4} a_s\sigma_{ij}^{(s)} + \pi_{ij}\,, \tag{4.6.113}$$

where

$$\pi_{ij} = C^{(\rho)}_{ijk\beta} v_{k,\beta} - \beta^{(\rho)}_{ij} f \, . \tag{4.6.114}$$

With the help of (4.6.109) and (4.6.113), the equations of equilibrium (4.6.2) reduce to

$$\pi_{\alpha i,\alpha} = 0 \qquad \text{on } \Sigma_\rho \, . \tag{4.6.115}$$

In view of (4.6.111) and (4.6.113), the conditions (4.6.3) become

$$[\pi_{\alpha i} n_\alpha]_\rho = 0 \qquad \text{on } L_\rho \, . \tag{4.6.116}$$

From (4.6.110), (4.6.112), (4.6.113), (4.6.83), (4.6.101) we obtain the conditions

$$[v_i]_1 = [v_i]_2 \, , \qquad [\pi_{\alpha i}]_1 \nu_\alpha = [\pi_{\alpha i}]_2 \nu_\alpha \qquad \text{on } \Gamma \, . \tag{4.6.117}$$

Thus, the functions v_i are the components of the displacement vector in the thermoelastic generalized plane strain problem (4.6.114)–(4.6.117) corresponding to the distribution temperature f. In what follows we assume that this problem is solved. The conditions (4.6.4) are satisfied on the basis of the equilibrium equations and the boundary conditions. From (4.6.5)–(4.6.7) we obtain the following system for the constants a_s

$$\sum_{s=1}^{4} M_{rs} a_s = A_r \, , \qquad (r = 1, 2, 3, 4) \, , \tag{4.6.118}$$

where

$$M_{\alpha\beta} = \sum_{\rho=1}^{2} \int_{\Sigma_\rho} x_\alpha \big[C^{(\rho)}_{3333} x_\beta + \sigma^{(\beta)}_{33} \big] \, da \, , \qquad M_{\alpha 3} = \sum_{\rho=1}^{2} \int_{\Sigma_\rho} x_\alpha \big[C^{(\rho)}_{3333} + \sigma^{(3)}_{33} \big] \, da \, ,$$

$$M_{\alpha 4} = \sum_{\rho=1}^{2} \int_{\Sigma_\rho} x_\alpha \big[C^{(\rho)}_{33\eta 3} \varepsilon_{\beta\eta 3} x_\beta + \sigma^{(4)}_{33} \big] \, da \, , \quad M_{3\alpha} = \sum_{\rho=1}^{2} \int_{\Sigma_\rho} \big[C^{(\rho)}_{3333} x_\alpha + \sigma^{(\alpha)}_{33} \big] \, da \, ,$$

$$M_{33} = \sum_{\rho=1}^{2} \int_{\Sigma_\rho} \big[C^{(\rho)}_{3333} + \sigma^{(3)}_{33} \big] \, da \, , \qquad M_{34} = \sum_{\rho=1}^{2} \int_{\Sigma_\rho} \big[C^{(\rho)}_{33\alpha 3} \varepsilon_{\beta\alpha 3} x_\beta + \sigma^{(4)}_{33} \big] \, da \, ,$$

$$M_{4\alpha} = \sum_{\rho=1}^{2} \int_{\Sigma_\rho} \varepsilon_{3\eta\beta} x_\eta \big[C^{(\rho)}_{\beta 333} x_\alpha + \sigma^{(\alpha)}_{\beta 3} \big] \, da \, , \tag{4.6.119}$$

$$M_{43} = \sum_{\rho=1}^{2} \int_{\Sigma_\rho} \varepsilon_{3\eta\beta} x_\eta \big[C^{(\rho)}_{\beta 333} + \sigma^{(\alpha)}_{\beta 3}\big]\, da\,,$$

$$M_{44} = \sum_{\rho=1}^{2} \int_{\Sigma_\rho} \varepsilon_{3\eta\beta} x_\eta \big[C^{(\rho)}_{\beta 3\nu 3} \varepsilon_{\lambda\nu 3} x_\lambda + \sigma^{(4)}_{\beta 3}\big]\, da\,,$$

$$A_\alpha = -\int_\Sigma x_\alpha \pi_{33}\, da\,, \quad A_3 = -\int_\Sigma \pi_{33}\, da\,, \quad A_4 = -\int_\Sigma \varepsilon_{3\alpha\beta} x_\alpha \pi_{\beta 3}\, da\,.$$

It is known (cf. Iesan [186]) that $\det(M_{rs}) \neq 0$, so that the system (4.6.118) determines the constants a_s. Thus, the problem $P^{(0)}$ is solved.

Let us study the recurrence process. We seek the solution of the problem $P^{(n+1)}$ in the form (4.6.90) where $v_i^{(s)}$ are the compnents of displacement field from the problem $Z^{(s)}$, w_i are unknown functions, and b_s $(s = 1, 2, 3, 4)$ are unknown constants. By (4.6.90),

$$\begin{aligned} t_{ij} =&(n+1)\bigg[\int_0^{x_3} t^*_{ij}\, dx_3 + C^{(\rho)}_{ij33}(b_1 x_1 + b_2 x_2 + b_3) - \\ &- C^{(\rho)}_{ij\alpha 3}\varepsilon_{\alpha\beta 3} b_4 x_\beta + \sum_{s=1}^{4} b_s \sigma^{(s)}_{ij} + \sigma_{ij} + k^{(\rho)}_{ij}\bigg]\,, \end{aligned} \tag{4.6.120}$$

where

$$\sigma_{ij} = C^{(\rho)}_{ijk\beta} w_{k,\beta}\,, \tag{4.6.121}$$

$$k^{(\rho)}_{ij} = C^{(\rho)}_{ijk3} u^*_k(x_1, x_2, 0)\,. \tag{4.6.122}$$

The equations of equilibrium reduce to

$$\sigma_{\alpha i,\alpha} + G^{(\rho)}_i = 0 \qquad \text{on } \Sigma_\rho\,, \tag{4.6.123}$$

where

$$G^{(\rho)}_i = t^*_{i3}(x_1, x_2, 0) + k^{(\rho)}_{\alpha i,\alpha}\,. \tag{4.6.124}$$

The conditions on the lateral boundary become

$$[\sigma_{\alpha i} n_\alpha]_\rho = p^{(\rho)}_i \qquad \text{on } L_\rho\,, \tag{4.6.125}$$

where

$$p^{(\rho)}_i = -k^{(\rho)}_{i\alpha} n_\alpha\,. \tag{4.6.126}$$

In view of (4.6.110), the conditions (4.6.101) reduce to

$$[w_i]_1 = [w_i]_2\,, \qquad [\sigma_{i\alpha}]_1 \nu_\alpha = [\sigma_{i\alpha}]_2 \nu_\alpha + q_i \quad \text{on } \Gamma\,, \tag{4.6.127}$$

where

$$q_i = \left[k_{i\alpha}^{(2)} - k_{i\alpha}^{(1)}\right]\nu_\alpha . \tag{4.6.128}$$

Thus, the functions w_i are the components of the displacement vector from the generalized isothermal plane strain problem (4.6.121), (4.6.123), (4.6.125), (4.6. 127). It is easy to see that the necessary and sufficient conditions for the existence of the solution to this problem are satisfied.

The conditions (4.6.4) are satisfied on the basis of the equations of equilibrium and the boundary conditions. The conditions (4.6.5)–(4.6.7) reduce to

$$\sum_{s=1}^{4} M_{rs} b_s = B_r , \qquad (r = 1, 2, 3, 4), \tag{4.6.129}$$

where M_{rs} are given by (4.6.119), and B_s are defined by

$$B_\alpha = -\sum_{\rho=1}^{2} \int_{\Sigma_\rho} x_\alpha (\sigma_{33} + k_{33}^{(\rho)})\, da , \qquad B_3 = -\sum_{\rho=1}^{2} \int_{\Sigma_\rho} (\sigma_{33} + k_{33}^{(\rho)})\, da ,$$

$$B_4 = -\sum_{\rho=1}^{2} \int_{\Sigma_\rho} \varepsilon_{3\alpha\beta} x_\alpha (\sigma_{\beta 3} + k_{\beta 3}^{(\rho)})\, da .$$

The system (4.6.129) determines the constants b_s. Thus, the problem is solved. The above results have been established by Iesan [181].

Remark. If the material which occupies the domain B_ρ is isotropic, then $u_3^{(s)} = 0$ $(s = 1, 2, 3)$ and the problems $Z^{(s)}$, $(s = 1, 2, 3)$, reduce to the following plane strain problems

$$\begin{aligned}
&\sigma_{\nu\alpha,\nu}^{(\beta)} + (\lambda^{(\rho)} x_\beta)_{,\alpha} = 0 , \qquad \sigma_{\nu\alpha,\nu}^{(3)} + \lambda_{,\alpha}^{(\rho)} = 0 ,\\
&\sigma_{\alpha\beta}^{(s)} = \lambda^{(\rho)} u_{\nu,\nu}^{(s)} \delta_{\alpha\beta} + \mu^{(\rho)} (u_{\alpha,\beta}^{(s)} + u_{\beta,\alpha}^{(s)}) \qquad \text{on } \Sigma_\rho ,\\
&[u_\alpha^{(s)}]_1 = [u_\alpha^{(s)}]_2 , \qquad [t_{\beta\alpha}^{(s)}]_1 \nu_\alpha = [t_{\beta\alpha}^{(s)}]_2 \nu_\alpha + p_\alpha^{(s)} \qquad \text{on } \Gamma ,\\
&[\sigma_{\nu\alpha}^{(\beta)} n_\nu]_\rho = -\lambda^{(\rho)} x_\beta n_\alpha , \qquad [\sigma_{\nu\alpha}^{(3)} n_\nu]_\rho = -\lambda^{(\rho)} n_\alpha \qquad \text{on } L_\rho ,
\end{aligned} \tag{4.6.130}$$

where $\lambda^{(\rho)}$ and $\mu^{(\rho)}$ are the Lamé moduli and

$$p_\alpha^{(\beta)} = (\lambda^{(2)} - \lambda^{(1)}) x_\beta \nu_\alpha , \qquad p_\alpha^{(3)} = (\lambda^{(2)} - \lambda^{(1)}) \nu_\alpha .$$

For the problem $Z^{(4)}$ we obtain $u_\alpha^{(4)} = 0$, $u_3^{(4)} = \varphi$, where φ is the solution of the boundary value problem

$$\begin{aligned}
&\left(\mu^{(\rho)}\varphi_{,\beta}\right)_{,\beta} = \varepsilon_{\alpha\beta3}\left(x_\beta\mu^{(\rho)}\right)_{,\alpha} \qquad \text{on } \Sigma_\rho\,,\\
&[\varphi]_1 = [\varphi]_2 \qquad \text{on } \Gamma\,,\\
&\mu^{(1)}\left[\frac{\partial\varphi}{\partial\nu}\right]_1 = \mu^{(2)}\left[\frac{\partial\varphi}{\partial\nu}\right]_2 + \left(\mu^{(1)} - \mu^{(2)}\right)\varepsilon_{\alpha\beta3}x_\beta\nu_\alpha \qquad \text{on } \Gamma\,,\\
&\left[\frac{\partial\varphi}{\partial n}\right]_\rho = \varepsilon_{\alpha\beta3}x_\beta n_\alpha \qquad \text{on } L_\rho\,.
\end{aligned} \tag{4.6.131}$$

In the case of isotropic solids, the solution of the problem $P^{(0)}$ has the form

$$\begin{aligned}
u_\alpha &= -\frac{1}{2}a_\alpha x_3^2 + \sum_{s=1}^{3} a_s u_\alpha^{(s)} + v_\alpha(x_1,\, x_2)\,,\\
u_3 &= (a_1x_1 + a_2x_2 + a_3)x_3\,,
\end{aligned} \tag{4.6.132}$$

where $u_\alpha^{(s)}$ are defined by (4.6.130) and v_α are the components of displacement vector from the thermoelastic plane strain problem

$$\begin{aligned}
&\pi_{\beta\alpha,\beta} = 0 \qquad \text{on } \Sigma_\rho\,,\\
&\pi_{\alpha\eta} = \lambda^{(\rho)}v_{\nu,\nu}\delta_{\alpha\eta} + \mu(v_{\alpha,\eta} + v_{\eta,\alpha}) - \beta f\delta_{\alpha\eta} \qquad \text{on } \Sigma_\rho\,,\\
&[v_\alpha]_1 = [v_\alpha]_2\,, \qquad [\pi_{\beta\alpha}]_1\nu_\beta = [\pi_{\beta\alpha}]_2\nu_\beta \qquad \text{on } \Gamma\,,\\
&[\pi_{\beta\alpha}n_\beta]_\rho = 0 \qquad \text{on } L_\rho\,.
\end{aligned} \tag{4.6.133}$$

The constants a_i $(i = 1,\, 2,\, 3)$ are given by

$$D_{ij}a_j = Q_i\,, \tag{4.6.134}$$

where

$$\begin{aligned}
D_{\alpha\beta} &= \sum_{\rho=1}^{2}\int_{\Sigma_\rho} x_\alpha\left[(\lambda^{(\rho)} + 2\mu^{(\rho)})x_\beta + \lambda^{(\rho)}u_{\nu,\nu}^{(\beta)}\right] da\,,\\
D_{\alpha3} &= \sum_{\rho=1}^{2}\int_{\Sigma_\rho} x_\alpha\left[\lambda^{(\rho)} + 2\mu^{(\rho)} + \lambda^{(\rho)}u_{\nu,\nu}^{(3)}\right] da\,,
\end{aligned}$$

$$D_{3\alpha} = \sum_{\rho=1}^{2} \int_{\Sigma_\rho} x_\alpha \big[\lambda^{(\rho)} + 2\mu^{(\rho)} + \lambda^{(\rho)} u^{(\alpha)}_{\nu,\nu}\big]\, da\,,$$

$$D_{33} = \sum_{\rho=1}^{2} \int_{\Sigma_\rho} \big[\lambda^{(\rho)} + 2\mu^{(\rho)} + \lambda^{(\rho)} u^{(3)}_{\nu,\nu}\big]\, da\,,$$

$$Q_\alpha = -\sum_{\rho=1}^{2} \int_{\Sigma_\rho} x_\alpha (\lambda^{(\rho)} v_{\nu,\nu} - \beta^{(\rho)} f)\, da\,,$$

$$Q_3 = -\sum_{\rho=1}^{2} \int_{\Sigma_\rho} (\lambda^{(\rho)} v_{\alpha,\alpha} - \beta^{(\rho)} f)\, da\,. \qquad (4.6.135)$$

e) **Deformation of a composed circular cylinder**. We now study the deformation of a heterogeneous circular cylinder which is subjected to a constant temperature variation $\theta = T^* =$ constant. We assume that the domain Σ_2 is bounded by two concentric circles L and Γ of radius r_2 and r_1, respectively, where $r_1 < r_2$. The domain Σ_1 is bounded by the circle Γ. We consider the case when the domains Σ_1 and Σ_2 are occupied by two different homogeneous and isotropic elastic materials. In this case the solutions of the problems (4.6.130) are

$$u_1^{(1)} = \begin{cases} \frac{1}{2}(A_1 - \nu^{(1)})(x_1^2 - x_2^2) + B_1 r^2 & \text{on } \Sigma_1\,, \\ \frac{1}{2}[A_2 - \nu^{(2)} + B_2 r^{-4}](x_1^2 - x_2^2) + C r^2 & \text{on } \Sigma_2\,, \end{cases}$$

$$u_2^{(1)} = \begin{cases} (A_1 - \nu^{(1)}) x_1 x_2 & \text{on } \Sigma_1\,, \\ [A_2 - \nu^{(2)} + B_2 r^{-4}] x_1 x_2 & \text{on } \Sigma_2\,, \end{cases}$$

$$u_1^{(2)} = u_2^{(1)}\,, \qquad (4.6.136)$$

$$u_2^{(2)} = \begin{cases} \frac{1}{2}(A_1 - \nu^{(1)})(x_2^2 - x_1^2) + B_1 r^2 & \text{on } \Sigma_1\,, \\ \frac{1}{2}[A_2 - \nu^{(2)} + B_2 r^{-4}](x_2^2 - x_1^2) + C r^2 & \text{on } \Sigma_2\,, \end{cases}$$

$$u_\alpha^{(3)} = \begin{cases} (D_1 - \nu^{(1)}) x_\alpha & \text{on } \Sigma_1\,, \\ [D_2 - \nu^{(2)} + D_3 r^{-2}] x_\alpha & \text{on } \Sigma_2\,, \end{cases}$$

where

$$r^2 = x_1^2 + x_2^2\,, \qquad A_1 = \eta^{(1)}(\nu^{(1)} - \nu^{(2)})(r_2^4 - r_1^4)d^{-1}\,,$$

$$A_2 = -\eta^{(2)}(\nu^{(1)} - \nu^{(2)})r_1^4 d^{-1}\,, \qquad B_1 = -\gamma^{(1)}(\nu^{(1)} - \nu^{(2)})(r_2^4 - r_1^4)d^{-1}\,,$$

$$B_2 = -\gamma^{(2)}(\nu^{(1)} - \nu^{(2)})r_1^4 r_2^4 d^{-1}\,, \qquad C = \gamma^{(2)}(\nu^{(1)} - \nu^{(2)})r_1^4 d^{-1}\,,$$

$$\nu^{(\alpha)} = \frac{\lambda^{(\alpha)}}{2(\lambda^{(\alpha)} + \mu^{(\alpha)})}\,, \qquad \eta^{(\alpha)} = \frac{\lambda^{(\alpha)} + 3\mu^{(\alpha)}}{2\mu^{(\alpha)}(\lambda^{(\alpha)} + \mu^{(\alpha)})}\,, \qquad \gamma^{(\alpha)} = \frac{1}{2\mu^{(\alpha)}}\,,$$

$$D_1 = (\eta^{(1)} - \gamma^{(1)})(\nu^{(1)} - \nu^{(2)})(r_2^2 - r_1^2)m\,,$$

$$D_2 = (\gamma^{(2)} - \eta^{(2)})(\nu^{(1)} - \nu^{(2)})r_1^2 m\,, \qquad D_3 = 2\gamma^{(2)}(\nu^{(2)} - \nu^{(1)})r_1^2 r_2^2 m\,,$$

$$d = \eta^{(2)} r_1^4 + \gamma^{(2)} r_2^4 + \eta^{(1)}(r_2^4 - r_1^4)\,,$$

$$m^{-1} = (\eta^{(1)} - \gamma^{(1)})(r_2^2 - r_1^2) + (\eta^{(2)} - \gamma^{(2)})r_1^2 + 2\gamma^{(2)} r_2^2\,.$$

The solution of the thermoelastic plane strain problem (4.6.133) is

$$v_\alpha = \begin{cases} K_1 x_\alpha & \text{on } \Sigma_1\,, \\ (K_2 + K_3 r^{-2})x_\alpha & \text{on } \Sigma_2\,, \end{cases} \tag{4.6.137}$$

where

$$\begin{aligned} K_1 =& \frac{1}{2}T^* m(\eta^{(1)} - \gamma^{(1)})\big[(\eta^{(2)} - \gamma^{(2)})\beta^{(2)}(r_2^2 - r_1^2) + \\ &+ (\eta^{(2)} - \gamma^{(2)})\beta^{(1)} r_1^2 + 2\beta^{(1)}\gamma^{(2)} r_2^2\big]\,, \\ K_2 =& \frac{1}{2}T^* m(\eta^{(2)} - \gamma^{(2)})\big[(\eta^{(1)} - \gamma^{(1)})\beta^{(2)}(r_2^2 - r_1^2) + \\ &+ (\eta^{(1)} - \gamma^{(1)})\beta^{(1)} r_1^2 + 2\beta^{(2)}\gamma^{(2)} r_2^2\big]\,, \\ K_3 =& T^* m\gamma^{(2)}\big[\beta^{(1)}(\eta^{(1)} - \gamma^{(1)}) - \beta^{(2)}(\eta^{(2)} - \gamma^{(2)})\big] r_1^2 r_2^2\,. \end{aligned}$$

From (4.6.134) we obtain

$$a_\alpha = 0\,, \qquad a_3 = \frac{(\beta^{(1)}S_1 + \beta^{(2)}S_2)T^* - 2\lambda^{(1)}K_1 S_1 - 2\lambda^{(2)}K_2 S_2}{E^{(1)}S_1 + E^{(2)}S_2 + 2\lambda^{(1)}D_1 S_1 + 2\lambda^{(2)}D_2 S_2}\,, \tag{4.6.138}$$

where

$$E^{(\alpha)} = \frac{\mu^{(\alpha)}(3\lambda^{(\alpha)} + 2\mu^{(\alpha)})}{\lambda^{(\alpha)} + \mu^{(\alpha)}}, \qquad S_1 = \pi r_1^2, \qquad S_2 = \pi(r_2^2 - r_1^2).$$

Thus, the solution of the problem is

$$u_\alpha = a_3 u_\alpha^{(3)} + v_\alpha, \qquad u_3 = a_3 x_3,$$

where $u_\alpha^{(3)}$ are given by (4.6.136), v_α are given by (4.6.137), and a_3 is given by (4.6.138).

The problem of the thermal stresses in bimetals has been studied in various papers (see, e.g. Blevins [25], Hsu, Reddy and Bert [172], Orthwein [289] and the literature cited therein).

Remark. The problem of thermal stresses in shells, is not considered here. The reader interested in this subject will find a full account in the books of Nowinski [282], and Podstrigach and Shvets [302]. A comprehensive bibliography on thermal stresses in shells has been presented by Keene and Hetnarski [218].

4.7 Exercises

1. *The surface of a spherical cavity of radius a in an unbounded homogeneous and isotropic body is free of tractions and is subject to a constant temperature T^*. Investigate the equilibrium of the body in the absence of body forces and heat sources.*

Solution. We assume that the body occupies the domain $B = \{\boldsymbol{x} : x_1^2+x_2^2+x_3^2 > a^2\}$. The basic equations are given by (4.3.4) and $\Delta\theta = 0$ on B. The thermal boundary conditions are

$$\theta = T^* \quad \text{for } r = a, \qquad \theta \to 0 \quad \text{as } r \to \infty, \tag{4.7.1}$$

where $r = (x_1^2 + x_2^2 + x_3^2)^{1/2}$. Since the boundary of the cavity is free of stress, we have the condition

$$t_{ji} x_j = 0 \quad \text{on } r = a. \tag{4.7.2}$$

Moreover, we have the condition

$$u_i = O(1) \qquad \text{as } r \to \infty. \tag{4.7.3}$$

The solution of the equation $\Delta\theta = 0$ which satisfies the conditions (4.7.1) is

$$\theta = aT^* r^{-1}. \tag{4.7.4}$$

We seek the components of the displacement vector in the form $u_i = x_i g(r)$, where g is an unknown function. The equations (4.3.4) reduce to

$$g'' + \frac{4}{r}g' = -\frac{\beta a}{(\lambda + 2\mu)r^3}T^* .$$

The solution of this equation which satisfies the condition (4.7.3) is

$$g = \frac{C}{r^3} + \frac{\beta a}{2(\lambda + 2\mu)r}T^* ,$$

where C is an arbitrary constant. From the constitutive equations we obtain

$$t_{ji}x_j = x_i\big[(3\lambda + 2\mu)g + (\lambda + 2\mu)rg' - \beta\theta\big] = -2\mu x_i\Big[2Cr^{-3} + \frac{\beta a}{(\lambda + 2\mu)r}T^*\Big].$$

From (4.6.140) we obtain

$$C = -\frac{\beta}{2(\lambda + 2\mu)a^3}T^* .$$

2. *A continuum body occupies the region* $\{\boldsymbol{x} : a_2^2 < x_1^2 + x_2^2 < a_1^2, \quad 0 < x_3 < h\}$, $a_1 > a_2 > 0$. *The generic cross-section* $\Sigma = \{\boldsymbol{x} : a_2^2 < x_1^2 + x_2^2 < a_1^2, \quad x_3 = s\}$ *of the tube is the assembly of the regions* Σ_1 *and* Σ_2, $\Sigma = \Sigma_1 \cup \Sigma_2$, *where* $\Sigma_1 = \{\boldsymbol{x} : b^2 < x_1^2 + x_2^2 < a_1^2, \quad x_3 = s\}$ *and* $\Sigma_2 = \{\boldsymbol{x} : a_2^2 < x_1^2 + x_2^2 < b^2, \quad x_3 = s\}$, $b > 0$. *The region* $B_1 = \{\boldsymbol{x} : (x_1, x_2) \in \Sigma_1, \quad 0 < x_3 < h\}$ *and* $B_2 = \{\boldsymbol{x} : (x_1, x_2) \in \Sigma_2, \quad 0 < x_3 < h\}$ *are occupied by different homogeneous and isotropic thermoelastic materials. The body is in equilibrium in the absence of body forces and heat sources. The lateral surfaces of the tube are subject to constant temperatures and are free of tractions. Investigate the thermoelastic plane strain parallel to the* x_1Ox_2*-plane.*

Solution. The tube is composed of two materials which are welded together along the surface $\Pi = \{\boldsymbol{x} : x_1^2 + x_2^2 = b^2, \quad 0 < x_3 < h\}$. We assume that the region B_ϱ is occupied by a thermoelastic material with the constitutive coefficients $\lambda^{(\varrho)}$, $\mu^{(\varrho)}$, $\beta^{(\varrho)}$, and $k^{(\varrho)}$, $(\varrho = 1, 2)$. The displacement vector, the temperature field, the stress vector and the heat flux vector must be continuous in passing from one medium to another, so that we have the conditions

$$\begin{aligned} &[u_i]_1 = [u_i]_2 , \qquad [t_{\beta i}]_1\nu_\beta = [t_{\beta i}]_2\nu_\beta , \\ &[\theta]_1 = [\theta]_2 , \qquad [q_\beta]_1\nu_\beta = [q_\beta]_2\nu_\beta , \qquad \text{on } \Pi , \end{aligned} \tag{4.7.5}$$

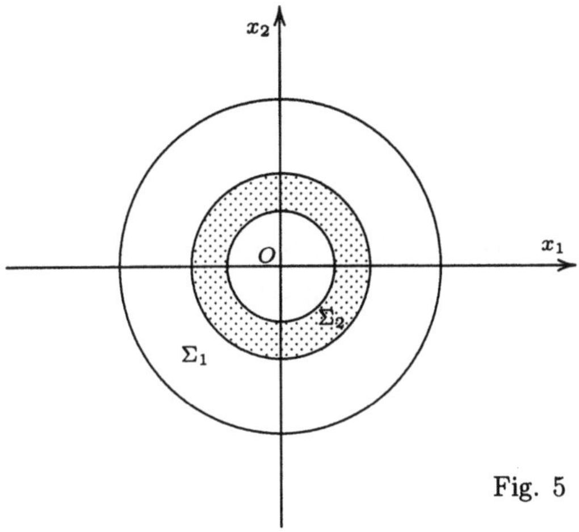

Fig. 5

where $(\nu_1,\, \nu_2,\, 0)$ are the direction cosines of the vector normal to Π, and we have indicated that the expressions in brackets are calculated for the material corresponding to the regions B_1 and B_2. The temperature field satisfies the equation

$$\Delta\theta = 0 \qquad \text{on } B_\varrho \quad (\varrho = 1,\, 2)\,, \tag{4.7.6}$$

and the boundary conditions

$$\begin{gathered} \theta = T_1 \qquad \text{for } r = a_1\,, \qquad \theta = T_2 \qquad \text{for } r = a_2\,, \\ [\theta]_1 = [\theta]_2\,, \qquad k^{(1)}\Big[\frac{\partial\theta}{\partial x_\alpha}\Big]_1 x_\alpha = k^{(2)}\Big[\frac{\partial\theta}{\partial x_\beta}\Big]_2 x_\beta\,, \qquad \text{for } r = b\,, \end{gathered} \tag{4.7.7}$$

$0 < x_3 < h$, where T_1 and T_2 are prescribed constants, and $r = (x_1^2 + x_2^2)^{1/2}$. We seek the function θ in the form $\theta = \theta(r)$. Clearly, from (4.7.6) we obtain

$$\theta = C_1 \ln r + C_2 \qquad \text{on } B_1\,, \qquad \theta = C_3 \ln r + C_4 \qquad \text{on } B_2\,, \tag{4.7.8}$$

where C_s $(s = 1,\, 2,\, 3,\, 4)$ are arbitrary constans. The conditions (4.7.7) reduce to

$$\begin{aligned} C_1 \ln a_1 + C_2 &= T_1\,, & C_3 \ln a_2 + C_4 &= T_2\,, \\ C_1 \ln b + C_2 &= C_3 \ln b + C_4\,, & k^{(1)} C_1 &= k^{(2)} C_3\,. \end{aligned} \tag{4.7.9}$$

It follows from

$$k^{(1)} > 0\,, \qquad k^{(2)} > 0\,, \qquad \ln(a_2/b) > 0\,, \qquad \ln(a_1/b) < 0\,,$$

that the determinant of the system (4.7.9) is different from zero. We obtain

$$\begin{aligned} C_1 &= Ak^{(2)}, \qquad & C_2 &= T_1 - Ak^{(2)} \ln a_1, \\ C_3 &= Ak^{(1)}, \qquad & C_4 &= T_2 - Ak^{(1)} \ln a_2, \end{aligned} \tag{4.7.10}$$

where

$$A = \frac{T_2 - T_1}{k^{(1)} \ln a_2 + (k^{(2)} - k^{(1)}) \ln b - k^{(2)} \ln a_1}.$$

Thus, the function θ is given by (4.7.8) where C_s are defined by (4.7.10). The displacement vector field satisfies the equations

$$\mu^{(\varrho)} \Delta u_i + \left(\lambda^{(\varrho)} + \mu^{(\varrho)}\right) u_{j,ji} - \beta^{(\varrho)} \theta_{,i} = 0 \qquad \text{on } B_\varrho \quad (\varrho = 1, 2). \tag{4.7.11}$$

The mechanical boundary conditions are

$$\begin{aligned} &[u_i]_1 = [u_i]_2, \qquad [t_{\beta i}]_1 x_\beta = [t_{\beta i}]_2 x_\beta \qquad \text{on } \Pi, \\ &[t_{\alpha i}]_\varrho x_\alpha = 0, \qquad r = a_\varrho \qquad (\varrho = 1, 2), \qquad 0 < x_3 < h. \end{aligned} \tag{4.7.12}$$

We seek the solution in the form

$$u_\alpha = x_\alpha \psi^{(\varrho)}(r), \qquad u_3 = 0 \qquad \text{on } B_\varrho, \qquad (\varrho = 1, 2), \tag{4.7.13}$$

where $\psi^{(1)}$ and $\psi^{(2)}$ are unknown functions. As in Section 4.4.a we find that the equations (4.7.11) are satisfied if and only if

$$\begin{aligned} \psi^{(1)} &= A_1 r^{-2} + A_2 - \frac{\beta C_1}{4(\lambda^{(1)} + 2\mu^{(1)}) r^2} (1 + 2 \ln r), \\ \psi^{(2)} &= A_3 r^{-2} + A_4 - \frac{\beta C_3}{4(\lambda^{(2)} + 2\mu^{(2)}) r^2} (1 + 2 \ln r), \end{aligned} \tag{4.7.14}$$

where A_s $(s = 1, 2, 3, 4)$ are arbitrary constants. Using the constitutive equations and (4.7.14) we obtain

$$\begin{aligned} [t_{\beta\alpha}]_1 x_\beta &= x_\alpha \left[2(\lambda^{(1)} + \mu^{(1)}) A_2 - 2\mu^{(1)} r^{-2} A_1 - G_1(r) \right], \\ [t_{\beta\alpha}]_2 x_\beta &= x_\alpha \left[2(\lambda^{(2)} + \mu^{(2)}) A_4 - 2\mu^{(2)} r^{-2} A_3 - G_2(r) \right], \qquad t_{\beta 3} = 0, \end{aligned}$$

where

$$\begin{aligned} G_1 &= \beta^{(1)} \theta(r) - 2\left(\lambda^{(1)} + \mu^{(1)}\right) F(r) - \left(\lambda^{(1)} + 2\mu^{(1)}\right) r F'(r), \qquad b < r < a_1, \\ G_2 &= \beta^{(2)} \theta(r) - 2\left(\lambda^{(2)} + \mu^{(2)}\right) H(r) - \left(\lambda^{(2)} + 2\mu^{(2)}\right) r H'(r), \qquad a_2 < r < b, \\ F &= -\frac{\beta^{(1)} C_1}{4\left(\lambda^{(1)} + 2\mu^{(1)}\right) r^2} (1 + 2 \ln r), \qquad b < r < a_1, \\ H &= -\frac{\beta^{(2)} C_3}{4\left(\lambda^{(2)} + 2\mu^{(2)}\right) r^2} (1 + 2 \ln r), \qquad a_2 < r < b. \end{aligned}$$

The conditions (4.7.12) reduce to the following system for the constants A_s ($s = 1, 2, 3, 4$)

$$\begin{aligned} &A_1 b^{-2} + A_2 - A_3 b^{-2} - A_4 = M\,,\\ &\mu^{(1)} b^{-2} A_1 - L^{(1)} A_2 - \mu^{(2)} b^{-2} A_3 - L^{(2)} A_4 = G_2(b) - G_1(b)\,,\\ &\mu^{(1)} a_1^{-2} A_1 - L^{(1)} A_2 = -\frac{1}{2} G_1(a_1)\,,\\ &\mu^{(2)} a_2^{-2} A_3 - L^{(2)} A_4 = -\frac{1}{2} G_2(a_2)\,, \end{aligned} \tag{4.7.15}$$

where

$$M = \frac{\beta(1+2\ln b)}{4b^2}\left(\frac{C_1}{\lambda^{(1)}+2\mu^{(1)}} - \frac{C_3}{\lambda^{(2)}+2\mu^{(2)}}\right), \quad L^{(\varrho)} = \lambda^{(\varrho)} + \mu^{(\varrho)}, \quad (\varrho = 1,2).$$

The determinant of the system (4.7.15) is

$$d = \mu^{(1)} L^{(1)}\left(\frac{1}{a_1^2} - \frac{1}{b^2}\right)\left(\frac{L^{(2)}}{b^2} + \frac{\mu^{(2)}}{a_2^{(2)}}\right) + \mu^{(2)} L^{(2)}\left(\frac{1}{b^2} - \frac{1}{a_2^2}\right)\left(\frac{\mu^{(1)}}{a_1^2} + \frac{L^{(1)}}{b^2}\right).$$

In view of the relations

$$L^{(\varrho)} > 0\,, \qquad \mu^{(\varrho)} > 0\,, \qquad (\varrho = 1,\, 2)\,, \qquad \frac{1}{a_2^2} > \frac{1}{b^2} > \frac{1}{a_1^2}\,,$$

we conclude that $d > 0$ so that the system (4.7.15) determines the constants A_s ($s = 1, 2, 3, 4$).

3. *A continuum body occupies the region* $\{\boldsymbol{x} \,:\, a_2^2 < x_1^2 + x_2^2 + x_3^2 < a_1^2\}$, $a_1 > a_2 > 0$. *The domain* $B_1 = \{\boldsymbol{x} \,:\, b^2 < x_1^2 + x_2^2 + x_3^2 < a_1^2\}$ *and* $B_2 = \{\boldsymbol{x} \,:\, a_2^2 < x_1^2 + x_2^2 + x_3^2 < b^2\}$, $b > 0$, *are occupied by different homogeneous and isotropic thermoelastic materials. The body is in equilibrium in the absence of body forces and heat sources. The outer surface is subject to the constant temperature* T_1, *and the inner surface is maintained at the constant temperature* T_2. *Investigate the deformation of the body when the surface of the shell is free of tractions.*

Solution. The shell is composed of two different materials which are welded together along the surface $S = \{\boldsymbol{x} \,:\, x_1^2 + x_2^2 + x_3^2 = b^2\}$, $b > 0$. We assume that the region B_ϱ is occupied by a thermoelastic material with the constitutive coefficients $\lambda^{(\varrho)}$, $\mu^{(\varrho)}$, $\beta^{(\varrho)}$ and $k^{(\varrho)}$ ($\varrho = 1, 2$). The displacement vector, the temperature field, the stress vector and the heat flux vector must be continuous in passing from one medium to another, so that we have the conditions

$$\begin{aligned} &[u_i]_1 = [u_i]_2\,, \qquad [t_{ji}]_1 \nu_j = [t_{ji}]_2 \nu_j\,,\\ &[\theta]_1 = [\theta]_2\,, \qquad [q_i]_1 x_i = [q_j]_2 x_j\,, \qquad \text{on } S\,, \end{aligned} \tag{4.7.16}$$

where ν_j are the direction cosines of the vector normal to S, and we have indicated that the expressions in brackets are calculated for the material corresponding to the regions B_1 and B_2. The temperature field satisfies the equation

$$\Delta\theta = 0 \qquad \text{on } B_\varrho \quad (\varrho = 1,\, 2)\,, \tag{4.7.17}$$

and the boundary conditions

$$\begin{aligned} &\theta = T_1 \quad \text{for } r = a_1\,, \qquad \theta = T_2 \quad \text{for } r = a_2\,, \\ &[\theta]_1 = [\theta]_2\,, \qquad k^{(1)}\left[\frac{\partial\theta}{\partial x_j}\right]_1 x_j = k^{(2)}\left[\frac{\partial\theta}{\partial x_i}\right]_2 x_i \qquad \text{for } r = b\,, \end{aligned} \tag{4.7.18}$$

where T_1 and T_2 are prescribed constants, and $r = \sqrt{x_1^2 + x_2^2 + x_3^2}$. We seek the function θ in the form $\theta = \theta(r)$. From (4.7.17) we obtain

$$\theta = A_1 r^{-1} + A_2 \qquad \text{on } B_1\,, \qquad \theta = A_3 r^{-1} + A_4 \qquad \text{on } B_2\,, \tag{4.7.19}$$

where A_s $(s = 1,\, 2,\, 3,\, 4)$ are arbitrary constants. The conditions (4.7.18) reduce to

$$\begin{aligned} &A_1 a_1^{-1} + A_2 = T_1\,, \qquad && A_3 a_2^{-1} + A_4 = T_2\,, \\ &A_1 b^{-1} + A_2 = A_3 b^{-1} + A_4\,, \qquad && k^{(1)} A_1 = k^{(2)} A_3\,. \end{aligned} \tag{4.7.20}$$

From (4.7.20) we find that

$$\begin{aligned} &A_1 = d^{-1} a_1 a_2 k^{(2)} b (T_2 - T_1)\,, \qquad && A_2 = T_1 - d^{-1} a_2 b k^{(2)} (T_2 - T_1)\,, \\ &A_3 = d^{-1} a_1 a_2 k^{(1)} b (T_2 - T_1)\,, \qquad && A_4 = T_2 - d^{-1} a_1 b k^{(1)} (T_2 - T_1)\,, \end{aligned} \tag{4.7.21}$$

where

$$d = a_2 k^{(2)} (a_1 - b) + a_1 k^{(1)} (b - a_2)\,.$$

Since $k^{(1)} > 0,\ k^{(2)} > 0,\ a_1 < b < a_2$, it follows that d is different from zero. Thus, the function θ is given by (4.7.19) where the constants A_s $(s = 1,\, 2,\, 3,\, 4)$ are defined by (4.7.21).

The displacement vector field satisfies the equations

$$\mu^{(\varrho)} \Delta u_i + \left(\lambda^{(\varrho)} + \mu^{(\varrho)}\right) u_{j,ji} - \beta^{(\varrho)} \theta_{,i} = 0 \qquad \text{on } B_\varrho\,, \quad (\varrho = 1,\, 2)\,. \tag{4.7.22}$$

The mechanical boundary conditions are

$$\begin{aligned} &[t_{ji}]_\varrho x_j = 0 \qquad \text{for } r = a_\varrho \qquad (\varrho = 1,\, 2)\,, \\ &[u_i]_1 = [u_i]_2\,, \qquad [t_{ji}]_1 x_j = [t_{ji}]_2 x_j \qquad \text{for } r = b\,. \end{aligned} \tag{4.7.23}$$

We seek the solution in the form

$$u_i = x_i \psi^{(\varrho)}(r) \qquad \text{on } B_\varrho, \quad (\varrho = 1, 2),$$

where $\psi^{(1)}$ and $\psi^{(2)}$ are unknown functions. The equilibrium equations (4.7.22) are satisfied if and only if

$$\begin{aligned} \psi^{(1)} &= C_1 + C_2 r^{-3} + \frac{\beta^{(1)} A_1}{2(\lambda^{(1)} + 2\mu^{(1)})} r^{-1}, \\ \psi^{(2)} &= C_3 + C_4 r^{-3} + \frac{\beta^{(2)} A_3}{2(\lambda^{(2)} + 2\mu^{(2)})} r^{-1}, \end{aligned} \tag{4.7.24}$$

where C_s $(s = 1, 2, 3, 4)$ are arbitrary constants. It follows from the constitutive equations and (4.7.24) that

$$\begin{aligned} [t_{ji}]_1 x_j &= x_i \Big[(3\lambda^{(1)} + 2\mu^{(1)}) C_1 - 4\mu^{(1)} r^{-3} C_2 - \frac{2\mu^{(1)} \beta^{(1)} A_1}{\lambda^{(1)} + 2\mu^{(1)}} r^{-1} - \beta^{(1)} A_2 \Big], \\ [t_{ji}]_2 x_j &= x_i \Big[(3\lambda^{(2)} + 2\mu^{(2)}) C_3 - 4\mu^{(2)} r^{-3} C_4 - \frac{2\mu^{(2)} \beta^{(2)} A_3}{\lambda^{(2)} + 2\mu^{(2)}} r^{-1} - \beta^{(2)} A_4 \Big]. \end{aligned}$$

The conditions (4.7.23) reduce to the following system for the constants C_1, C_2, C_3 and C_4,

$$\begin{aligned} &(3\lambda^{(1)} + 2\mu^{(1)}) C_1 - 4\mu^{(1)} a_1^{-3} C_2 = M_1, \\ &(3\lambda^{(2)} + 2\mu^{(2)}) C_3 - 4\mu^{(2)} a_2^{-3} C_4 = M_2, \\ &C_1 + C_2 b^{-3} - C_3 - C_4 b^{-3} = M_3, \\ &(3\lambda^{(1)} + 2\mu^{(1)}) C_1 - 4\mu^{(1)} b^{-3} C_2 - (3\lambda^{(2)} + 2\mu^{(2)}) C_3 + 4\mu^{(2)} b^{-3} C_4 = M_4, \end{aligned} \tag{4.7.25}$$

where

$$\begin{aligned} M_1 &= \frac{2\mu^{(1)} \beta^{(1)} A_1}{\lambda^{(1)} + 2\mu^{(1)}} a_1^{-1} + \beta^{(1)} A_2, \\ M_2 &= \frac{2\mu^{(2)} \beta^{(2)} A_3}{\lambda^{(2)} + 2\mu^{(2)}} a_2^{-1} + \beta^{(2)} A_4, \\ M_3 &= \frac{\beta^{(2)} A_3}{2(\lambda^{(2)} + 2\mu^{(2)})} b^{-1} - \frac{\beta^{(1)} A_1}{2(\lambda^{(1)} + 2\mu^{(1)})} b^{-1}, \\ M_4 &= \frac{2\mu^{(1)} \beta^{(1)} A_1}{\lambda^{(1)} + 2\mu^{(1)}} b^{-1} + \beta^{(1)} A_2 - \frac{2\mu^{(2)} \beta^{(2)} A_3}{\lambda^{(2)} + 2\mu^{(2)}} b^{-1} - \beta^{(2)} A_4. \end{aligned}$$

The determinant of the system (4.7.25) is given by

$$D = 4\mu^{(2)}L^{(2)}\left(L^{(1)}b^{-3} + 4\mu^{(1)}a_1^{-3}\right)\left(a_2^{-3} - b^{-3}\right) + \\ + 4\mu^{(1)}L^{(1)}\left(L^{(2)}b^{-3} + 4\mu^{(2)}a_2^{-3}\right)\left(b^{-3} - a_1^{-3}\right),$$

where $L^{(\varrho)} = 3\lambda^{(\varrho)} + 2\mu^{(\varrho)}$, $(\varrho = 1, 2)$. If we take into account that

$$3\lambda^{(\varrho)} + 2\mu^{(\varrho)} > 0, \qquad \mu^{(\varrho)} > 0, \quad (\varrho = 1, 2), \qquad a_2 < b < a_1,$$

then we conclude that $D \neq 0$. Thus the system (4.7.25) determines the constants A_s $(s = 1, 2, 3, 4)$.

4. *Determine the equilibrium solution corresponding to a concentrated heat source in a homogeneous and isotropic space.*

Solution. We assume that the body forces are absent and that the heat source is concentrated at the point y. Thus, $f_i = 0$ and $S = A\delta(x - y)$, where δ is the Dirac delta and A is a given constant. The temperature field θ satisfies the equation

$$\Delta\theta = -\frac{A}{k}\delta(x - y).$$

The solution of this equation which vanishes at infinity is

$$\theta = \frac{A}{4\pi k r}, \tag{4.7.26}$$

where $r^2 = (x_i - y_i)(x_i - y_i)$. It follows from (4.3.6) that the displacement potential satisfies the equation

$$\Delta\Phi = \frac{\beta A}{4\pi k(\lambda + 2\mu)r}. \tag{4.7.27}$$

Clearly, the function

$$\Phi = \frac{\beta A r}{8\pi k(\lambda + 2\mu)},$$

is a solution of the equation (4.7.27). From (4.3.5) we obtain

$$\begin{aligned} u_i &= \frac{\beta A(x_i - y_i)}{8\pi k(\lambda + 2\mu)r}, \\ t_{ij} &= -\frac{\mu\beta A}{4\pi k(\lambda + 2\mu)r}\left[\delta_{ij} + (x_i - y_i)(x_i - y_i)r^{-2}\right]. \end{aligned} \tag{4.7.28}$$

The functions u_i and t_{ij} satisfy the conditions at infinity required by the uniqueness theorem. Thus (4.7.28) is the solution of the problem.

5. *A homogeneous and isotropic thermoelastic material occupies the region* $\{\boldsymbol{x} : a_2^2 < x_1^2 + x_2^2 < a_1^2, \quad 0 < x_3 < h\}$, *where* $a_1 > a_2 > 0$. *The inner surface of the tube is perfectly insulated. The temperature gradient on the outer surface is* $M\mathbf{e}_2$ *where* M *is a prescribed constant. The lateral surface of the tube is free of tractions. The body is in equilibrium in the absence of body forces and heat sources. Investigate the thermoelastic plane strain parallel to the* x_1Ox_2 *plane.*

Solution. The temperature field θ satisfies the following boundary value problem

$$\begin{aligned} &\Delta\theta = 0 \quad \text{on } \Sigma\,, \qquad \frac{\partial\theta}{\partial n} = 0 \quad \text{for } r = a_2\,, \\ &\frac{\partial\theta}{\partial x_1} = 0\,, \qquad \frac{\partial\theta}{\partial x_2} = M \quad \text{for } r = a_1\,, \end{aligned} \tag{4.7.29}$$

where $\Sigma = \{\boldsymbol{x} : a_2^2 < x_1^2 + x_2^2 < a_1^2, \quad x_3 = 0\}$ and $r = (x_1^2 + x_2^2)^{1/2}$. We shall use the polar coordinates (r, φ) so that $x_1 = r\cos\varphi, \quad x_2 = r\sin\varphi$. Since the lateral surface is free of tractions, we have the conditions

$$\sigma_{rr} = 0\,, \qquad \sigma_{r\varphi} = 0 \quad \text{for } r = a_1 \quad \text{and } r = a_2\,. \tag{4.7.30}$$

From (4.6.167) we obtain

$$\begin{aligned} &\frac{1}{r}\frac{\partial}{\partial r}\left(r\frac{\partial\theta}{\partial r}\right) + \frac{1}{r^2}\frac{\partial^2\theta}{\partial\varphi^2} = 0 \quad \text{on } \Sigma\,, \\ &\frac{\partial\theta}{\partial r} = 0 \quad \text{for } r = a_2\,, \qquad \frac{\partial\theta}{\partial r} = M\sin\varphi \quad \text{for } r = a_1\,. \end{aligned} \tag{4.7.31}$$

We seek the solution in the form $\theta = F(r)\sin\varphi$, where F is an unknown function. Then, from (4.7.31) we obtain the equation

$$\frac{1}{r}\frac{d}{dr}\left(r\frac{dF}{dr}\right) - \frac{1}{r^2}F = 0 \tag{4.7.32}$$

and the conditions

$$\frac{dF}{dr} = 0 \quad \text{for } r = a_2\,, \qquad \frac{dF}{dr} = M \quad \text{for } r = a_1\,. \tag{4.7.33}$$

The solution of the equation (4.7.32) is

$$F = C_1 r^{-1} + C_2 r\,,$$

where C_1 and C_2 are arbitrary constants. From (4.7.33) we get

$$C_1 = \frac{Ma_1^2a_2^2}{a_1^2 - a_2^2}\,, \qquad C_2 = \frac{Ma_1^2}{a_1^2 - a_2^2}\,.$$

Thus, we have

$$\theta = (C_1 r^{-1} + C_2 r)\sin\varphi\,. \tag{4.7.34}$$

We can reduce the thermoelastic problem to one for unheated body with zero body forces by using the thermoelastic potential Φ. The equation (4.5.30) becomes

$$\Delta\Phi = \frac{\beta}{\lambda + 2\mu}(C_1 r^{-1} + C_2 r)\sin\varphi \qquad \text{on } \Sigma\,. \tag{4.7.35}$$

A solution of this equation is

$$\Phi = \frac{\beta}{2(\lambda + 2\mu)}\Big(C_1 r\ln r + \frac{1}{4}C_2 r^3\Big)\sin\varphi\,.$$

Thus, we obtain the following components of the displacement vector field $\boldsymbol{u}^*$ corresponding to the thermoelastic potential Φ,

$$u_r^* = \frac{\partial\Phi}{\partial r}\,, \qquad u_\varphi^* = \frac{1}{r}\frac{\partial\Phi}{\partial\varphi}\,.$$

The components of the stress tensor corresponding to the displacement field $\boldsymbol{u}^*$ and the temperature field θ are given by

$$\begin{aligned}
\sigma_{rr} &= \lambda(\varepsilon_{rr}^* + \varepsilon_{\varphi\varphi}^*) + 2\mu\varepsilon_{rr}^* - \beta\theta = \lambda\Delta\Phi + 2\mu\frac{\partial^2\Phi}{\partial r^2} - (\lambda + 2\mu)\Delta\Phi = \\
&= -2\mu\Big(\frac{1}{r}\frac{\partial\Phi}{\partial r} + \frac{1}{r^2}\frac{\partial^2\Phi}{\partial\varphi^2}\Big) = -\frac{\beta\mu}{2(\lambda + 2\mu)}(2C_1 r^{-1} + C_2 r)\sin\varphi\,, \\
\sigma_{\varphi\varphi}^* &= -2\mu\frac{\partial^2\Phi}{\partial r^2} = -\frac{\beta\mu}{2(\lambda + 2\mu)}(2C_1 r^{-1} + 3C_2 r)\sin\varphi\,, \\
\sigma_{r\varphi}^* &= 2\mu\Big(\frac{1}{r}\frac{\partial^2\Phi}{\partial r\partial\varphi} - \frac{1}{r^2}\frac{\partial\Phi}{\partial\varphi}\Big) = -\frac{\beta\mu}{2(\lambda + 2\mu)}(2C_1 r^{-1} + C_2 r)\cos\varphi\,.
\end{aligned} \tag{4.7.36}$$

The functions σ_{rr}^* and $\sigma_{r\varphi}^*$ do not satisfy boundary conditions (4.7.30). The solution of the problem has the form $u_\alpha = u_\alpha^* + u_\alpha^\circ$, where the function u_α satisfy the equations (4.5.31), (4.5.33) and the boundary conditions

$$\sigma_{rr}^\circ = -\sigma_{rr}^*\,, \qquad \sigma_{r\varphi}^\circ = -\sigma_{r\varphi}^* \quad \text{for } r = a_1 \quad \text{and } r = a_2\,. \tag{4.7.37}$$

Here, σ_{rr}° and $\sigma_{r\varphi}^\circ$ are stresses corresponding to $\boldsymbol{u}^\circ$. The new problem can be solved with the aid of the Airy function. Clearly, the function

$$A = (A_1 r^{-1} + A_2 r^3 + A_3 r\ln r)\sin\varphi\,,$$

where A_1, A_2 and A_3 are arbitrary constants, satisfies the equation (4.5.37) on Σ. We obtain

$$\sigma^{\circ}_{rr} = \frac{1}{r}\frac{\partial A}{\partial r} + \frac{1}{r^2}\frac{\partial^2 A}{\partial \varphi^2} = (-2A_1 r^{-3} + 2A_2 r + A_3 r^{-1})\sin\varphi\,,$$

$$\sigma^{\circ}_{\varphi\varphi} = \frac{\partial^2 A}{\partial r^2} = (2A_1 r^{-3} + 6A_2 r + A_3 r^{-1})\sin\varphi\,,$$

$$\sigma^{\circ}_{r\varphi} = -\frac{\partial}{\partial r}\Big(\frac{1}{r}\frac{\partial A}{\partial \varphi}\Big) = (2A_1 r^{-3} - 2A_2 r - A_3 r^{-1})\cos\varphi\,.$$

The displacement vector field must be single-valued. This restriction on the conditions (4.7.37) implies that

$$A_1 = -\frac{\beta\mu a_1^2 a_2^2 C_1}{2(\lambda+2\mu)(a_1^2+a_2^2)}\,, \qquad A_2 = \frac{\beta\mu}{2(\lambda+2\mu)}\Big(\frac{C_1}{a_1^2+a_2^2} + \frac{1}{2}C_2\Big), \qquad A_3 = 0\,.$$

The solution is given by

$$\sigma_{rr} = \sigma^*_{rr} + \sigma^{\circ}_{rr} = \frac{\mu\beta C_1 r}{(\lambda+2\mu)(a_1^2+a_2^2)}(1 - a_1^2 r^{-2})(1 - a_2^2 r^{-2})\sin\varphi\,,$$

$$\sigma_{\varphi\varphi} = \sigma^*_{\varphi\varphi} + \sigma^{\circ}_{\varphi\varphi} = \frac{\mu\beta C_1 r}{(\lambda+2\mu)(a_1^2+a_2^2)}\left[3 - (a_1^2+a_2^2)r^{-2} - a_1^2 a_2^2 r^{-4}\right]\sin\varphi\,,$$

$$\sigma_{r\varphi} = \sigma^*_{r\varphi} + \sigma^{\circ}_{r\varphi} = -\frac{\mu\beta C_1 r}{(\lambda+2\mu)(a_1^2+a_2^2)}(1 - a_1^2 r^{-2})(1 - a_2^2 r^{-2})\cos\varphi\,.$$

6. *A homogeneous and isotropic thermoelastic material occupies the region* $B = \{\boldsymbol{x} : x_1^2 + x_2^2 + x_3^2 > a^2\}$, *where* $a > 0$. *The surface of the cavity is perfectly insulated and free of tractions. At infinity, the temperature gradient is* $H\boldsymbol{e}_3$ *where* H *is a prescribed constant. Investigate the equilibrium of the body in the absence of body forces and heat sources.*

Solution. The temperature field is the solution of the equation $\Delta\theta = 0$ on B. We shall use the spherical coordinates (R, φ, ψ) such that $x_1 = R\sin\varphi\cos\psi$, $x_2 = R\sin\varphi\sin\psi$, $x_3 = R\cos\varphi$. The thermal boundary conditions are

$$\frac{\partial\theta}{\partial n} = 0 \quad \text{for } R = a_2\,, \qquad \frac{\partial\theta}{\partial n} = H\cos\varphi \quad \text{as } R \to \infty\,. \tag{4.7.38}$$

The mechanical boundary conditions are

$$\sigma_{RR} = 0\,, \qquad \sigma_{R\varphi} = 0\,, \qquad \sigma_{R\psi} = 0 \qquad \text{for } R = a\,, \tag{4.7.39}$$

and the conditions at infinity. We seek the temperature field in the form $\theta = G(R)\cos\varphi$, where G is an unknown function. For the function G we obtain the equation

$$\frac{d}{dR}\left(R^2\frac{dG}{dR}\right) - 2G = 0\,,$$

and the conditions

$$\frac{dG}{dR} = 0 \quad \text{for } R = a\,, \qquad \frac{dG}{dR} = H \quad \text{as } R \to \infty\,.$$

We find that

$$G = A_1 R + A_2 R^{-2}\,,$$

where A_1 and A_2 are given by

$$A_1 = H\,, \qquad A_2 = \frac{1}{2}Ha^3\,. \tag{4.7.40}$$

Thus,

$$\theta = (A_1 R + A_2 R^{-2})\cos\varphi\,, \tag{4.7.41}$$

where A_1 and A_2 are given by (4.7.40). We note that the term $A_1 R\cos\varphi = A_1 x_3$ causes no stresses. Thus, the equation (4.3.6) becomes

$$\Delta\Phi = \frac{\beta A_2}{(\lambda + 2\mu)R^2}\cos\varphi\,.$$

A particular solution of this equation is

$$\Phi = -\frac{\beta A_2}{2(\lambda + 2\mu)}\cos\varphi\,. \tag{4.7.42}$$

The particular solution corresponding to the thermoelastic potential (4.7.42) is characterized by

$$u_R^* = \frac{\partial\Phi}{\partial R} = 0\,, \qquad u_\varphi^* = \frac{1}{R}\frac{\partial\Phi}{\partial\varphi} = \frac{\beta A_2}{2(\lambda + 2\mu)R}\sin\varphi\,, \qquad u_\psi^* = \frac{1}{R\sin\varphi}\frac{\partial\Phi}{\partial\psi} = 0\,,$$

$$\varepsilon_{RR}^* = \varepsilon_{R\psi}^* = \varepsilon_{\varphi\psi}^* = 0, \qquad \varepsilon_{\varphi\varphi}^* = \varepsilon_{\psi\psi}^* = \frac{\beta A_2\cos\varphi}{2(\lambda + 2\mu)R^2}, \qquad \varepsilon_{R\varphi}^* = -\frac{\beta A_2\sin\varphi}{2(\lambda + 2\mu)R^2},$$

$$\sigma_{RR}^* = -\frac{2\mu\beta A_2}{(\lambda + 2\mu)R^2}\cos\varphi\,, \qquad \sigma_{\varphi\varphi}^* = \sigma_{\psi\psi}^* = -\frac{\mu\beta A_2}{(\lambda + 2\mu)R^2}\cos\varphi\,,$$

$$\sigma_{R\varphi}^* = -\frac{\mu\beta A_2}{(\lambda + 2\mu)R^2}\sin\varphi\,, \qquad \sigma_{R\psi}^* = \sigma_{\varphi\psi}^* = 0\,.$$

Clearly,

$$\overset{*}{\sigma}_{RR} = -2P\cos\varphi\,, \qquad \overset{*}{\sigma}_{R\varphi} = -P\sin\varphi\,, \qquad \overset{*}{\sigma}_{R\psi} = 0 \qquad \text{for } R = a\,, \tag{4.7.43}$$

where

$$P = \frac{\mu\beta A_2}{(\lambda + 2\mu)a^2}\,. \tag{4.7.44}$$

We seek the solution of the problem in the form $\boldsymbol{u} = \boldsymbol{u}^* + \boldsymbol{u}^\circ$ where $\boldsymbol{u}^\circ$ is the solution of the equations (4.3.3) with $f_i = 0$ and $\theta = 0$. Clearly, if we take $\boldsymbol{u}^\circ = \operatorname{grad}\boldsymbol{L}$, where the function L satisfies the equation $\Delta L = 0$ on B, then $\boldsymbol{u}^\circ$ satisfies the equations of elastostatics with zero body forces. We take

$$L = CR^{-2}\cos\varphi\,,$$

where C is an arbitrary constant. The components of the stress tensor corresponding to $\boldsymbol{u}^\circ$ are given by

$$\begin{aligned} \overset{\circ}{\sigma}_{RR} &= \frac{12\mu C}{R^4}\cos\varphi\,, \qquad & \overset{\circ}{\sigma}_{\varphi\varphi} = \overset{\circ}{\sigma}_{\psi\psi} &= -\frac{6\mu C}{R^4}\cos\varphi\,, \\ \overset{\circ}{\sigma}_{R\varphi} &= \frac{6\mu C}{R^4}\sin\varphi\,, \qquad & \overset{\circ}{\sigma}_{R\psi} = \overset{\circ}{\sigma}_{\varphi\psi} &= 0\,. \end{aligned} \tag{4.7.45}$$

In view of (4.7.43) and (4.7.45), the conditions (4.7.39) are satisfied if

$$C = \frac{Pa^4}{6\mu}\,.$$

The thermal stresses on B are given by

$$\begin{aligned} \sigma_{RR} &= \overset{*}{\sigma}_{RR} + \overset{\circ}{\sigma}_{RR} = 2P\Big(\frac{a^4}{R^4} - \frac{a^2}{R^2}\Big)\cos\varphi\,, \\ \sigma_{\varphi\varphi} &= \sigma_{\psi\psi} = -P\Big(\frac{a^2}{R^2} + \frac{a^4}{R^4}\Big)\cos\varphi\,, \\ \sigma_{R\varphi} &= P\Big(\frac{a^4}{R^4} - \frac{a^2}{R^2}\Big)\sin\varphi\,, \qquad \sigma_{R\psi} = \sigma_{\varphi\psi} = 0\,. \end{aligned}$$

5 Nonlinear Thermoelasticity

5.1 Continuous Dependence Results

In this section we consider the equations of the nonlinear theory of thermoelastodynamics. In [92], Dafermos has established the continuous dependence of smooth thermodynamic processes upon the initial state and supply terms for nonconductors of heat. The results of Dafermos have been extended by Chirita [69] to heat-conducting elastic bodies. This section is devoted to the continuous dependence results given in [92], [69].

In what follows we assume that the density ϱ_0 is smooth and strictly positive,

$$\varrho_0(\boldsymbol{X}) \geq \varrho_1 > 0, \qquad \boldsymbol{X} \in \bar{B}. \tag{5.1.1}$$

Moreover, we assume that the response functions (1.3.1) are smooth functions on $\mathcal{C}$. In particular we suppose that the partial derivatives of $\widetilde{\psi}$, $\widetilde{T}_{KL}$, $\widetilde{\eta}$ and $\widetilde{Q}_L$, at any fixed state $(x_{i,A},\ T,\ T_{,K},\ X_M) \in \mathcal{C}$, are bounded functions on $\bar{B}_0 \times [0, t_1]$.

We shall say that $U = (\boldsymbol{x},\ T)$ is a *smooth thermodynamic process* if $\dot{x}_i$, $x_{i,K}$, T and $T_{,K}$ are Lipschitz-continuous, uniformly on bounded subsets of their domain, and satisfy the equations of motion (1.2.3) and the balance law of energy

$$\frac{d}{dt}\Big(\frac{1}{2}\varrho_0 \dot{x}_i \dot{x}_i + \psi + \varrho_0 T \eta\Big) = \varrho_0 f_i \dot{x}_i + (T_{Ki}\dot{x}_i + Q_K)_{,K} + \varrho_0 s, \tag{5.1.2}$$

on $B_0 \times [0, t_1]$. A smooth thermodynamic process will be called admissible if it satisfies the Clausius-Duhem inequality (1.2.18). It follows that a smooth admissible thermodynamic process satisfies the relations (1.3.5)-(1.3.8). In what follows it will be convenient to denote the components of the deformation gradient $\boldsymbol{F} = (x_{i,A})$ by F_{iA}. Moreover, the symbol $|\cdot|$ denotes a norm in a tensor space, while $\|\cdot\|$ denotes an L_2-norm.

Let $U = (x_i, T)$ and $\bar{U} = (\bar{x}_i, \bar{T})$ be two smooth admissible thermodynamic processes on $\bar{B} \times [0, t_1]$ corresponding to the supply terms $(\boldsymbol{f},\ s)$ and $(\bar{\boldsymbol{f}},\ \bar{s})$, respectively. We introduce the function W on $[0, t_1]$ by

$$\begin{aligned} W = \int_B \Big\{&\frac{1}{2}\varrho_0 (v_i - \bar{v}_i)(v_i - \bar{v}_i) + \psi - \bar{\psi} - \\ &- \bar{T}_{Ki}(F_{iK} - \bar{F}_{iK}) + \varrho_0 (T - \bar{T})\eta\Big\}\, dV, \end{aligned} \tag{5.1.3}$$

where

$$
\begin{aligned}
&v_i = \dot{x}_i\,, && \bar{v}_i = \dot{\bar{x}}_i\,, && \bar{F}_{iK} = \bar{x}_{i,K}\,,\\
&\bar{\psi} = \widetilde{\psi}(\bar{x}_{i,A}, \bar{T}, \bar{T}_{,B}, X_M), && && \bar{T}_{Ai} = \widetilde{T}_{Ai}(\bar{x}_{i,A}, \bar{T}, \bar{T}_{,B}, X_M)\,.
\end{aligned}
\tag{5.1.4}
$$

It follows from (1.3.6) that W is of quadratic order in

$$\|(\boldsymbol{v} - \bar{\boldsymbol{v}}, \boldsymbol{F} - \bar{\boldsymbol{F}}, T - \bar{T})\|_{L^2(B_0)}.$$

We now establish an evolutionary relation for W. Let E be the function on $\partial B_0 \times [0, t_1]$ defined by

$$E = \big[(T_{Ki} - \bar{T}_{Ki})(v_i - \bar{v}_i) + \frac{1}{T}(T - \bar{T})(Q_K - \bar{Q}_K)\big] N_K\,. \tag{5.1.5}$$

Theorem 5.1.1. *Let U and $\bar{U}$ be two smooth admissible processes corresponding to the supply terms $(\boldsymbol{f}, s)$ and $(\bar{\boldsymbol{f}}, \bar{s})$, respectively. Then*

$$
\begin{aligned}
\dot{W} = \int_{B_0} \{&\varrho_0(f_i - \bar{f}_i)(v_i - \bar{v}_i) + \frac{1}{T}\varrho_0(s - \bar{s})(T - \bar{T}) -\\
&- \frac{1}{T\bar{T}}(\varrho_0\bar{s} + \bar{Q}_{K,K})(T - \bar{T})^2 + \dot{\bar{F}}_{iK}\big[T_{Ki} - \bar{T}_{Ki} -\\
&- \frac{\partial \bar{T}_{Ki}}{\partial \bar{F}_{jA}}(F_{jA} - \bar{F}_{jA}) - \frac{\partial \bar{T}_{Ki}}{\partial \bar{T}}(T - \bar{T})\big] -\\
&- \varrho_0\dot{\bar{T}}\big[\eta - \bar{\eta} - \frac{\partial \bar{\eta}}{\partial \bar{F}_{iK}}(F_{iK} - \bar{F}_{iK}) - \frac{\partial \bar{\eta}}{\partial \bar{T}}(T - \bar{T})\big] -\\
&- (Q_K - \bar{Q}_K)(\frac{T - \bar{T}}{T})_{,K}\}\, dV + \int_{\partial B_0} E\, dA\,.
\end{aligned}
\tag{5.1.6}
$$

Proof. By (5.1.3),

$$
\begin{aligned}
\dot{W} = \int_{B_0} \{&\frac{d}{dt}(\frac{1}{2}\varrho_0 \boldsymbol{v}^2 + \psi + \varrho_0 T\eta) - \frac{d}{dt}(\frac{1}{2}\varrho_0\bar{\boldsymbol{v}}^2 + \bar{\psi} + \varrho_0\bar{T}\bar{\eta}) -\\
&-\varrho_0\dot{v}_i\bar{v}_i - \varrho_0 v_i\dot{\bar{v}}_i + 2\varrho_0\bar{v}_i\dot{\bar{v}}_i - \dot{\bar{T}}_{Ki}(F_{iK} - \bar{F}_{iK}) -\\
&-\bar{T}_{Ki}(\dot{F}_{iK} - \dot{\bar{F}}_{iK}) - \varrho_0\dot{\bar{T}}(\eta - \bar{\eta}) - \varrho_0\bar{T}(\dot{\eta} - \dot{\bar{\eta}})\}\, dV.
\end{aligned}
\tag{5.1.7}
$$

With the helps of (1.2.3), (5.1.2) and the divergence theorem we can write

$$\begin{aligned}\dot{W} = &\int_{\partial B_0} (T_{Ki} - \bar{T}_{Ki})(v_i - \bar{v}_i)N_K\, dA + \int_{B_0} \{\varrho_0(f_i - \bar{f}_i)(v_i - \bar{v}_i) + \\ &+ \dot{\bar{F}}_{iK}(T_{Ki} - \bar{T}_{Ki}) - \dot{\bar{T}}_{Kj}(F_{jK} - \bar{F}_{jK}) + \varrho_0 s + \\ &+ Q_{K,K} - \varrho_0\bar{s} - \bar{Q}_{K,K} - \varrho_0\dot{\bar{T}}(\eta - \bar{\eta}) - \varrho_0\bar{T}(\dot{\eta} - \dot{\bar{\eta}})\}\, dV.\end{aligned} \tag{5.1.8}$$

We note that

$$\dot{\bar{T}}_{Kj} = \frac{\partial \bar{T}_{Mi}}{\partial \bar{F}_{jK}}\dot{\bar{F}}_{iM} - \varrho_0\frac{\partial \bar{\eta}}{\partial \bar{F}_{jK}}\dot{\bar{T}},$$

$$\varrho_0\dot{\bar{\eta}} = -\frac{\partial \bar{T}_{Kj}}{\partial \bar{T}}\dot{\bar{F}}_{jK} + \varrho_0\frac{\partial \bar{\eta}}{\partial \bar{T}}\dot{\bar{T}},$$

so that

$$\begin{aligned}\dot{W} = &\int_{\partial B_0} (T_{ki} - \bar{T}_{Ki})(v_i - \bar{v}_i)N_K\, dA + \\ &+ \int_{B_0} \{\varrho_0(f_i - \bar{f}_i)(v_i - \bar{v}_i) + \dot{\bar{F}}_{iK}\big[T_{Ki} - \bar{T}_{Ki} - \frac{\partial \bar{T}_{Ki}}{\partial \bar{F}_{jB}}(F_{jB} - \\ &- \bar{F}_{jB}) - \frac{\partial \bar{T}_{Ki}}{\partial \bar{T}}(T - \bar{T})\big] - \varrho_0\dot{\bar{T}}\big[\eta - \bar{\eta} - \frac{\partial \bar{\eta}}{\partial \bar{F}_{iK}}(F_{iK} - \bar{F}_{iK}) - \\ &- \frac{\partial \bar{\eta}}{\partial \bar{T}}(T - \bar{T})\big] + \varrho_0 s + Q_{L,L} - \varrho_0\bar{s} - \bar{Q}_{K,K} - \\ &- \varrho_0\dot{\bar{\eta}}(T - \bar{T}) - \varrho_0\bar{T}(\dot{\eta} - \dot{\bar{\eta}})\}\, dV\,.\end{aligned} \tag{5.1.9}$$

By (1.3.9),

$$\begin{aligned}&\varrho_0 s + Q_{L,L} - \varrho_0\bar{s} - \bar{Q}_{L,L} - \varrho_0\dot{\bar{\eta}}(T - \bar{T}) - \varrho_0\bar{T}(\dot{\eta} - \dot{\bar{\eta}}) = \\ &= \varrho_0 s + Q_{K,K} - \varrho_0\bar{s} - \bar{Q}_{L,L} + \frac{1}{T}\bar{T}(\varrho_0\bar{s} + \bar{Q}_{L,L} - \\ &- \varrho_0 s - Q_{K,K}) - \frac{1}{T\bar{T}}(\varrho_0\bar{s} + \bar{Q}_{K,K})(\bar{T} - T)^2 = \\ &= \frac{1}{T}\varrho_0(s - \bar{s})(T - \bar{T}) + \big[\frac{1}{T}(Q_K - \bar{Q}_K)(T - \bar{T})\big]_{,K} - \\ &- (Q_K - \bar{Q}_K)\big(1 - \frac{\bar{T}}{T}\big)_{,K} - \frac{1}{T\bar{T}}(\varrho_0\bar{s} + \bar{Q}_{K,K})(T - \bar{T})^2.\end{aligned} \tag{5.1.10}$$

It follows from (5.1.8)-(5.1.10) that (5.1.6) holds. □
In view of $(1.3.1)_4$ we can write

$$\begin{aligned} Q_M - \bar{Q}_M &= \bar{R}_{MiN}(F_{iN} - \bar{F}_{iN}) + \bar{D}_M(T - \bar{T}) + \\ &+ \bar{K}_{MN}(T_{,N} - \bar{T}_{,N}) + o_M\Big(|T - \bar{T}| + |\boldsymbol{F} - \bar{\boldsymbol{F}}| + |\boldsymbol{G} - \bar{\boldsymbol{G}}|\Big), \end{aligned} \tag{5.1.11}$$

where $\boldsymbol{G} = \text{grad}_X T$ and

$$\bar{R}_{MiN} = \frac{\partial \bar{Q}_M}{\partial \bar{F}_{iN}}, \qquad \bar{D}_M = \frac{\partial \bar{Q}_M}{\partial \bar{T}}, \qquad \bar{K}_{MN} = \frac{\partial \bar{Q}_M}{\partial \bar{T}_{,N}}. \tag{5.1.12}$$

We say that a thermodynamic process $(\boldsymbol{x}, T)$ resides in the region of state space where the material is a definite heat conductor, if K_{MN} is positive definite.

Theorem 5.1.2. *Let $\overline{U} = (\bar{\boldsymbol{x}}, \bar{T})$ be a smooth admissible process on $B_0 \times [0, t_1]$ and residing in the region of state space where the material is a definite heat conductor. Then there exist positive constants δ_0 and m_0 with the property: if $U = (\boldsymbol{x}, T)$ is any smooth admissible process on $B_0 \times [0, t_1]$ such that*

$$|\boldsymbol{F}(\boldsymbol{X},t) - \bar{\boldsymbol{F}}(\boldsymbol{X},t)| + |T(\boldsymbol{X},t) - \bar{T}(\boldsymbol{X},t)| + |\boldsymbol{G}(\boldsymbol{X},t) - \bar{\boldsymbol{G}}(\boldsymbol{X},t)| < \delta_0, \tag{5.1.13}$$

$(\boldsymbol{X}, t) \in B_0 \times [0, t_1]$, then we have

$$-\int_{B_0} (Q_K - \bar{Q}_K)\left(1 - \bar{T}T^{-1}\right)_{,K} dV \le m_0 \|(\boldsymbol{F} - \bar{\boldsymbol{F}}, T - \bar{T})(\cdot, t)\|^2_{L^2(B_0)}, \tag{5.1.14}$$

for any $t \in [0, t_1]$.

Proof. Since K_{MN} is positive definite, there exists a positive constant λ such that

$$\int_{B_0} \frac{1}{T} \bar{K}_{AB}(G_A - \bar{G}_A)(G_B - \bar{G}_B)\, dV \ge \lambda \int_{B_0} (G_A - \bar{G}_A)(G_A - \bar{G}_A)\, dV. \tag{5.1.15}$$

In view of (5.1.11) and (5.1.15), it follows that there exists a positive constant δ_0 such that whenever (5.1.13) holds, we have

$$\begin{aligned} -\int_{B_0} (Q_A - \bar{Q}_A)\left(1 - \bar{T}T^{-1}\right)_{,A} dV \le \int_{B_0} \Big[& e(T - \bar{T})^2 + \\ & + g_K(G_K - \bar{G}_K)(T - \bar{T}) + h_{iK}(F_{iK} - \bar{F}_{iK})(T - \bar{T}) + \end{aligned} \tag{5.1.16}$$

$$+J_{AiK}\big(G_A-\bar{G}_A\big)\big(F_{iK}-\bar{F}_{iK}\big)\Big]\,dV-\lambda\int_{B_0}\big(G_K-\bar{G}_K\big)\big(G_K-\bar{G}_K\big)\,dV,$$

where

$$e=\bar{D}_A T_{,A}T^{-2}, \qquad g_B=\bar{K}_{BA}T_{,A}T^{-2}-\bar{D}_B T^{-1},$$

$$h_{iK}=\bar{R}_{AiK}T_{,A}T^{-2}, \qquad J_{AiK}=-\bar{R}_{AiK}T^{-1}.$$

By means of Schwarz's inequality combined with the arithmetic-geometric mean inequality, we obtain

$$\begin{aligned}&-2\int_{B_0}\big(Q_K-\bar{Q}_K\big)\big(1-\bar{T}T^{-1}\big)_{,K}\,dV\le(\alpha_1^2+\alpha_2^2-\\&-2\lambda)\|\big(\boldsymbol{G}-\bar{\boldsymbol{G}}\big)(\cdot,t)\|^2_{L^2(B_0)}+\big(M_1^2\alpha_1^{-2}+M_4^2+\\&+1)\|\big(T-\bar{T}\big)(\cdot,t)\|^2_{L^2(B_0)}+\big(M_2^2\alpha_2^2+M_3^2\big)\|\big(\boldsymbol{F}-\bar{\boldsymbol{F}}\big)(\cdot,t)\|^2_{L^2(B_0)},\end{aligned}\tag{5.1.17}$$

where α_1 and α_2 are arbitrary positive constants, and

$$M_1=\max|\boldsymbol{g}|,\quad M_2=\max|\boldsymbol{J}|,\quad M_3=\max|\boldsymbol{h}|,\quad M_4^2=2\max|e|,\quad \text{on } \bar{B}_0\times[0,t_1].$$

We now take α_1 and α_2 such that

$$\alpha_1^2+\alpha_2^2-2\lambda<0.$$

Then, from (5.1.17) we obtain (5.1.14) where

$$m_0=\frac{1}{2}\max\big(M_1^2\alpha_1^{-2}+M_4^2+1,\ M_2^2\alpha_2^{-2}+M_3^2\big).\ \square$$

Following Dafermos [92], we say that a smooth admissible process $\bar{U}=(\bar{\boldsymbol{x}},\bar{T})$ resides in the convexity region of internal energy if:
(α) for each $(\boldsymbol{X},t)\in\bar{B}_0\times[0,t_1]$ there exists a positive constant μ_0 such that

$$\frac{\partial^2\bar{\psi}}{\partial\bar{F}_{iK}\partial\bar{F}_{iL}}\Lambda_{iK}\Lambda_{iL}\ge\mu_0\Lambda_{iK}\Lambda_{iK},\tag{5.1.18}$$

for all Λ_{iK};

$$(\beta)\qquad \frac{\partial\bar{\eta}}{\partial\bar{T}}>0\qquad \text{on } B_0\times[0,\,t_1].\tag{5.1.19}$$

Theorem 5.1.3. *Let $\bar{U} = (\bar{\boldsymbol{x}}, \bar{T})$ be a smooth admissible process on $\bar{B}_0 \times [0, t_1]$, residing in the region of state space where the internal energy is a convex function and material is a definite conductor of heat and corresponding to the supply terms $(\bar{\boldsymbol{f}}, \bar{s}) \in L^\infty(B_0 \times [0, t_1])$. Then there exist the positive constants δ, α, M and N with the following property: if $U = (\boldsymbol{x}, T)$ is any smooth admissible process on $B_0 \times [0, t_1]$ corresponding to the supply terms $(\boldsymbol{f}, s) \in L^\infty(B_0 \times [0, t_1])$ such that*

$$|\boldsymbol{F}(\boldsymbol{X}, t) - \bar{\boldsymbol{F}}(\boldsymbol{X}, t)| + |T(\boldsymbol{X}, t) - \bar{T}(\boldsymbol{X}, t)| + \\ + |\boldsymbol{G}(\boldsymbol{X}, t) - \bar{\boldsymbol{G}}(\boldsymbol{X}, t)| < \delta, \qquad (\boldsymbol{X}, t) \in B_0 \times [0, t_1], \tag{5.1.20}$$

$$E = 0 \qquad \text{on } \partial B_0 \times [0, t_1],$$

then we have

$$\|(\boldsymbol{v} - \bar{\boldsymbol{v}}, \boldsymbol{F} - \bar{\boldsymbol{F}}, T - \bar{T})(\cdot, s)\|_{L^2(B_0)} \leq \{M\|(\boldsymbol{v} - \bar{\boldsymbol{v}}, \boldsymbol{F} - \bar{\boldsymbol{F}}, T - \\ - \bar{T})(\cdot, 0)\|_{L^2(B_0)} + N\int_0^s \|(\boldsymbol{f} - \bar{\boldsymbol{f}}, s - \bar{s})(\cdot, t)\|_{L^2(B_0)}\, dt\} \exp(\alpha s), \tag{5.1.21}$$

for any $s \in [0, t_1]$.

Proof. By (1.3.6), (5.1.18) and (5.1.19), we conclude that there exist positive constants δ_1 and μ such that whenever

$$|\boldsymbol{F}(\boldsymbol{X}, t) - \bar{\boldsymbol{F}}(\boldsymbol{X}, t)| + |T(\boldsymbol{X}, t) - \bar{T}(\boldsymbol{X}, t)| < \delta_1, \quad (\boldsymbol{X}, t) \in B_0 \times [0, t_1]$$

holds, then we have

$$[\psi - \bar{\psi} - \bar{T}_{Ki}(F_{iK} - \bar{F}_{iK}) + \varrho_0(T - \bar{T})\eta](\boldsymbol{X}, t) \geq \\ \geq \mu\Big(|\boldsymbol{F}(\boldsymbol{X}, t) - \bar{\boldsymbol{F}}(\boldsymbol{X}, t)|^2 + |T(\boldsymbol{X}, t) - \bar{T}(\boldsymbol{X}, t)|^2\Big), \tag{5.1.22}$$

for all $(\boldsymbol{X}, t) \in B_0 \times [0, t_1]$. Since $\bar{U}$ resides in the region where the material is a definite conductor of heat, from Theorems 5.1.1 and 5.1.2 we conclude that there exist the positive constants δ_0, m_1 and ν_1 such that whenever (5.1.13) holds we have the following estimate

$$\dot{W}(t) \leq m_1\|(\boldsymbol{F} - \bar{\boldsymbol{F}}, T - \bar{T})(\cdot, t)\|^2_{L^2(B_0)} + \\ + \nu_1\|(\boldsymbol{f} - \bar{\boldsymbol{f}}, s - \bar{s})(\cdot, t)\|_{L^2(B_0)}\|(\boldsymbol{v} - \bar{\boldsymbol{v}}, T - \bar{T})(\cdot, t)\|_{L^2(B_0)}, \tag{5.1.23}$$

for all $t \in [0, t_1]$. Let

$$z(t) = \|(\boldsymbol{v} - \bar{\boldsymbol{v}}, \boldsymbol{F} - \bar{\boldsymbol{F}}, T - \bar{T})(\cdot, t)\|^2_{L^2(B_0)},$$

$t \in [0, t_1]$. We choose $\delta = \min(\delta_0, \delta_1)$. Then we fix $s \in [0, t_1]$ and integrate the inequality (5.1.23) over $[0, \sigma]$, $\sigma \in [0, s]$.We obtain

$$\begin{aligned} W(t) \leq W(0) + m_1 \int_0^\sigma z^2(t)\, dt + \\ + \nu_1 \int_0^\sigma \|(\boldsymbol{f} - \bar{\boldsymbol{f}}, s - \bar{s})(\cdot, t)\|_{L^2(B_0)} z(t)\, dt . \end{aligned} \tag{5.1.24}$$

With the aid of (5.1.3) and (5.1.22) we conclude that there exists a positive constant m such that

$$W(t) \geq m z(t), \qquad t \in [0, t_1]. \tag{5.1.25}$$

If we use the estimate

$$W(0) \leq c_1 z(0),$$

where c_1 is a positive constant, then from (5.1.24) and (5.1.25) we obtain an inequality of the form (2.4.6) with

$$f(t) = z(t), \qquad g(t) = \|(\boldsymbol{f} - \bar{\boldsymbol{f}}, s - \bar{s})(\cdot, t)\|_{L^2(B_0)}, \qquad t \in [0, t_1].$$

An application of Lemma 2.4.3 yields (5.1.21). □

The next uniqueness result is a direct consequence of Theorem 5.1.3 .

Theorem 5.1.4. *Let U and $\bar{U}$ be as in Theorem 5.1.3. Assume that the corresponding supply terms $(\boldsymbol{f}, s)$ and $(\bar{\boldsymbol{f}}, \bar{s})$ coincide on $B_0 \times [0, t_1]$ and that both processes originate from the same state. Then $\boldsymbol{x} = \bar{\boldsymbol{x}}$ and $T = \bar{T}$* on $B_0 \times [0, t_1]$.

We now establish a continuous dependence result by using the strong ellipticity condition.

We say that a process $\bar{U} = (\bar{\boldsymbol{x}}, \bar{T})$ resides in the strong ellipticity region if there exists a positive constant γ with the property that, for any ξ_i, ζ_K, $\chi \in \mathbb{R}$ and for all $(\boldsymbol{X}, t) \in B_0 \times [0, t_1]$, we have

$$\frac{\partial^2 \bar{\psi}}{\partial \bar{F}_{iK} \partial \bar{F}_{jL}} \xi_i \xi_j \zeta_K \zeta_L - \frac{\partial^2 \bar{\psi}}{\partial \bar{T}^2} \chi^2 \geq \gamma \left(|\xi|^2 |\zeta|^2 + \chi^2 \right). \tag{5.1.26}$$

Lemma 5.1.1. *Let $\bar{U} = (\bar{\boldsymbol{x}}, \bar{T})$ be a smooth admissible process on $B_0 \times [0, t_1]$ residing in the strong ellipticity region. Let $U = (\boldsymbol{x}, T)$ be any admissible smooth process on $B_0 \times [0, t_1]$ such that*

$$\boldsymbol{x} = \bar{\boldsymbol{x}} \qquad \text{on} \quad \partial B_0 \times [0, t_1]. \tag{5.1.27}$$

Then there exist the constants $\mu_1 > 0$ and κ with the property that

$$\int_{B_0}\Big[\frac{\partial^2\bar{\psi}}{\partial\bar{F}_{iK}\partial\bar{F}_{jL}}(F_{iK}-\bar{F}_{iK})(F_{jL}-\bar{F}_{jL})-\frac{\partial^2\bar{\psi}}{\partial\bar{T}^2}(T-\bar{T})^2\Big]\,dV \geq \\ \geq \mu_1\|(\boldsymbol{F}-\bar{\boldsymbol{F}}, T-\bar{T})(\cdot,\,t)\|^2_{L^2(B_0)}-\kappa\|(\boldsymbol{x}-\bar{\boldsymbol{x}})(\cdot,\,t)\|^2_{L^2(B_0)}\,. \tag{5.1.28}$$

The proof of this result is a straightforward generalization of the proof of the standard Garding inequality [128], p.366.

Theorem 5.1.5. *Let $\bar{U} = (\bar{\boldsymbol{x}}, \bar{T})$ be a smooth admissible process on $B_0 \times [0,\,t_1]$ residing in the region where the strong ellipticity condition (5.1.26) holds and the material is a definite conductor of heat. We assume that U corresponds to the supply terms $(\bar{\boldsymbol{f}},\,\bar{s}) \in L^\infty(B_0 \times [0,\,t_1])$. Then there exist the positive constants δ, α, β, M and N with the following property: if $U = (\boldsymbol{x},\,T)$ is a smooth admissible process on $B_0 \times [0,\,t_1]$ corresponding to the supply terms $(\boldsymbol{f},\,s) \in L^\infty(B_0 \times [0,\,t_1])$ and such that*

$$|\boldsymbol{F}(\boldsymbol{X},\,t)-\bar{\boldsymbol{F}}(\boldsymbol{X},\,t)|+|T(\boldsymbol{X},\,t)-\bar{T}(\boldsymbol{X},\,t)|+ \\ +|\boldsymbol{G}(\boldsymbol{X},t)-\bar{\boldsymbol{G}}(\boldsymbol{X},\,t)|<\delta\,, \qquad (\boldsymbol{X},\,t)\in B_0\times[0,\,t_1]\,, \tag{5.1.29}$$

$$\boldsymbol{x}=\bar{\boldsymbol{x}}\,, \qquad (T-\bar{T})(Q_K-\bar{Q}_K)N_K=0 \qquad \text{on} \quad \partial B_0\times[0,\,t_1]\,,$$

then

$$\|(\boldsymbol{v}-\boldsymbol{v},\,\boldsymbol{F}-\bar{\boldsymbol{F}}, T-\bar{T})(\cdot,\,s)\|_{L^2(B_0)} \leq \\ \leq \big\{M\|(\boldsymbol{v}-\bar{\boldsymbol{v}},\,\boldsymbol{F}-\bar{\boldsymbol{F}}, T-\bar{T})(\cdot,\,0)\|_{L^2(B_0)}+ \\ +N\int_0^s \|\boldsymbol{f}-\bar{\boldsymbol{f}},\,s-\bar{s})(\cdot,\,t)\|_{L^2(B_0}\,dt\big\}\exp{(\alpha s+\beta s^2)}\,. \tag{5.1.30}$$

Proof. By (1.3.6) and Lemma 5.1.1, there exists a positive constant δ_2 such that whenever

$$|\boldsymbol{F}(\boldsymbol{X},\,t)-\bar{\boldsymbol{F}}(\boldsymbol{X},\,t)|+|T(\boldsymbol{X},\,t)-\bar{T}(\boldsymbol{X},\,t)|<\delta_2\,, \quad (\boldsymbol{X},\,t)\in B_0\times[0,\,t_1]\,, \tag{5.1.31}$$

holds, we have

$$\int_{B_0}[\psi-\bar{\psi}-\bar{T}_{Ki}(F_{iK}-\bar{F}_{iK})+\varrho(T-\bar{T})\eta]\,dV \geq \\ \geq \mu_1\|(\boldsymbol{F}-\bar{\boldsymbol{F}}, T-\bar{T})(\cdot,\,t)\|^2_{L^2(B_0)}-\kappa\|(\mathbf{x}-\bar{\mathbf{x}})(\cdot,\,t)\|^2_{L^2(B_0)}\,. \tag{5.1.32}$$

Clearly, there exist the positive constants δ_0, m_1 and ν_1 such that whenever (5.1.13) holds, the inequality (5.1.23) holds. Now we put $\delta = \min(\delta_0, \delta_2)$. By means of Schwarz inequality and Poincaré inequality we find that

$$\begin{aligned}\int_{B_0} |\boldsymbol{x}(\boldsymbol{X}, \tau) - \bar{\boldsymbol{x}}(\boldsymbol{X}, \tau)|^2 \, dV = \int_{B_0} |\boldsymbol{x}(\boldsymbol{X}, 0) - \bar{\boldsymbol{x}}(\boldsymbol{X}, 0) + \\ + \int_0^\tau [\boldsymbol{v}(\boldsymbol{X}, t) - \bar{\boldsymbol{v}}(\boldsymbol{X}, t)] \, dt|^2 \, dV \leq 2c \int_{B_0} |\boldsymbol{F}(\boldsymbol{X}, 0) - \\ - \bar{\boldsymbol{F}}(\boldsymbol{X}, 0)|^2 \, dV + 2\tau \int_0^\tau \int_{B_0} |\boldsymbol{v}(\boldsymbol{X}, t) - \bar{\boldsymbol{v}}(\boldsymbol{X}, t)|^2 \, dV \, dt \,, \qquad c = \text{const.}.\end{aligned} \tag{5.1.33}$$

From (5.1.3), (5.1.32) and (5.1.33) we conclude that there exist the positive constants b_1, b_2 and b_3 such that

$$\begin{aligned} z(s) \leq & b_1 W(s) + b_2 \int_{B_0} |\boldsymbol{F}(\boldsymbol{X}, 0) - \bar{\boldsymbol{F}}(\boldsymbol{X}, 0)|^2 \, dV + \\ & + b_3 s \int_0^s \int_{B_0} |\boldsymbol{v}(\boldsymbol{X}, t) - \bar{\boldsymbol{v}}(\boldsymbol{X}, t)|^2 \, dV \, dt \,. \end{aligned} \tag{5.1.34}$$

It follows from (5.1.24), (5.1.34) and Lemma 2.4.3 that (5.1.30) holds. □

5.2 Acceleration waves

In Section 1.7 we have considered a moving surface which propagates through a continuous body. A propagating singular surface is said to be an acceleration wave if $\boldsymbol{x}$, $\dot{\boldsymbol{x}}$, $\boldsymbol{x}_{,K}$, T and η are continuous functions everywhere, while $\ddot{\boldsymbol{x}}$, $\boldsymbol{x}_{,KL}$, $\dot{\boldsymbol{x}}_{,K}$, $\boldsymbol{x}_{,KLM}$, $\dot{\boldsymbol{x}}_{,LM}$, $\ddot{\boldsymbol{x}}_{,K}$, $\dddot{\boldsymbol{x}}$, $T_{,K}$, $\dot{T}$, $\eta_{,K}$ and $\dot{\eta}$ may exhibit jump discontinuities across the singular surface. There is a vast literature on the theory of acceleration waves in heat-conducting materials (see, for example,Coleman and Gurtin [85], Truesdell [358], Chen [64], Mc Carthy [249]). Here we follow the work of Chadwick and Currie [54]. For a thermoelastic material it follows from the constitutive equations that the stress, entropy and internal energy are continuous and the heat flux vector is discontinuous on an acceleration wave $\Sigma(t)$. It is known (see, for example, Mc Carthy [249]) that the energy equation implies

$$[Q_K]N_K = 0 \,, \tag{5.2.1}$$

on $\Sigma(t)$. Chadwick and Currie [54] have considered the equation

$$G_A \widehat{Q}_A(E_{MN}, T, G_K) = 0 \,, \tag{5.2.2}$$

where $G_A = T_{,A}$, as a condition on the temperature gradient $\boldsymbol{G}$ in a given deformation-temperature state (E_{MN}, T). When the heat-flux functions $\widehat{Q}_A$ vanish identically in the state (E_{MN}, T), the material is said to be a non-conductor. For a non-conductor, the equation (5.2.2) holds for all choices of $\boldsymbol{G}$. A material for which (5.2.2) has no non-zero solution is said to be a normal conductor. An anomalous conductor is a material which is neither a non-conductor nor a normal conductor. This classification of materials into normal conductors, anomalous conductors and non-conductors is due to Chadwick and Currie [54]. In this section we examine some properties of acceleration waves travelling through homogeneous normal conductors. A detailed treatment of acceleration waves in the three types of materials previously defined is given in [54].

A region in which the material is homogeneously deformed, at rest, and at constant temperature is said to be a region of uniform state (cf. Chadwick and Currie [54]). We confine our attention to waves entering a region of uniform state.

Let

$$\boldsymbol{a} = [\ddot{\boldsymbol{x}}], \qquad g = [T_{,K}]N_K, \qquad h = [\eta_{,K}]N_K. \tag{5.2.3}$$

Here g and h are respectively thermal and entropic amplitudes of the wave. It follows from the compatibility conditions (1.7.73) that

$$\begin{aligned}
&[x_{i,KL}] = U_N^{-2} a_i N_K N_L, \qquad &&[\dot{x}_{i,K}] = -U_N^{-1} a_i N_K, \\
&[T_{,K}] = [G_K] = gN_K, &&[\dot{T}] = -U_N g, \\
&[\dot{\eta}] = -U_N h, &&[\eta_{,K}] = hN_K.
\end{aligned} \tag{5.2.4}$$

An acceleration wave of zero thermal amplitude is called an *isothermal wave* and a wave of zero entropic amplitude an *isentropic wave.* We have the following theorem due to Chadwick and Currie [54].

Theorem 5.2.1. *An acceleration wave propagating into a region of uniform state of a homogeneous normal conductor is necessarily isothermal.*

Proof. Since $\boldsymbol{G}$ vanishes ahead of $\Sigma(t)$, it follows from (1.3.7) that the heat flux is also identically zero in this region. The value of $\boldsymbol{G}$ on the rearward of $\Sigma(t)$ is $-[\boldsymbol{G}]$ and the relations (5.2.1) and (5.2.4) imply

$$N_K \widehat{Q}_K(E_{AB}, T, -gN_L) = 0.$$

The above relation leads to

$$-[G_K]\widehat{Q}_K(E_{AB}, T, -[G_L]) = 0. \tag{5.2.5}$$

Since the considered material is a normal conductor, the equation (5.2.5) holds only for $[\boldsymbol{G}] = 0$. From (5.2.4) we conclude that the acceleration wave is an isothermal wave. □

We consider the equations of motion in the form (1.2.3). In view of (1.3.1) and (1.3.6) we obtain

$$T_{Ai} = \frac{\partial \widetilde{\psi}}{\partial x_{i,A}}\,, \qquad \varrho_0 \eta = -\frac{\partial \widetilde{\psi}}{\partial T}\,. \tag{5.2.6}$$

Since the considered material is homogeneous, the equations of motion can be written in the form

$$B_{MiKj} x_{j,MK} - \beta_{Mi} T_{,M} + \varrho_0 f_i = \varrho_0 \ddot{x}_i\,, \tag{5.2.7}$$

where

$$B_{MiKj} = \frac{\partial^2 \widetilde{\psi}}{\partial x_{i,M} \partial x_{j,K}}\,, \qquad \beta_{Ai} = -\frac{\partial^2 \widetilde{\psi}}{\partial x_{i,A} \partial T}\,. \tag{5.2.8}$$

We introduce the notation

$$D_{MiKjRs} = \frac{\partial B_{MiKj}}{\partial x_{s,R}}\,. \tag{5.2.9}$$

We note that B_{AiKj}, β_{Ai} and D_{MiKjRs} are determined by the deformation-temperature state.

We assume that the body force is continuous. On calculating the jump in (5.2.7) across the wave front, making use of the conditions (5.2.4) and the fact that the wave is isothermal, we obtain

$$(Q^0_{ij}(\boldsymbol{N}) - \varrho_0 U_N^2 \delta_{ij}) a_j = 0\,, \tag{5.2.10}$$

where

$$Q^0_{ij}(\boldsymbol{N}) = B_{MiKj} N_M N_K\,. \tag{5.2.11}$$

The tensor $Q^0_{ij}(\boldsymbol{N})$ is called the isothermal acoustic tensor for the direction $\boldsymbol{N}$. In view of the symmetry of $Q^0_{ij}(\boldsymbol{N})$ we deduce from (5.2.10) the following result (Truesdell [353]): the amplitude vector of an isothermal acceleration wave travelling in the direction of the unit vector $\boldsymbol{N}$ is a proper vector of the isothermal acoustic tensor $Q^0_{ij}(\boldsymbol{N})$ and the speed of propagation U_N is such that $\varrho_0 U_N^2$ is the corresponding proper number. Conversely, any proper vector of $Q^0_{ij}(\boldsymbol{N})$ is a possible amplitude of an isothermal acceleration wave provided that the corresponding proper number is non-negative.

For the remainder of this section we confine our attention to plane acceleration waves. In order to derive the growth equation governing the variation of the acceleration amplitude, we first differentiate the equations (5.2.7) with respect to time,

$$\begin{aligned} &D_{MiKjRs} \dot{x}_{s,R} x_{j,MK} + B_{MiKj} \dot{x}_{j,MK} - \\ &- \Big(\frac{\partial \beta_{Mi}}{\partial x_{j,B}} \dot{x}_{j,B} + \frac{\partial \beta_{Mi}}{\partial T} \dot{T}\Big) T_{,M} - \beta_{Mi} \dot{T}_{,M} + \varrho_0 \dot{f}_i = \varrho_0 \dddot{x}_i\,. \end{aligned} \tag{5.2.12}$$

Since the temperature and its first partial derivatives are continuous on an isothermal acceleration wave, we obtain from (1.7.2) the following compatibility conditions

$$[T_{,AB}] = PN_A N_B\,, \qquad [\dot{T}_{,A}] = -U_N P N_A\,, \qquad P = [T_{,KL}]N_K N_L\,. \qquad (5.2.13)$$

The jump in the third-order derivatives may be evaluted from (1.7.73) by recalling that $\dot{x}_i$ and $x_{i,K}$ are continuous across $\Sigma(t)$. Thus we get

$$\begin{aligned}&[\dot{x}_{i,KL}] = c_i N_K N_L\,, \qquad [\dddot{x}_i] = c_i U_N^2 + 2\frac{\delta_D a_i}{\delta t}\,,\\ &c_i = [\dot{x}_{i,KL} N_K N_L]\,.\end{aligned} \qquad (5.2.14)$$

We note that

$$[fg] = f^+[g] + g^+[f] - [f][g]\,.$$

If the material ahead of the wave front is at rest, then

$$[\dot{x}_{s,R}x_{j,MK}] = -[\dot{x}_{s,R}][x_{j,MK}]\,.$$

It follows from (5.2.12)-(5.2.14) that

$$\begin{aligned}&2\varrho_0\frac{\delta_D a_i}{\delta t} - U_N^{-3} d_{isj} a_s a_j - \beta_{Mi} P U_N U_M =\\ &= \left(Q^0_{ij}(\boldsymbol{N}) - \varrho_0 U_N^2 \delta_{ij}\right)c_j\,,\end{aligned} \qquad (5.2.15)$$

where

$$d_{isj} = D_{MiKjRs} N_M N_K N_R\,. \qquad (5.2.16)$$

We now consider the energy equation (1.3.8), rewritten as

$$\varrho_0 T\dot{\eta} = \frac{\partial Q_A}{\partial x_{i,M}} x_{i,MA} + \frac{\partial Q_A}{\partial T} T_{,A} + \frac{\partial Q_A}{\partial T_{,B}} T_{,AB} + \varrho_0 s\,. \qquad (5.2.17)$$

We assume that the heat supply s is continuous. In view of the uniform state assumption each term of equation (5.2.17) is zero ahead of $\Sigma(t)$. Evaluation on the rearward side of $\Sigma(t)$ therefore gives

$$\begin{aligned}&-\varrho_0 T U_N h = [x_{i,MA}]\frac{\partial \widetilde{Q}_A}{\partial x_{i,M}}(x_{r,L},\, T,\, -gN_K) +\\ &+ [T_{,A}]\frac{\partial \widetilde{Q}_A}{\partial T}(x_{r,M},\, T,\, -gN_K) + PN_A N_B \frac{\partial \widetilde{Q}_A}{\partial T_{,B}}(x_{r,M},\, T,\, -gN_K) = 0\,.\end{aligned} \qquad (5.2.18)$$

It follows from (1.3.7) that

$$\frac{\partial \widetilde{Q}_A}{\partial x_{i,M}} = 0\,, \qquad \frac{\partial \widetilde{Q}_A}{\partial T} = 0 \qquad \text{if} \quad T_{,K} = 0\,. \tag{5.2.19}$$

Moreover, we have (Chadwick and Set [52])

$$\frac{\partial \widetilde{Q}_A}{\partial T_{,B}}(x_{r,M},\, T,\, 0)Y_A Y_B \geq 0 \qquad \text{for all} \quad Y_M \in \mathbb{R}\,. \tag{5.2.20}$$

For an isothermal acceleration wave, $g = 0$. The result of setting $g = 0$ in equation (5.2.18) and using (5.2.19) is

$$\varrho_0 T U_N h = -\tilde{k} P\,, \tag{5.2.21}$$

where

$$\tilde{k} = \frac{\partial \widetilde{Q}_A}{\partial T_{,B}}(x_{r,M},\, T,\, 0) N_A N_B\,. \tag{5.2.22}$$

From the constitutive equations (1.3.6) we obtain

$$\varrho_0 \dot{\eta} = -\frac{\partial^2 \psi}{\partial T \partial x_{i,K}} \dot{x}_{i,K} - \frac{\partial^2 \psi}{\partial T^2} \dot{T}\,. \tag{5.2.23}$$

Forming the jump of each term on $\Sigma(t)$ and utilising the compatibility conditions (5.2.4) we obtain, for an isothermal acceleration wave,

$$h = (\varrho_0 U_N^2)^{-1} \beta_{Ki} N_K a_i\,. \tag{5.2.24}$$

It follows from (5.2.20) and (5.2.22) that $\tilde{k} \geq 0$. We assume that $\tilde{k} \neq 0$. Then (5.2.21) and (5.2.24) implies that

$$P = -(T/\tilde{k} U_N) \beta_{Ki} N_K a_i\,. \tag{5.2.25}$$

When the equation (5.2.15) is multiplied by a_i and use is made of the propagation condition (5.2.10), the symmetry of $Q^0_{ij}(\boldsymbol{N})$ and (5.2.25), we arrive at the following equation

$$2\varrho_0 a_i \frac{\delta_D a_i}{\delta t} - U_N^{-3} d_{isj} a_i a_s a_j + (T/\tilde{k})(\beta_{Ki} N_K a_i)^2 = 0\,. \tag{5.2.26}$$

It is convenient to write

$$a_i = a m_i\,, \tag{5.2.27}$$

where $\boldsymbol{m}$ is a unit proper vector of the isothermal acoustical tensor $Q^0_{ij}(\boldsymbol{N})$ associated with the proper number $\varrho_0 U_N^2$ and a is called the strength of the wave. We note that

$$\frac{\delta_D}{\delta t} = U_N \frac{d}{dN}\,, \tag{5.2.28}$$

where N is distance measured along the normal to $\Sigma(t)$. In view of (5.2.27) and (5.2.28) the equation (5.2.26) reduces to

$$\frac{da}{dN} = \alpha a^2 - \beta a\,, \tag{5.2.29}$$

where

$$\alpha = (1/2\varrho_0 U_N^4) d_{isj} m_i m_s m_j\,,$$

$$\beta = (T/2\varrho_0 \tilde{k} U_N)(\beta_{Ki} N_K m_i)^2\,.$$

We note that $\beta > 0$. The solution of the equation (5.2.29) is

$$a = a_0 e^{-\beta N}\{1 - (\alpha a_0/\beta)(1 - e^{-\beta N})\}^{-1}\,,$$

where $a = a_0$ when $N = 0$. If $\alpha = 0$, then $a = a_0 \exp(-\beta N)$. We now assume that $\alpha \neq 0$. Then we have

i) if $a_0 \text{sgn}\alpha > \beta/|\alpha|$, a/a_0 increases monotonically, becoming infinite at time $-(1/\beta U_N)\log\{1 - (\beta/\alpha a_0)\}$;

ii) if $a_0 \text{sgn}\alpha < \beta/|\alpha|$, a/a_0 decreases monotonically, approaching zero exponentially as $t \to \infty$;

iii) if $a_0 \text{sgn}\alpha = \beta/|\alpha|$, $a = a_0$ for $t \geq 0$.

Detailed anlyses of the growth of the acceleration waves in heat-conducting elastic materials have been given by Chadwick and Currie [54],[56].

In the theory of elastic materials there are close parallels between acceleration waves and plane harmonic disturbances of a uniform equilibrium configuration. The relationship between harmonic and acceleration plane waves in the context of thermoelasticity has been studied by Chadwick [58]. It is shown that it is no longer the case that for a result referring to one kind of waves there is an associated result for waves of the other kind.

5.3 Second-order theory

In the theory of elasticity, the problem of second-order effects has been a subject of many investigations. In the context of thermoelasticity Herrmann [163] and Chadwick and Set [52] were the first to consider the second-order theory. In this section we use the method of Chadwick and Set [52] to derive from the nonlinear equations a consistent approximation correct to terms which are of the second order of smallness. First, we give an alternative form of the basic equations of nonlinear thermoelasticity. We introduce the notations

$$X_i = \delta_{iK} X_K\,, \qquad E_{ij} = \delta_{iK}\delta_{jL}E_{KL}\,, \qquad T_{ji} = \delta_{jK}T_{Ki}\,, \qquad Q_i = \delta_{iK}Q_K\,,$$

where δ_{iK} is the Kronecker delta. Throughout this section the comma notation is used for partial derivatives with respect to the referential coordinates X_i. The equations of motion (1.2.3) can be written in the form

$$T_{ji,j} + \varrho_0 f_i = \varrho_0 \ddot{u}_i\,. \tag{5.3.1}$$

The energy equation (1.3.8) reduces to

$$\varrho_0 T\dot{\eta} = Q_{j,j} + \varrho_0 s\,. \tag{5.3.2}$$

The constitutive equations become

$$\begin{aligned} &\psi = \hat{\psi}(E_{ij},\, T,\, X_s)\,,\\ &T_{ji} = \frac{1}{2}(u_{i,s} + \delta_{is})\Big(\frac{\partial\psi}{\partial E_{js}} + \frac{\partial\psi}{\partial E_{sj}}\Big)\,,\\ &\varrho_0\eta = -\frac{\partial\psi}{\partial T}\,,\\ &Q_i = \hat{Q}_i\big(E_{rs},\, T,\, T_{,k},\, X_s\big)\,, \end{aligned} \tag{5.3.3}$$

where

$$2E_{ij} = u_{i,j} + u_{j,i} + u_{s,i}u_{s,j}\,. \tag{5.3.4}$$

Let $\mathcal{M}$ be the set of all admissible thermodynamic processes on B_0. The trivial process in which the body rests permanently in the reference configuration B_0 is represented in $\mathcal{M}$ by the element $\big(\boldsymbol{X},\, T_0\big)$. From Section 3.4 we can see that we can construct a length l_0 and a time t_0 which together with reference density ϱ_0 and the reference temperature T_0 enable us to use dimensionless quantities. We denote

$$\begin{aligned} a_1 &= \sup\{|u_{i,j}| + l_0T_0^{-1}|T_{,r}| + T_0^{-1}|\theta|\}\,,\\ a_2 &= \sup\{|u_{i,j}| + l_0T_0^{-1}|T_{,r}| + T_0^{-1}|\theta| + l_0|u_{i,jr}| +\\ &\quad + l_0^2T_0^{-1}|T_{,ij}| + t_0|\dot{u}_{i,j}| + t_0T_0^{-1}|\dot{\theta}|\}\,, \end{aligned} \tag{5.3.5}$$

where

$$|A_{ij\dots s}| = (A_{ij\dots s}A_{ij\dots s})^{1/2}.$$

Following [52] we refer to a_1 as the primary amplitude and to a_2 as the secondary amplitude of the admissible process $U = (\boldsymbol{x}, T)$. Let $\mathcal{M}_0$ be the subset of $\mathcal{M}$ defined as the intersection of the domains of the functionals a_1 and a_2. Clearly, $a_1 \leq a_2$. In what follows we consider processes from $\mathcal{M}_0$ which satisfy the conditions $a_1^3 \ll 1$, $a_2^3 \ll 1$. The restriction $a_1^3 \ll 1$ suffices for the derivation of second-order constitutive equations. The stronger assumption $a_2^3 \ll 1$ is needed to justify the substitution of the resulting expressions into the equations of motion (5.3.1), and energy equation (5.3.2).

We assume that $\hat{\psi}$ is of class C^4 with respect to E_{ij} and T in the neighbourhood

$$|E_{ij}| + T_0^{-1}|\theta| < \delta,$$

and that $\widehat{Q}_i$ are of class C^3 with respect to E_{rs}, $T_{,k}$ and T in the neighbourhood

$$|E_{ij}| + l_0 T_0^{-1}|T_{,r}| + T_0^{-1}|\theta| < \delta,$$

where the positive constant δ is such that

$$\{u_{i,j}\,;\, |u_{i,j}| \leq a_1\} \subseteq \{u_{i,j}\,;\, |E_{ij}| < \delta\}.$$

We assume that $\hat{\psi}$, T_{ji} and η vanish in the reference configuration. We have

$$\begin{aligned}\hat{\psi} =& \frac{1}{2}C_{ijrs}E_{ij}E_{rs} - \beta_{ij}E_{ij}\theta - \frac{1}{2}a\,\theta^2 + \frac{1}{6}C_{ijkmnr}E_{ij}E_{km}E_{nr} + \\ &+ \frac{1}{2}C^{(1)}_{ijrs}E_{ij}E_{rs}\theta - \frac{1}{2}\beta^{(1)}_{ij}E_{ij}\theta^2 - \frac{1}{6}b\,\theta^3 + O(a_1^4), \qquad (5.3.6)\\ \widehat{Q}_i =& (k_{ij} + m_{ijrs}E_{rs} + \frac{1}{2}k_{irj}\theta_{,r} + k^{(1)}_{ij}\theta)\theta_{,j} + O(a_1^3),\end{aligned}$$

where

$$C_{ijrs} = (\partial^2\hat{\psi}/\partial E_{ij}\partial E_{rs})_0, \qquad C_{ijkmnr} = (\partial^3\hat{\psi}/\partial E_{ij}\partial E_{mn}\partial E_{rs})_0,$$

$$C^{(1)}_{ijrs} = (\partial^3\hat{\psi}/\partial E_{ij}\partial E_{rs}\partial T)_0, \qquad \beta_{ij} = -(\partial^2\hat{\psi}/\partial E_{ij}\partial T)_0,$$

$$\beta_{ij}^{(1)} = -(\partial^3\hat{\psi}/\partial E_{ij}\partial T^2)_0\,, \qquad a = -(\partial^2\hat{\psi}/\partial T^2)\,,$$

$$b = -(\partial^3\hat{\psi}/\partial T^3)\,, \qquad k_{ij} = (\partial\widehat{Q}_i/\partial T)_0\,, \tag{5.3.7}$$

$$k_{ijr} = (\partial^2\widehat{Q}_i/\partial T_{,j}\partial T_{,r})_0\,, \qquad k_{ij}^{(1)} = (\partial^2\widehat{Q}_i/\partial T_{,j}\partial T)_0\,,$$

$$m_{ijr} = (\partial^2\widehat{Q}_i/\partial T_{,j}\partial E_{rs})_0\,.$$

The above coefficients have the following symmetries

$$C_{ijrs} = C_{rsij} = C_{jirs} \qquad C_{ijrsmn} = C_{jirsmn} = C_{ijmnrs} = C_{mnrsij}\,,$$

$$C_{ijrs}^{(1)} = C_{jirs}^{(1)} = C_{rsij}^{(1)}\,, \qquad \beta_{ij} = \beta_{ji}\,, \tag{5.3.8}$$

$$\beta_{ij}^{(1)} = \beta_{ji}^{(1)}\,, \qquad k_{ijr} = k_{jir}\,, \qquad m_{ijrs} = m_{ijsr}\,.$$

By (5.3.4),

$$E_{ij} = e_{ij} + \frac{1}{2}u_{s,i}u_{s,j}\,, \qquad e_{ij} = \frac{1}{2}(u_{i,j} + u_{j,i})\,. \tag{5.3.9}$$

It follows from (5.3.3), (5.3.6) and (5.3.9) that

$$T_{ij} = C_{ijrs}e_{rs} - \beta_{ij}\theta + \frac{1}{2}C_{ijrsmn}e_{rs}e_{mn} + C_{irmn}u_{m,n}u_{j,r} +$$

$$+ \frac{1}{2}C_{ijmn}u_{r,m}u_{r,n} + (C_{ijrs}^{(1)}e_{rs} - \beta_{ir}u_{j,r})\theta - \frac{1}{2}\beta_{ij}^{(1)}\theta^2 + O(a_1^3)\,, \tag{5.3.10}$$

$$\varrho_0\eta = \beta_{ij}e_{ij} + a\theta - \frac{1}{2}C_{ijrs}^{(1)}e_{ij}e_{rs} + \frac{1}{2}\beta_{ij}u_{m,i}u_{m,j} + \beta_{ij}^{(1)}e_{ij}\theta +$$

$$+ \frac{1}{2}b\theta^2 + O(a_1^3)\,.$$

From $(5.3.6)_2$ we get

$$\hat{Q}_i = (k_{ij} + m_{ijrs}e_{rs} + \frac{1}{2}k_{irj}\theta_{,r} + k_{ij}^{(1)}\theta)\theta_{,j} + O(a_1^3)\,. \tag{5.3.11}$$

Second-order approximations to the constitutive equations follow from (5.3.10) and (5.3.11) on placing the restriction $a_1^3 \ll 1$ on the class of admissible thermodynamic processes under consideration and taking this inequality to justify the neglect of the remainder terms. When the expressions (5.3.10) and (5.3.11) are introduced into the equations (5.3.1) and (5.3.2) the terms of the resulting equations contain not only the dependent variables $u_{i,j}$, $\theta_{,k}$ and θ appearing in the constitutive equations but the additional variables $u_{i,rs}$, $\theta_{,rs}$, $\dot{u}_{i,j}$ and $\dot{\theta}$. This means that the remainder terms in (5.3.10) and (5.3.11) do not in general give rise to terms in the field equations which are $O(a_1^3)$ and that the smallness of the primary amplitude, as specified by the condition $a_1^3 \ll 1$, does not, therefore, justify the derivation of second-order equations of thermoelasticity. It is for this reason that we replace the restriction $a_1^3 \ll 1$ by the stronger requirement $a_2^3 \ll 1$. From (5.3.5) we see that any variable which is $O(a_1^n)$ is a variable which is $O(a_2^n)$. Order relations with respect to a_2 are hence preserved when the operators $\partial/\partial t$ and $\partial/\partial X_j$ are applied to (5.3.10), (5.3.11), and it follows that, under the strengthened condition $a_2^3 \ll 1$ on the class of admissible processes, second-order equations of thermoelasticity are obtained by direct substitution of these expressions into the field equations with neglect of terms which are $O(a_2^3)$. In the case of homogeneous and anisotropic materials we obtain

$$
\begin{aligned}
&C_{ijrs}u_{r,sj} - \beta_{ij}\theta_{,j} + C_{ijrsmn}u_{r,s}u_{m,nj} + C_{jrmn}(u_{m,n}u_{i,rj} + \\
&+ u_{m,nj}u_{i,r}) + C_{ijmn}u_{r,m}u_{r,nj} + C^{(1)}_{ijrs}(u_{r,js}\theta + u_{r,s}\theta_{,j}) - \\
&- \beta_{jr}(u_{i,rj}\theta + u_{i,r}\theta_{,j}) - \beta^{(1)}_{ij}\theta\theta_{,j} + \varrho_0 f_i = \varrho_0 \ddot{u}_i\,, \\
&k_{ij}\theta_{,ij} + m_{ijrs}(u_{r,si}\theta_{,j} + u_{r,s}\theta_{,ij}) + k_{irj}\theta_{,r}\theta_{,ij} + k^{(1)}_{ij}(\theta\theta_{,j})_{,i} - \\
&- T_0 a\dot{\theta} - T_0(\beta_{ij}\dot{u}_{i,j} - C^{(1)}_{ijrs}u_{i,j}\dot{u}_{r,s} + \beta_{ij}u_{m,i}\dot{u}_{m,j} + \beta^{(1)}_{ij}u_{i,j}\dot{\theta}) - \\
&- (\beta_{ij} + T_0\beta^{(1)}_{ij})\dot{u}_{i,j}\theta - (a + bT_0)\theta\dot{\theta} = -\varrho_0 s\,.
\end{aligned}
\tag{5.3.12}
$$

The constitutive equations for isotropic bodies are

$$
\begin{aligned}
T_{ij} =&\lambda e_{rr}\delta_{ij} + 2\mu e_{ij} - \beta\theta\delta_{ij} + \{\frac{1}{2}\nu_1 e_{rr}e_{ss} + \nu_2 e_{rs}e_{rs} + \\
&+ \frac{1}{2}\lambda u_{m,n}u_{m,n} + \lambda^{(1)}e_{rr}\theta - \frac{1}{2}\beta^{(1)}\theta^2\}\delta_{ij} + 2(\nu_2 e_{pp} + \mu^{(1)}\theta)e_{ij} + \\
&+ (\lambda e_{pp} - \beta\theta)u_{j,i} + 4(\mu + \nu_3)e_{ir}e_{jr} - \mu u_{p,i}u_{j,p}\,, \\
\varrho_0\eta =&\beta e_{rr} + a\theta - \frac{1}{2}\lambda^{(1)}e_{rr}e_{ss} - \mu^{(1)}e_{ij}e_{ij} +
\end{aligned}
\tag{5.3.13}
$$

$$+\frac{1}{2}\beta u_{p,r}u_{p,r}+\beta^{(1)}e_{rr}\theta+\frac{1}{2}b\theta^2\,,$$

$$Q_i = k\theta_{,i}+m_1\theta_{,i}e_{rr}+2m_2e_{ir}\theta_{,r}+k^{(1)}\theta\theta_{,i}\,.$$

In the case of homogeneous and isotropic bodies the equations (5.3.12) become

$$\begin{aligned}
&\{\lambda+\mu+(\nu_1+\nu_2)u_{s,s}+(\lambda^{(1)}+\mu^{(1)})\theta\}u_{r,ir}+\\
&+\{\mu+(\lambda+\nu_2)u_{s,s}+(\mu^{(1)}-\beta)\theta\}u_{i,rr}+\{(\lambda+\mu)u_{r,r}+\\
&+2(\nu_2+\nu_3)u_{(r,s)}\}u_{r,is}+2(\mu+\nu_3)u_{(r,s)}u_{i,rs}+\{(\lambda+\mu)u_{i,r}+\\
&+2(\nu_2+\nu_3)u_{(i,r)}\}u_{j,jr}+2(\mu+\nu_3)u_{(i,q)}u_{q,rr}-(\beta+\beta^{(1)}\theta-\\
&-\lambda^{(1)}u_{r,r})\theta_{,i}+(2\mu^{(1)}u_{(i,r)}-\beta u_{i,r})\theta_{,r}+\varrho_0 f_i=\varrho_0\ddot{u}_i\,,\\
&(k+k^{(1)}\theta+m_1u_{r,r})\theta_{,ss}+2m_2u_{r,s}\theta_{,rs}+(m_1u_{j,jr}+2m_2u_{(r,s),s}+\\
&+k^{(1)}\theta_{,r})\theta_{,r}-T_0a\dot{\theta}-(a+bT_0)\theta\dot{\theta}-T_0\beta^{(1)}u_{r,r}\dot{\theta}-\{T_0\beta+\\
&+(\beta+T_0\beta^{(1)})\theta-T_0\lambda^{(1)}u_{r,r}\}\dot{u}_{s,s}-T_0(\beta u_{r,s}-2\mu^{(1)}u_{(r,s)})\dot{u}_{r,s}=-\varrho_0 s\,,
\end{aligned}\tag{5.3.14}$$

where

$$u_{(i,j)}=\frac{1}{2}(u_{i,j}+u_{j,i})\,.$$

Application . Let us consider plane wave solutions for homogeneous and isotropic bodies with zero body force and heat supply. We assume that

$$u_i=u_i(X_1,\,t)\,,\qquad\qquad \theta=\theta(X_1,\,t)\,.$$

From (5.3.4) we obtain

$$\begin{aligned}
&\{\lambda+2\mu+(3\lambda+6\mu+\nu_1+6\nu_2+8\nu_3)u_{1,1}+(\lambda^{(1)}+\\
&+2\mu^{(1)}-\beta)\theta\}u_{1,11}+(\lambda+2\mu+\nu_2+2\nu_3)u_{\Gamma,1}u_{\Gamma,11}-\\
&-\{\beta-(\lambda^{(1)}+2\mu^{(1)}-\beta)u_{1,1}+\beta^{(1)}\}\theta_{,1}=\varrho_0\ddot{u}_1\,,\\
&\{\mu+(\lambda+2\mu+\nu_2+2\nu_3)u_{1,1}+(\mu^{(1)}-\beta)\theta\}u_{\Gamma,11}+\\
&+(\lambda+2\mu+\nu_2+2\nu_3)u_{\Gamma,1}u_{1,11}+(\mu^{(1)}-\beta)u_{\Gamma,1}\theta_{,1}=\varrho_0\ddot{u}_\Gamma\,,
\end{aligned}\tag{5.3.15}$$

$$\{k + k^{(1)}\theta + (m_1 + 2m_2)u_{1,1}\}\theta_{,11} + \{(m_1 + 2m_2)u_{1,11} +$$

$$+ k^{(1)}\theta_{,1}\}\theta_{,1} - T_0 a\dot{\theta} - (a + bT_0)\theta\dot{\theta} - T_0\beta^{(1)}u_{1,1}\dot{\theta} -$$

$$- \{T_0\beta - T_0(\lambda^{(1)} + 2\mu^{(1)} - \beta)u_{1,1} + (\beta + T_0\beta^{(1)})\theta\}\dot{u}_{1,1} -$$

$$- T_0(\mu^{(1)} - \beta)u_{\Lambda,1}\dot{u}_{\Lambda,1} = 0,$$

where Γ, $\Lambda = 2,3$ and the summation convention applies to the repeated index Λ. It is clear that the equations (5.3.15) permit the existence of purely longitudinal waves to the extent that when $u_2 = u_3 = 0$ there remains a pair of coupled partial differential equations in u_1 and θ. A detailed study of the longitudinal thermoelastic waves has been presented by Johnson [211]. The transverse waves have been investigated by Chadwick and Set [52]. In the context of second-order theory these waves are influenced by the thermoelastic interaction.

5.4 Plane strain

In this section we study the plane strain problem for thermoelastic bodies. The results we give here are due to Green and Adkins [148], p.96, Chaudhry [63] and Iesan [175].

a) **The governing equations.** We consider the equilibrium equations of the nonlinear thermoelasticity in curvilinear coordinates for a homogeneous and isotropic body. We use the notation of Section 1.6 and choose the curvilinear coordinate system so that $\theta^3 = X_3$, $X_\alpha = X_\alpha(\theta^1, \theta^2)$. We define the state of thermoelastic plane strain to be that state in which

$$x_\alpha = x_\alpha(\theta^1, \theta^2), \qquad x^3 = \theta^3, \qquad T = T(\theta^1, \theta^2). \tag{5.4.1}$$

We can write

$$\boldsymbol{X} = \boldsymbol{R}^*(\theta^1, \theta^2) + \theta^3\boldsymbol{a}_3, \qquad \mathbf{x} = \boldsymbol{r}^*(\theta^1, \theta^2) + \theta^3\boldsymbol{A}_3. \tag{5.4.2}$$

It follows from (1.6.3)-(1.6.6), (5.4.2) that

$$\begin{aligned} &\boldsymbol{g}_i = \boldsymbol{a}_i, \quad \boldsymbol{G}_i = \boldsymbol{A}_i, \quad g_{\alpha\beta} = a_{\alpha\beta}, \quad g_{33} = 1, \quad g_{\alpha 3} = 0, \\ &G_{\alpha\beta} = A_{\alpha\beta}, \quad G_{\alpha 3} = 0, \quad G_{33} = 1, \quad g = a, \quad G = A. \end{aligned} \tag{5.4.3}$$

The strain tensor (1.6.8) becomes

$$2\gamma_{\alpha\beta} = A_{\alpha\beta} - a_{\alpha\beta}, \qquad \gamma_{i3} = 0. \tag{5.4.4}$$

The strain invariants defined in (1.6.3) reduce to

$$I_1 = 1 + a^{\alpha\beta} A_{\alpha\beta}\,, \qquad I_2 = \frac{A}{a}(a_{\alpha\beta} A^{\alpha\beta} + 1)\,, \qquad I_3 = \frac{A}{a}\,. \tag{5.4.5}$$

We note that

$$a_{\alpha\beta} A^{\alpha\beta} A = a^{\alpha\beta} A_{\alpha\beta} a\,,$$

so that the invariants I_s are not independent. We therefore have

$$I_3 - I_2 + I_1 - 1 = 0\,. \tag{5.4.6}$$

We introduce the notations

$$I = I_1\,, \quad J = I_3\,, \quad J_1^* = I_1 - 3\,, \quad J_2^* = I_2 - 2I_1 + 3\,.$$

The free energy can be expressed in the forms

$$\psi = \hat{\psi}(I_1,\, I_2,\, I_3,\, T) = \psi_1(I,\, J,\, \theta) = \sigma_2(J_1^*,\, J_2^*,\, \theta)\,, \qquad \theta = T - T_0\,, \tag{5.4.7}$$

so that the relations (1.6.49) become

$$\begin{aligned} &\tau^{\alpha\beta} = \mathcal{H} a^{\alpha\beta} + \mathcal{L} A^{\alpha\beta}\,, \\ &\tau^{\alpha 3} = 0\,, \qquad \tau^{33} = \Phi + (I-1)\Psi + p\,, \end{aligned} \tag{5.4.8}$$

where

$$\begin{aligned} &\mathcal{H} = \frac{2}{\sqrt{J}}\frac{\partial\psi_1}{\partial I} = \frac{2}{\sqrt{J}}\Big(\frac{\partial\psi_2}{\partial J_1^*} - \frac{\partial\psi_2}{\partial J_2^*}\Big)\,, \\ &\mathcal{L} = 2\sqrt{J}\frac{\partial\psi_1}{\partial J} = 2\sqrt{J}\frac{\partial\psi_2}{\partial J_2^*}\,, \qquad \Phi = \frac{2}{\sqrt{J}}\frac{\partial\hat{\psi}}{\partial I_1}\,, \\ &\Psi = \frac{2}{\sqrt{J}}\frac{\partial\hat{\psi}}{\partial I_2}\,, \qquad p = 2\sqrt{J}\frac{\partial\hat{\psi}}{\partial I_3}\,. \end{aligned} \tag{5.4.9}$$

It follows from (1.6.39), (1.6.52) and (5.4.1) that

$$q^\alpha = (C_1 \delta^\alpha_\beta + C_2 \gamma^\alpha_\beta + C_3 \gamma^\alpha_\varrho \gamma^\varrho_\beta) T\|^\beta\,, \qquad q^3 = o\,, \tag{5.4.10}$$

where $T\|^\beta = A^{\alpha\beta} T\|_\alpha$, $\gamma^\alpha_\beta = a^{\alpha\varrho}\gamma_{\varrho\beta}$, and C_i are functions of the invariants I, J, T, $T\|^\alpha T\|_\alpha$, $T\|^\alpha T\|_\beta \gamma^\beta_\alpha$, $T\|^\alpha T\|_\beta \gamma^\beta_\varrho \gamma^\varrho_\alpha$. In the absence of body forces the equilibrium equations reduce to

$$\tau^{\alpha\beta}\|_{\beta} = 0\,. \tag{5.4.11}$$

We assume that $s = 0$, so that the energy equation becomes

$$q^{\alpha}\|_{\alpha} = 0\,. \tag{5.4.12}$$

The equations of equilibrium (5.4.11) may be satisfied by putting

$$\tau^{\alpha\beta} = \epsilon^{\alpha\nu}\epsilon^{\beta\varrho}\varphi\|_{\nu\varrho}\,, \tag{5.4.13}$$

where φ is Airy's stress function, and $\epsilon^{\alpha\beta}$ is the alternating symbol,

$$\epsilon^{\alpha\beta}\sqrt{A} = e_{\alpha\beta 3}\,.$$

From (5.4.8) and (5.4.13) we get

$$\varphi\|_{\alpha\beta} = \frac{A}{a}\mathcal{H}a_{\alpha\beta} + \mathcal{L}A_{\alpha\beta}\,. \tag{5.4.14}$$

The total force $\boldsymbol{F}$ exerted across a part P_1P of the curve P_1P_2, measured per unit length of the x_3-axis, is (cf. Green and Adkins [148])

$$\boldsymbol{P} = P^1\boldsymbol{A}_1 + P^2\boldsymbol{A}_2 = \epsilon^{\alpha\beta}\varphi\|_{\alpha}\boldsymbol{A}_{\beta}\,. \tag{5.4.15}$$

The solution of plane strain probleme is often greatly simplified by the introduction of complex variables. We introduce complex coordinates $(\zeta,\,\bar{\zeta})$ and $(z,\,\bar{z})$ by means of the relations

$$\begin{aligned} \zeta &= X_1 + iX_2\,, & \bar{\zeta} &= X_1 - iX_2\,, \\ z &= x_1 + ix_2\,, & \bar{z} &= x_1 - ix_2\,. \end{aligned} \tag{5.4.16}$$

If the components of displacement referred to the X_α-axes are $(u_1,\,u_2,\,0)$, the complex displacement function D is defined by

$$D = u_1 + iu_2\,. \tag{5.4.17}$$

From (5.4.16) and (5.4.17) we obtain

$$z = \zeta + D\,, \qquad \bar{z} = \bar{\zeta} + \bar{D}\,. \tag{5.4.18}$$

We now take $\theta^1 = z$ and $\theta^2 = \bar{z}$. Then, we get

$$A_{12} = \frac{1}{2}, \quad A^{12} = 2, \quad A = -\frac{1}{4}, \quad A_{11} = A_{22} = A^{11} = A^{22} = 0,$$

$$\begin{aligned}
\sqrt{a} &= \frac{i}{2}\frac{\partial(\zeta,\bar{\zeta})}{\partial(z,\bar{z})} = \frac{i}{2}\Big(1 - \frac{\partial D}{\partial z} - \frac{\partial \bar{D}}{\partial \bar{z}} + \frac{\partial D}{\partial z}\frac{\partial \bar{D}}{\partial \bar{z}} - \frac{\partial D}{\partial \bar{z}}\frac{\partial \bar{D}}{\partial z}\Big), \\
a_{11} &= \bar{a}_{22} = \frac{\partial \bar{D}}{\partial z}\Big(\frac{\partial D}{\partial z} - 1\Big), \qquad a_{12} = \frac{1}{2\sqrt{J}} + \frac{\partial D}{\partial \bar{z}}\frac{\partial \bar{D}}{\partial z}, \\
a^{11} &= \overline{a^{22}} = \frac{1}{a}a_{22}, \qquad\qquad a^{12} = -\frac{1}{a}a_{12}.
\end{aligned} \tag{5.4.19}$$

It follows from (5.4.4) and (5.4.19) that

$$\begin{aligned}
\gamma_{11} &= \bar{\gamma}_{22} = \frac{1}{2}\frac{\partial \bar{D}}{\partial z}\Big(1 - \frac{\partial D}{\partial z}\Big), \\
\gamma_{12} &= \frac{1}{4}\Big(\frac{\partial D}{\partial z} + \frac{\partial \bar{D}}{\partial \bar{z}} - \frac{\partial D}{\partial z}\frac{\partial \bar{D}}{\partial \bar{z}} - \frac{\partial D}{\partial \bar{z}}\frac{\partial \bar{D}}{\partial z}\Big).
\end{aligned} \tag{5.4.20}$$

From (5.4.5) and (5.4.19) we get

$$\begin{aligned}
I_1 &= 1 + 2\sqrt{J} + 4J\frac{\partial D}{\partial \bar{z}}\frac{\partial \bar{D}}{\partial z}, \\
I_2 &= 2\sqrt{J} + 4J\frac{\partial D}{\partial \bar{z}}\frac{\partial \bar{D}}{\partial z} + J, \qquad J = -\frac{1}{4a}.
\end{aligned} \tag{5.4.21}$$

If we denote by T^{ij} the contravariant components of the stress tensor referred to the coordinates z, $\bar{z}$, x_3 then from (5.4.13) we find that

$$T^{11} = \overline{T^{22}} = -4\frac{\partial^2 \varphi}{\partial \bar{z}^2}, \qquad T^{12} = 4\frac{\partial^2 \phi}{\partial z \partial \bar{z}}. \tag{5.4.22}$$

In view of (5.4.14),

$$\frac{\partial^2 \phi}{\partial z^2} = J\frac{\partial \bar{D}}{\partial z}\Big(\frac{\partial D}{\partial z} - 1\Big)\mathcal{H}, \qquad \frac{\partial^2 \phi}{\partial z \partial \bar{z}} = \Big(\frac{1}{2}\sqrt{J} + J\frac{\partial D}{\partial \bar{z}}\frac{\partial \bar{D}}{\partial z}\Big)\mathcal{H} + \frac{1}{2}\mathcal{L}. \tag{5.4.23}$$

It follows from (5.4.15) that the components of the resultant force $\boldsymbol{F}$ can be expressed in the form

$$P^1 = \bar{P}^2 = 2i\frac{\partial \phi}{\partial \bar{z}}. \tag{5.4.24}$$

b) **Successive approximations.** We assume that the complex displacement function D and the temperature T are functions of a real parameter ε which can be expanded as absolutely convergent series

$$D = \varepsilon D_1(z,\,\bar{z}) + \varepsilon^2 D_2(z,\,\bar{z}) + \cdots\,, \qquad \theta = \varepsilon T_1(z,\,\bar{z}) + \varepsilon^2 T_2(z,\,\bar{z}) + \cdots\,, \tag{5.4.25}$$

in the range $\varepsilon \in [0, \varepsilon_0]$. Here ε_0 is a value of the parameter which depends upon the problem under consideration. We assume that the first and second partial derivatives of D and θ with respect to z and $\bar{z}$ exist and can be obtained by term-by-term differentiation of the series (5.4.25), and that the resulting series are absolutely convergent in the range $\varepsilon \in [0, \varepsilon_0]$ and for relevant values of z. In what follows the calculations are limited to the first two terms of the power series expansion in terms of ε.

From (5.4.19), (5.4.21) and (5.4.25) we obtain

$$\sqrt{J} = 1 + \varepsilon a_1 + \varepsilon^2 a_2 + \cdots\,, \tag{5.4.26}$$

where

$$\begin{aligned} a_1 &= \frac{\partial D_1}{\partial z} + \frac{\partial \bar{D}_1}{\partial \bar{z}}\,, \\ a_2 &= \frac{\partial D_2}{\partial z} + \frac{\partial \bar{D}_2}{\partial \bar{z}} + \frac{\partial D_1}{\partial z}\frac{\partial \bar{D}_1}{\partial \bar{z}} + \frac{\partial D_1}{\partial \bar{z}}\frac{\partial \bar{D}_1}{\partial z} + \Big(\frac{\partial D_1}{\partial z}\Big)^2 + \Big(\frac{\partial \bar{D}_1}{\partial \bar{z}}\Big)^2 . \end{aligned} \tag{5.4.27}$$

Clearly

$$J = 1 + 2a_1\varepsilon + (a_1^2 + 2a_2)\varepsilon^2 + \cdots\,. \tag{5.4.28}$$

By (5.4.21), (5.4.26) and (5.4.28),

$$\begin{aligned} J_1^* &= 2a_1\varepsilon + 2\Big(a_2 + 2\frac{\partial D_1}{\partial \bar{z}}\frac{\partial \bar{D}_1}{\partial z}\Big)\varepsilon^2 + \cdots\,, \\ J_2^* &= \Big(a_1^2 - 4\frac{\partial D_1}{\partial \bar{z}}\frac{\partial \bar{D}_1}{\partial z}\Big)\varepsilon^2 + \cdots\,. \end{aligned} \tag{5.4.29}$$

We introduce the notations

$$\begin{aligned} &A = \frac{\partial^2 \psi_2}{\partial J_1^{*2}}\,, \qquad B = \frac{\partial \psi_2}{\partial J_2^*}\,, \qquad C = \frac{\partial^2 \psi_2}{\partial J_1^* \partial J_2^*}\,, \qquad G = -\frac{\partial^2 \psi_2}{\partial J_1^* \partial \theta}\,, \\ &H = \frac{\partial^3 \psi_2}{\partial J_1^{*3}}\,, \qquad M = -\frac{1}{2}\frac{\partial^3 \psi_2}{\partial J_1^* \partial \theta^2}\,, \qquad K = -\frac{\partial^2 \psi_2}{\partial J_2^* \partial \theta}\,, \\ &P = -\frac{1}{2}\frac{\partial^3 \psi_2}{\partial \theta \partial \theta \partial J_1^{*2}} \qquad \text{for} \quad \varepsilon = 0\,. \end{aligned} \tag{5.4.30}$$

Then, we have

$$\frac{\partial \psi_2}{\partial J_1^*} = \varepsilon(2Aa_1 - GT_1) + \varepsilon^2[2Aa_2 - GT_2 + (2H + C)a_1^2 + $$
$$+ 4(A - C)\frac{\partial D_1}{\partial \bar{z}}\frac{\partial \bar{D}_1}{\partial z} - MT_1^2 - 4Pa_1T_1], \tag{5.4.31}$$
$$\frac{\partial \psi_2}{\partial J_2^*} = B + \varepsilon(2Ca_1 - KT_1) + \cdots .$$

From (5.4.9), (5.4.26) and (5.4.31) we find that

$$\mathcal{H} = \mathcal{H}_0 + \varepsilon\mathcal{H}_1 + \cdots , \tag{5.4.32}$$

where

$$\mathcal{H}_0 = 2B, \qquad \mathcal{H}_1 = 2[(2A - 2C + B)a_1 - (G - K)T_1]. \tag{5.4.33}$$

Similarly, we get

$$\frac{1}{2}(\mathcal{H}\sqrt{J} + \mathcal{L}) = \varepsilon[(2A + B)a_1 - GT_1] + \varepsilon^2[(2A + B)a^2 - $$
$$- GT_2 + 4(A - C)\frac{\partial D_1}{\partial \bar{z}}\frac{\partial \bar{D}_1}{\partial z} + (2H + 3C)a_1^2 - \tag{5.4.34}$$
$$- Ka_1T_1 - MT_1^2 - 4Pa_1T_1] + \cdots .$$

If we now put

$$\varphi = \mathcal{H}_0[\varepsilon\varphi_1(z, \bar{z}) + \varepsilon^2\varphi_2(z, \bar{z}) + \cdots], \tag{5.4.35}$$

then from (5.4.28), (5.4.32) and $(5.4.23)_1$ we obtain

$$\frac{\partial^2 \varphi_1}{\partial z^2} = -\frac{\partial \bar{D}_1}{\partial z},$$
$$\frac{\partial^2 \varphi_2}{\partial z^2} = -\frac{\partial \bar{D}_2}{\partial z} + \frac{2(A - C)}{B}\left(\frac{\partial D_1}{\partial z} + \frac{\partial \bar{D}_1}{\partial \bar{z}}\right)\frac{\partial \bar{D}_1}{\partial z} - \tag{5.4.36}$$
$$- \frac{\partial \bar{D}_1}{\partial \bar{z}}\frac{\partial \bar{D}_1}{\partial z} - \frac{G - K}{B}T_1\frac{\partial D_1}{\partial z}.$$

In the view of (5.4.34), (5.4.28), (5.4.33), (5.4.23) we get

$$\begin{aligned}
\frac{\partial^2\varphi_1}{\partial z\partial\bar z} &= -\frac{2A+B}{2B}\Big(\frac{\partial D_1}{\partial z}+\frac{\partial\bar D_1}{\partial\bar z}\Big)+\frac{G}{2B}T_1\,,\\
\frac{\partial^2\varphi_2}{\partial z\partial\bar z} &= -\frac{2A+B}{2B}\Big(\frac{\partial D_2}{\partial z}+\frac{\partial\bar D_2}{\partial\bar z}\Big)+\frac{G}{2B}T_2-\\
&-\frac{s_1}{2B}\Big(\frac{\partial D_1}{\partial z}+\frac{\partial\bar D_1}{\partial\bar z}\Big)^2-\frac{s_2}{2B}\frac{\partial D_1}{\partial\bar z}\frac{\partial\bar D_1}{\partial z}+\frac{2A+B}{2B}\frac{\partial D_1}{\partial z}\frac{\partial\bar D_1}{\partial\bar z}+\\
&+\frac{K}{2B}T_1\Big(\frac{\partial D_1}{\partial z}+\frac{\partial\bar D_1}{\partial\bar z}\Big)+\frac{M}{2B}T_1^2+\frac{2P}{B}\Big(\frac{\partial D_1}{\partial z}+\frac{\partial\bar D_1}{\partial\bar z}\Big)T_1\,,
\end{aligned}\tag{5.4.37}$$

where

$$s_1=2A+3C+B+2H\,,\qquad s_2=6A-4C-B\,.$$

If we introduce the notations

$$\begin{aligned}
&c_1=1+A/B\,,\qquad c_2=C/B\,,\qquad c_4=H/B\,,\\
&\beta_1=-G/B\,,\qquad \beta_2=-K/B\,,\qquad \beta_3=-M/B\,,\\
&\beta_4=-4P/B\,,\qquad \alpha_1=c_1-c_2-1\,,\\
&\alpha_2=2c_1+3c_2+2c_4-1\,,\qquad \alpha_3=6c_1-4c_2-7\,,\\
&\nu=\frac{A+B}{2A+B}\,,\qquad E=-\frac{4B(3A+2B)}{2A+B}\,,
\end{aligned}\tag{5.4.38}$$

then from (5.4.36) and (5.4.37) we obtain

$$\begin{aligned}
\frac{\partial^2\varphi_1}{\partial z^2} &= -\frac{\partial\bar D_1}{\partial z}\,,\\
\frac{\partial^2\varphi_1}{\partial z\partial\bar z} &= \frac{1}{2(1-2\nu)}\Big(\frac{\partial D_1}{\partial z}+\frac{\partial\bar D_1}{\partial\bar z}\Big)-\frac{1}{2}\beta_1T_1\,,
\end{aligned}\tag{5.4.39}$$

and

$$\begin{aligned}
\frac{\partial^2\varphi_2}{\partial z^2} &= -\frac{\partial\bar D_2}{\partial z}+2\alpha_1\Big(\frac{\partial D_1}{\partial z}+\frac{\partial\bar D_1}{\partial\bar z}\Big)\frac{\partial\bar D_1}{\partial z}-\\
&-\frac{\partial\bar D_1}{\partial\bar z}\frac{\partial\bar D_1}{\partial z}+(\beta_1-\beta_2)T_1\frac{\partial D_1}{\partial z}\,,
\end{aligned}$$

$$\frac{\partial^2 \varphi_2}{\partial z \partial \bar{z}} = \frac{1}{2(1-2\nu)}\Big(\frac{\partial D_2}{\partial z} + \frac{\partial \bar{D}_2}{\partial \bar{z}}\Big) - \frac{1}{2}\beta_1 T_2 - \tag{5.4.40}$$

$$- \frac{1}{2(1-2\nu)} \frac{\partial D_1}{\partial z}\frac{\partial \bar{D}_1}{\partial \bar{z}} - \frac{1}{2}\alpha_2\Big(\frac{\partial D_1}{\partial z} + \frac{\partial \bar{D}_1}{\partial \bar{z}}\Big)^2 - \frac{1}{2}\alpha_3 \frac{\partial D_1}{\partial \bar{z}}\frac{\partial \bar{D}_1}{\partial z} -$$

$$- \frac{1}{2}(\beta_2 + \beta_4) T_1 \Big(\frac{\partial D_1}{\partial z} + \frac{\partial \bar{D}_1}{\partial \bar{z}}\Big) - \frac{1}{2}\beta_3 T_1^2 .$$

The functions C_i which appear in (5.4.10) can be expressed in the form

$$C_1 = k + \varepsilon\big[2k_1\Big(\frac{\partial D_1}{\partial z} + \frac{\partial \bar{D}_1}{\partial \bar{z}}\Big) + k_2 T_1\big] + \cdots, \quad C_2 = k_3 + \cdots \tag{5.4.41}$$

where

$$k = C_1 , \qquad k_1 = \frac{\partial C_1}{\partial J_3^*} , \qquad k_2 = \frac{\partial C_1}{\partial \theta} , \qquad k_3 = C_2 \quad \text{for } \varepsilon = 0 .$$

By (5.4.19) and (5.4.20),

$$\begin{aligned} a^{11} = \overline{a^{22}} = 4\varepsilon \frac{\partial D_1}{\partial \bar{z}} + \cdots , \qquad & a^{12} = 2a + \cdots , \\ \gamma_{11} = \bar{\gamma}_{22} = \frac{1}{2}\frac{\partial \bar{D}_1}{\partial z} + \cdots , \qquad & \gamma_{12} = \frac{1}{4}\varepsilon\Big(\frac{\partial D_1}{\partial z} + \frac{\partial \bar{D}_1}{\partial \bar{z}}\Big) + \cdots . \end{aligned} \tag{5.4.42}$$

Since

$$T\|^1 = 2\frac{\partial T}{\partial \bar{z}} , \qquad T\|^2 = 2\frac{\partial T}{\partial z} ,$$

from (5.4.10) and (5.4.41) we obtain

$$q^\alpha = \varepsilon q^\alpha_{(1)} + \varepsilon^2 q^\alpha_{(2)} + \cdots , \tag{5.4.43}$$

where

$$q^1_{(1)} = 2k\frac{\partial T_1}{\partial z} , \qquad q^2_{(1)} = 2k\frac{\partial T_1}{\partial z} ,$$

$$\begin{aligned} q^1_{(2)} =& 2k\frac{\partial T_2}{\partial z} + (4k_1 + k_3)\Big(\frac{\partial D_1}{\partial z} + \frac{\partial \bar{D}_1}{\partial \bar{z}}\Big)\frac{\partial T_1}{\partial \bar{z}} + \\ & + 2k_2 T_1 \frac{\partial T_1}{\partial \bar{z}} + 2k_3 \frac{\partial D_1}{\partial \bar{z}}\frac{\partial T_1}{\partial z} , \\ q^2_{(2)} =& 2k\frac{\partial T_2}{\partial z} + (4k_1 + k_3)\Big(\frac{\partial D_1}{\partial z} + \frac{\partial \bar{D}_1}{\partial \bar{z}}\Big)\frac{\partial T_1}{\partial z} + \\ & + 2k_2 T_1 \frac{\partial T_1}{\partial z} + 2k_3 \frac{\partial \bar{D}_1}{\partial z}\frac{\partial T_1}{\partial \bar{z}} . \end{aligned} \tag{5.4.44}$$

The equation (5.4.12) yields

$$q^{\alpha}_{(1)}\|_{\alpha} = 0\,, \qquad\qquad q^{\alpha}_{(2)}\|_{\alpha}\,. \tag{5.4.45}$$

In what follows we assume that the plane domain under consideration is a simply-connected region and that the mechanical loadings are zero. In Section 4.5 we have shown that in the context of linear thermoelastostatics the stresses $t_{\alpha\beta}$ are zero. Let us study the second order approximation.

From (5.4.44) and (5.4.45) we obtain

$$\begin{aligned}
&\frac{\partial^2 T_1}{\partial z \partial \bar{z}} =0\,,\\
-4k&\frac{\partial^2 T_2}{\partial z \partial \bar{z}} =4k_2\frac{\partial T_1}{\partial z}\frac{\partial T_1}{\partial \bar{z}} + (4k_1 + k_3)\frac{\partial T_1}{\partial \bar{z}}\frac{\partial}{\partial z}\Big(\frac{\partial D_1}{\partial z} +\\
&+\frac{\partial \bar{D}_1}{\partial \bar{z}}\Big) + \frac{\partial T_1}{\partial z}\frac{\partial}{\partial \bar{z}}\Big(\frac{\partial D_1}{\partial z} + \frac{\partial \bar{D}_1}{\partial \bar{z}}\Big) +\\
&+ 2k_3\Big[\frac{\partial}{\partial z}\Big(\frac{\partial D_1}{\partial \bar{z}}\frac{\partial T_1}{\partial z}\Big) + \frac{\partial}{\partial \bar{z}}\Big(\frac{\partial \bar{D}_1}{\partial z}\frac{\partial T_1}{\partial \bar{z}}\Big)\Big]\,.
\end{aligned} \tag{5.4.46}$$

By $(5.4.46)_1$,

$$T_1(z,\,\bar{z}) = F'(z) + \overline{F'}(\bar{z})\,, \tag{5.4.47}$$

where $F'(z) = dF/dz$. We introduce the functions u^0_α by

$$D_1 = D_0 + \beta_1(1-2\nu)F(z)\,, \tag{5.4.48}$$

where $D_0 = u^0_1 + iu^0_2$. From (5.4.40) and (5.4.48) we get

$$\frac{\partial^2 \varphi_1}{\partial z^2} = -\frac{\partial D_0}{\partial z}\,, \qquad \frac{\partial^2 \varphi_1}{\partial z \partial \bar{z}} = \frac{1}{2(1-2\nu)}\Big(\frac{\partial D_0}{\partial z} + \frac{\partial \bar{D}_0}{\partial \bar{z}}\Big)\,. \tag{5.4.49}$$

If we put

$$T^{11} = \overline{T^{22}} = \varepsilon T^{11}_{(1)} + \varepsilon^2 T^{11}_{(2)} + \cdots\,, \qquad T^{12} = \varepsilon T^{12}_{(1)} + \varepsilon^2 T^{12}_{(2)} + \cdots\,, \tag{5.4.50}$$

then $T^{11}_{(1)}$, $T^{12}_{(1)}$ and D_0 correspond to the classical theory of elasticity in the absence of mechanical loadings. We conclude that $D_0 = 0$ *modulo* a rigid displacement, and $T^{11}_{(1)} = T^{12}_{(1)} = 0$. From (5.4.48) we conclude that

$$D_1 = \beta_1(1-2\nu)F(z)\,. \tag{5.4.51}$$

In view of (5.4.47) and (5.4.51) the equation $(5.4.46)_2$ becomes

$$\frac{\partial^2 T_2}{\partial z \partial \bar{z}} = \gamma F''(z)\bar{F}''(\bar{z}), \tag{5.4.52}$$

where γ is defined by

$$-4k\gamma = 4k_2 + \frac{2\beta_1}{1-2c_1}(4k_1 + k_3). \tag{5.4.53}$$

From (5.4.52) we obtain

$$T_2(z, \bar{z}) = \gamma F'(z)\bar{F}'(\bar{z}) + G'(z) + \bar{G}'(\bar{z}). \tag{5.4.54}$$

With the help of (5.4.51) and (5.4.54), from the equations (5.4.40) we find that

$$\begin{aligned} \frac{\partial^2 \varphi_2}{\partial z^2} &= -\frac{\partial \bar{D}_2}{\partial z}, \\ \frac{\partial^2 \varphi_2}{\partial z \partial \bar{z}} &= \frac{1}{2(1-2\nu)}\Big(\frac{\partial D_2}{\partial z} + \frac{\partial \bar{D}_2}{\partial \bar{z}}\Big) - \frac{1}{2}\beta_1[G'(z) + \bar{G}'(\bar{z})] - \\ &\quad - \frac{1}{2}[\beta_1^2(1-2\nu) + \gamma\beta_1 + 2b]F'(z)\bar{F}'(\bar{z}) - \frac{1}{2}b\big[\big(F'(z)\big)^2 + \big(\bar{F}'(\bar{z})\big)^2\big], \end{aligned} \tag{5.4.55}$$

where

$$b = \beta_1^2(1-2\nu)^2\alpha_2 + \beta_1(1-2\nu)(\beta_2 + \beta_4) + \beta_3. \tag{5.4.56}$$

Next, we introduce the function $D_2^*(z, \bar{z})$ by

$$D_2(z, \bar{z}) = D_2^*(z, \bar{z}) + h_1 F(z)\bar{F}'(\bar{z}) + h_2 G(z) + h_3 \int^z \big(F'(z)\big)^2 dz, \tag{5.4.57}$$

where

$$\begin{aligned} h_1 &= \frac{1-2\nu}{2}[\beta_1^2(1-2\nu) + \gamma\beta_1 + 2b], \\ h_2 &= \beta_1(1-2\nu), \qquad\qquad h_3 = b(1-2\nu). \end{aligned} \tag{5.4.58}$$

It follows from (5.4.55) that

$$\begin{aligned} \frac{\partial^2 \varphi_2}{\partial z^2} &= -\frac{\partial \bar{D}_2^*}{\partial z} - h_1\bar{F}(\bar{z})F''(z), \\ \frac{\partial^2 \varphi_2}{\partial z \partial \bar{z}} &= \frac{1}{2(1-2\nu)}\Big(\frac{\partial D_2^*}{\partial z} + \frac{\partial \bar{D}_2^*}{\partial \bar{z}}\Big). \end{aligned} \tag{5.4.59}$$

From $(5.4.59)_1$ we get

$$\frac{\partial \varphi_2}{\partial z} = -\bar{D}_2^* - h_1 \bar{F}(\bar{z}) F'(z) + 4(1-\nu)\bar{\Omega}(\bar{z}) , \tag{5.4.60}$$

where $\bar{\Omega}(\bar{z})$ is an arbitrary function of $\bar{z}$. This relation implies that

$$\begin{aligned} \frac{\partial D_2^*}{\partial z} + \frac{\partial \bar{D}_2^*}{\partial \bar{z}} = &- 2\frac{\partial^2 \varphi_2}{\partial z \partial \bar{z}} - 2h_1 F'(z)\bar{F}'(\bar{z}) + \\ &+ 4(1-\nu)[\Omega'(z) + \bar{\Omega}'(\bar{z})] . \end{aligned} \tag{5.4.61}$$

With the help of (5.4.61), the equation $(5.4.59)_2$ becomes

$$\frac{\partial^2 \varphi_2}{\partial z \partial \bar{z}} = \Omega'(z) + \bar{\Omega}'(\bar{z}) - \frac{h_1}{2(1-\nu)} F'(z)\bar{F}'(\bar{z}) , \tag{5.4.62}$$

so that

$$\varphi_2(z, \bar{z}) = z\bar{\Omega}(\bar{z}) + \bar{z}\Omega(z) + \omega(z) + \bar{\omega}(\bar{z}) - \frac{h_1}{2(1-\nu)} F(z)\bar{F}(\bar{z}) , \tag{5.4.63}$$

where $\bar{\omega}(\bar{z})$ is a further arbitrary function of $\bar{z}$.

It follows from (5.4.60) and (5.4.63) that

$$D_2^*(z, \bar{z}) = (3-4\nu)\Omega(z) - z\bar{\Omega}'(\bar{z}) - \bar{\omega}'(\bar{z}) - \frac{1-2\nu}{2(1-\nu)} h_1 F(z)\bar{F}'(\bar{z}) . \tag{5.4.64}$$

We denote

$$\begin{aligned} D_2^0(z, \bar{z}) &= (3-4\nu)\Omega(z) - z\bar{\Omega}'(\bar{z}) - \bar{\omega}'(\bar{z}) , \\ \varphi_2^0(z, \bar{z}) &= z\bar{\Omega}(\bar{z}) + z\Omega(z) + \omega(z) + \bar{\omega}(\bar{z}) . \end{aligned} \tag{5.4.65}$$

Then from (5.4.57), (5.4.63) and (5.4.64) we get

$$\begin{aligned} D_2(z, \bar{z}) = &D_2^0(z, \bar{z}) + \frac{h_1}{2(1-\nu)} F(z)\bar{F}(\bar{z}) + \\ &+ h_2 G(z) + h_3 \int^z \left[F'(z)\right]^2 dz , \\ \varphi_2(z, \bar{z}) = &\varphi_2^0(z, \bar{z}) - \frac{1}{2(1-\nu)} F(z)\bar{F}(\bar{z}) . \end{aligned} \tag{5.4.66}$$

These results show that in this case the temperature field can induce non zero stresses $T^{11}_{(2)}, T^{12}_{(2)}$. We assume that the boundary L is free from applied stress. Then we have the condition

$$\Omega(\eta) + \eta\bar{\Omega}'(\bar{\eta}) + \bar{\omega}'(\bar{\eta}) = \frac{h_1}{2(1-\nu)} F(\eta)\bar{F}'(\eta), \qquad \eta \in L. \tag{5.4.67}$$

The complex potential functions $\Omega(z)$ and $\omega(z)$ can be determined by the methods of classical elasticity (see, e.g. Mushelishvili [264], p.108, Green and Zerna [152], p.192).

c) **Application.** To illustrate the foregoing theory we consider the plane strain problem for a circular cylinder of radius R. We assume that L is the circle $|z| = R$. We suppose that the body is subject to a temperature field in which

$$T_1 = dx_1,$$

where d is a given constant. Then,

$$F'(z) = \frac{1}{2} dz. \tag{5.4.68}$$

From (5.4.67) and (5.4.68) we obtain

$$\Omega(\eta) + \eta\bar{\Omega}'(\bar{\eta}) + \bar{\omega}'(\bar{\eta}) = \frac{h_1 d^2 R^2 \eta}{16(1-\nu)}, \qquad \eta \in L. \tag{5.4.69}$$

Clearly, the complex potentials

$$\Omega(z) = \frac{h_1 d^2 R^2}{32(1-\nu)} z, \qquad \omega(z) = 0, \tag{5.4.70}$$

satisfy (5.4.69). It follows from (5.4.65), (5.4.66) and (5.4.70) that

$$\varphi_2(z, \bar{z}) = \frac{h_1 d^2 R^2}{16(1-\nu)} \left(z\bar{z} - \frac{1}{2R^2} z^2 \bar{z}^2\right). \tag{5.4.71}$$

In view of (5.4.22), (5.4.34), (5.4.35) and (5.4.71) we find that

$$T^{11}_{(2)} = T^{22}_{(2)} = \frac{h_1 \mathcal{H}_0 d^2 z^2}{4(1-\nu)}, \qquad T^{12}_{(2)} = \frac{h_1 \mathcal{H}_0 d^2 z^2}{4(1-\nu)} \left(1 - \frac{2}{R^2} z\bar{z}\right).$$

5.5 Special solutions

Controlable states of a class of materials satisfy the pertinent field equations identically in the response material functions which characterize the class. Such states are of interest not only in their being exact solutions to generally non-linear boundary-value problems, but also for their immediate application to the experimental determination of material response functions. Ericksen [121] has established that no less than five families of deformations can be supported by surface tractions alone in all homogeneous, isotropic and incompressible elastic bodies. The class of non-linear thermoelastic materials has been analyzed by Petroski and Carlson [297], [298], Hayes *et al.* [162], Laws [231], Chen and Petroski [65]. It was established that for compressible media there are no essential controllable states. In this case only the trivial states in which the deformation is homogeneous and the temperature is constant are controllable. For the incompressible thermoelastic materials there exist essential controllable states. In this section we present some families of controllable states.

We consider the non linear theory of thermoelastic equilibrium for homogeneous and incompressible materials in the absence of body forces and heat supply. The condition of incompressibility is

$$I_3 = 1\,. \tag{5.5.1}$$

In this case the equations (1.6.23) and (1.6.48)reduce to

$$\tau^{ij}\|_i = 0\,, \tag{5.5.2}$$

$$q^i\|_i = 0\,, \tag{5.5.3}$$

respectively. The constitutive equations are

$$\begin{aligned} \tau^{ij} &= pG^{ij} + \Phi g^{ij} + \Psi B^{ij}\,, \\ q^i &= \left(\mathscr{C}_1^* \delta^i_j + \mathscr{C}_2^* \gamma^i_j + \mathscr{C}_3^* \gamma^i_m \gamma^m_j\right) T\|^j\,, \end{aligned} \tag{5.5.4}$$

where Φ and Ψ are given by (1.6.50), p is an unknown function, and $\mathscr{C}_j^*$ are functions of the invariants

$$\begin{aligned} &I_1\,, I_2\,, \quad I_4^* = T\|_i T\|^i\,, \quad I_5^* = T\|^i T\|_j \gamma^j_i\,, \\ &I_6^* = T\|^i T\|_j \gamma^j_s \gamma^s_i\,, \qquad I_7^* = T\,. \end{aligned} \tag{5.5.5}$$

First, we consider the class of materials given by (see [298])

$$\begin{aligned} \psi &= \hat{\psi}(I_1,\, I_2,\, T)\,, \\ \mathscr{C}_i^* &= \hat{\mathscr{C}}_i(I_1,\, I_2,\, I_4^*,\, I_5^*,\, I_6^*)\,. \end{aligned} \tag{5.5.6}$$

$i)$ Let us study the thermomechanical deformation defined by

$$x_1 = \frac{C}{\sqrt{A}} X_1 - \frac{D}{\sqrt{B}} X_2 \,, \qquad x_2 = \frac{D}{\sqrt{A}} X_1 + \frac{C}{\sqrt{B}} X_2 \,,$$
$$x_3 = \sqrt{AB} X_3 \,, \qquad T = T_0 + T_1 X_3 \,, \tag{5.5.7}$$

where A, B, C, D, T_0 and T_1 are constants which satisfy the conditions

$$C^2 + D^2 = 1 \,, \qquad A,\, B > 0 \,. \tag{5.5.8}$$

If we take $\theta^i = X_i$, then we obtain

$$G_{11} = \frac{1}{A} \,, \qquad G_{22} = \frac{1}{B} \,, \qquad G_{33} = AB \,, \qquad g_{ij} = \delta_{ij} \,,$$
$$G^{11} = A \,, \quad G^{22} = B \,, \quad G^{33} = \frac{1}{AB} \,, \quad G_{ij} = G^{ij} = 0 \quad (i \neq j) \,, \quad G = 1 \,.$$
$$I_1 = \frac{1}{A} + \frac{1}{B} + AB \,, \qquad I_2 = A + B + \frac{1}{AB} \,, \tag{5.5.9}$$
$$I_4^* = T_1^2 \,, \quad I_5^* = \frac{1}{2}(AB-1)T_1^2 \,, \quad I_6^* = \frac{1}{4}(AB-1)^2 T_1^2 \,,$$
$$B^{11} = \frac{1}{B} + AB \,, \quad B^{22} = \frac{1}{A} + AB \,, \quad B^{33} = \frac{1}{A} + \frac{1}{B} \,, \quad B^{ij} = 0 \quad (i \neq j) \,.$$

In this case $\hat{\psi}$ does not depend on X_1 and X_2, and $\hat{\mathscr{C}}_i^*$ are constants. It follows from (5.5.2), (5.5.4), (5.5.9) that p is independent of X_1 and X_2. Moreover, from the third equations (5.5.2) we find

$$p = -p_0 - 2AB\left(\frac{\partial \psi}{\partial I_1} + \frac{A+B}{AB}\frac{\partial \psi}{\partial I_2}\right) , \tag{5.5.10}$$

where p_0 is an arbitrary constant. In view of (5.5.4), (5.5.9) and (5.5.10) we can determine the stress tensor. We note that τ^{ij} are independent of X_1 and X_2, and

$$\tau^{ij} = 0 \,, \quad (i \neq j) \,. \tag{5.5.11}$$

It follows from (5.5.4) and (5.5.9) that

$$q^1 = q^2 = 0 \,,$$
$$q^3 = \left[\mathscr{C}_1^* + \frac{1}{2}\mathscr{C}_2^*(AB-1) + \frac{1}{4}\mathscr{C}_3^*(AB-1)^2\right] T_1 \,. \tag{5.5.12}$$

The equations (5.5.3) are satisfied identically. Thus, we conclude that for any medium from the class under consideration, the thermomechanical deformation (5.5.7) satisfies the equations of equilibrium and the equation of energy in the absence of body force and heat supply.

ii) We now assume that

$$\theta^1 = r\,, \qquad \theta^2 = \varphi\,, \qquad \theta^3 = z\,, \tag{5.5.13}$$

where $(r,\ \varphi,\ z)$ are cylindrical coordinates in the deformed body, i. e.

$$x_1 = r\cos\varphi\,, \qquad x_2 = r\sin\varphi\,, \qquad x_3 = z\,. \tag{5.5.14}$$

We denote by $(\varrho,\ \vartheta,\ Z)$ the cylindrical coordinates in the reference configuration,

$$X_1 = \varrho\cos\vartheta\,, \qquad X_2 = \varrho\sin\vartheta\,, \qquad X_3 = Z\,. \tag{5.5.15}$$

Let us consider the thermomechanical deformation defined by

$$r = \frac{1}{\sqrt{\lambda}}\varrho\,, \qquad \varphi = \vartheta\,, \qquad z = \lambda Z\,, \qquad T = T_0 + T_1\vartheta\,, \tag{5.5.16}$$

where $\lambda(>0)$, T_0 and T_1 are constants.

If for definiteness we take the reference configuration of the body to be that bounded by the surfaces $X_3 = \pm h\,,\ \vartheta = 0\,,\ \vartheta = \alpha\,,\ \varrho = \varrho_1\,,\ \varrho = \varrho_2\,,$ then we can interpret (5.5.16) as a thermomechanical deformation in which the deformed body occupies the domain limited by the surfaces $z = \pm\lambda h\,,\ \varphi = 0\,,\ \varphi = \alpha\,,\ r_1 = \varrho_1/\sqrt{\lambda}\,,\ r_2 = \varrho_2/\sqrt{\lambda}\,.$ In the deformed body the plane $\varphi = 0$ is at temperature $T_0\,,$ and the plane $\varphi = \alpha$ is at temperature $T_0 + \alpha T_1\,.$

It follows from (5.5.14)-(5.5.16), (1.6.3), (1.6.4) that

$$\begin{gathered}
G_{11} = G^{11} = G_{33} = G^{33} = 1\,, \quad G_{22} = r^2\,, \quad G^{22} = r^{-2}\,, \\
G = r^2 \qquad G_{ij} = G^{ij} = 0\,, \quad (i \neq j)\,, \\
g_{11} = \lambda\,, \quad g_{22} = \lambda r^2\,, \quad g_{33} = \lambda^{-2}\,, \quad g = r^2\,, \\
g^{11} = \frac{1}{\lambda}\,, \quad g^{22} = \frac{1}{\lambda r^2}\,, \quad g^{33} = \lambda^2\,, \quad g_{ij} = g^{ij} = 0\,, \quad (i \neq j)\,.
\end{gathered} \tag{5.5.17}$$

The invariants I_s are given by

$$I_1 = \frac{2}{\lambda} + \lambda^2\,, \qquad I_2 = 2\lambda + \frac{1}{\lambda^2}\,, \qquad I_3 = 1\,. \tag{5.5.18}$$

With the help of the relations

$$T\|^1 = T\|^3 = 0\,, \qquad T\|^2 = \frac{1}{\lambda r^2} T_1\,, \qquad \gamma_1^1 = \frac{1}{2\lambda}(1-\lambda)\,,$$
$$\gamma_2^2 = \frac{1}{2\lambda}(1-\lambda)\,, \qquad \gamma_3^3 = \frac{1}{2}(\lambda^2 - 1)\,, \qquad \gamma_j^i = 0 \; (i \neq j)\,, \tag{5.5.19}$$

we find that

$$I_4^* = \frac{1}{\lambda r^2} T_1^2\,, \quad I_5^* = \frac{1}{2\lambda^2 r^2}(1-\lambda) T_1^2\,, \quad I_6^* = \frac{1}{4\lambda^3 r^2}(1-\lambda)^2 T_1^2\,. \tag{5.5.20}$$

Thus, we conclude that the free energy is independent of r and z, and $\mathscr{C}_s^*$ are independent of φ and z. It follows from (1.6.50), (5.5.17) and (5.5.18) that

$$B^{11} = \frac{1+\lambda^3}{\lambda^2}\,, \quad B^{22} = \frac{1+\lambda^3}{\lambda^2 r^2}\,, \quad B^{33} = 2\lambda\,, \quad B^{ij} = 0 \; (i \neq j)\,. \tag{5.5.21}$$

Thus, we obtain

$$\tau^{11} = p + \frac{1}{\lambda}\Phi + \frac{1+\lambda^3}{\lambda^2}\Psi\,, \qquad \tau^{22} = \frac{1}{r^2} p + \frac{1}{\lambda r^2}\Phi + \frac{1+\lambda^3}{\lambda^2 r^2}\Psi\,,$$
$$\tau^{33} = p + \lambda^2 \Phi + 2\lambda \Psi\,, \qquad \tau^{23} = \tau^{31} = \tau^{12} = 0\,. \tag{5.5.22}$$

The equilibrium equations become

$$\frac{\partial \tau^{11}}{\partial r} + \frac{\tau^{11} - r^2 \tau^{22}}{r} = 0\,, \qquad \frac{\partial \tau^{22}}{\partial \varphi} = 0\,, \qquad \frac{\partial \tau^{33}}{\partial z} = 0\,. \tag{5.5.23}$$

These equations are satisfied if and only if

$$p = -p_0 - \frac{1}{\lambda}\Phi - \frac{1+\lambda^3}{\lambda^2}\Psi\,, \tag{5.5.24}$$

where p_0 is an arbitrary constant. From (5.5.22) and (5.5.24) we get

$$\tau^{11} = r^2 \tau^{22} = -p_0\,, \qquad \tau^{23} = \tau^{31} = \tau^{12} = 0\,,$$
$$\tau^{33} = -p_0 + (\lambda^2 - \frac{1}{\lambda})\Phi + \frac{\lambda^3 - 1}{\lambda^2}\Psi\,. \tag{5.5.25}$$

By (5.5.19) and (5.5.4),

$$q^1 = q^3 = 0\,,$$

$$q^2 = \frac{1}{\lambda r^2}\left[\mathscr{C}_1^* + \frac{1}{2\lambda}(1-\lambda)\mathscr{C}_2^* + \frac{1}{4\lambda^2}(1-\lambda)^2\mathscr{C}_3^*\right]T_1\,. \tag{5.5.26}$$

Clearly the equation (5.5.3) is satisfied. Thus, (5.5.16) is a controllable state.

It is known (cf. Truesdell and Noll [354]) that the constitutive equations $(5.5.4)_2$ can be expressed in the form

$$T\|^i = (B_1\delta^i_j + B_2\gamma^i_j + B_3\gamma^i_m\gamma^m_j)q^j\,, \tag{5.5.27}$$

where B_s are prescribed functions of the invariants

$$I_1\,,\quad I_2\,,\quad J_4 = q^i q_i\,,\quad J_5 = q^i q_j \gamma^j_i\,,$$
$$J_6 = q^i q_j \gamma^j_r \gamma^r_i\,,\quad T\,. \tag{5.5.28}$$

In [231], Laws has studied controllable states of thermoelastic materials defined by the constitutive equations $(5.5.4)_1$ and (5.5.27). In what follows we present some results established in [231].

iii) Let us consider the thermomechanical deformation defined by

$$r^2 = 2AX_1 + B\,,\qquad \varphi = CX_2 + DX_3 + G_1\,,$$
$$z = EX_2 + FX_3 + G_2\,,\qquad q_r = \frac{1}{r}K\,,\qquad q_\varphi = q_z = 0\,, \tag{5.5.29}$$

where q_r, q_φ, q_z are the physical components of the heat flux vector $\boldsymbol{q}$, and A, B, C, D, E, F, G_1, G_2 and K are constants which satisfy the relation

$$A(CF - DE) = 1\,. \tag{5.5.30}$$

We choose the curvilinear coordinates θ^i as in (5.5.13). The metric tensors G_{ij} and G^{ij} are given by (5.5.17). It follows from (5.5.29) and (5.5.30) that

$$g_{11} = \frac{r^2}{A^2}\,,\quad g_{22} = A^2(F^2+E^2)\,,\quad g_{33} = A^2(D^2+C^2)\,,$$
$$g_{12} = g_{13} = 0\,,\quad g_{23} = -A^2(FD+EC)\,,\quad g = r^2\,,$$
$$g^{11} = \frac{A^2}{r^2}\,,\qquad g^{22} = D^2 + C^2\,,\qquad g^{33} = F^2 + E^2\,,$$
$$g^{12} = g^{13} = 0\,,\qquad g^{23} = FD + EC\,. \tag{5.5.31}$$

Thus, we obtain

$$I_1 = \frac{A^2}{r^2} + (D^2 + C^2)r^2 + F^2 + E^2 \,,$$
$$I_2 = \frac{r^2}{A^2} + \frac{A^2}{r^2}(F^2 + E^2) + A^2(D^2 + C^2) \,, \quad I_3 = 1 \,. \tag{5.5.32}$$

By (5.5.17), (5.5.31) we find the nonzero components of the tensor γ_i^j,

$$\gamma_1^1 = \frac{1}{2}\left(\frac{A^2}{r^2} - 1\right), \qquad \gamma_2^2 = \frac{1}{2}\left[(D^2 + C^2)r^2 - 1\right],$$
$$\gamma_3^3 = \frac{1}{2}(F^2 + E^2 - 1)\,, \quad r^2\gamma_3^2 = \gamma_2^3 = \frac{1}{2}r^2(FD + EC)\,. \tag{5.5.33}$$

We have $q^1 = q_1 = q_r$, $q^2 = q_2 = q^3 = q_3 = 0$, so that

$$J_4 = \frac{K^2}{r^2}\,, \quad J_5 = \frac{K^2}{2r^2}\left(\frac{A^2}{r^2} - 1\right), \quad J_6 = \frac{K^2}{4r^2}\left(\frac{A^2}{r^2} - 1\right)^2. \tag{5.5.34}$$

It follows from (3.1.34) and (5.5.29) that the equation (5.5.3) is satisfied. From (5.5.27), (5.5.29) and (5.5.32) we find that

$$\frac{\partial T}{\partial r} = \left[B_1 + \frac{1}{2}\left(\frac{A^2}{r^2} - 1\right)B_2 + \frac{1}{4}\left(\frac{A^2}{r^2} - 1\right)^2 B_3\right]\frac{K}{r}\,,$$
$$\frac{\partial T}{\partial \varphi} = \frac{\partial T}{\partial z} = 0\,. \tag{5.5.35}$$

We note that the functions B_s depend only on T and r. Thus the temperature T can be determined from the equation $(5.5.35)_1$. Let us now study the stress tensor. It follows from (1.6.50), (5.5.17) and (5.5.31) that

$$B^{11} = I_1\frac{A^2}{r^2} - \frac{A^4}{r^4}\,, \quad B^{22} = I_1(D^2 + C^2) - r^2(D^2 + C^2) - (FD + EC)^2\,,$$
$$B^{33} = I_1(F^2 + E^2) - (FD + EC)^2 r^2 - (F^2 + E^2)^2\,, \tag{5.5.36}$$
$$B^{12} = B^{13} = 0\,, \quad B^{23} = (FD + EC)\left[I_1 - F^2 - E^2 - r^2(D^2 + C^2)\right].$$

The stress tensor is given by

$$
\begin{aligned}
\tau^{11} =& p+\Phi\frac{A^2}{r^2}+\frac{A^2}{r^2}(I_1-\frac{A^2}{r^2})\Psi\,,\\
\tau^{22} =& \frac{1}{r^2}p+\Phi(D^2+C^2)+\big[I_1(D^2+C^2)-r^2(D^2+C^2)-\\
&-(FD+EC)^2\big]\Psi\,,\\
\tau^{33} =& p+\Phi(F^2+E^2)+\big[I_1(F^2+E^2)-(FD+EC)^2r^2-\\
&-(F^2+E^2)^2\big]\Psi\,,\\
\tau^{23} =& (FD+EC)\{\Phi+\big[I_1-F^2-E^2-r^2(D^2+C^2)\big]\Psi\}\,,\\
\tau^{12} =& \tau^{21}=0\,.
\end{aligned}
\tag{5.5.37}
$$

In view of (5.5.32) and (5.5.35) we conclude that Φ and Ψ depend only on r. The equilibrium equations can be expressed in the form (5.5.23). It follows that the equilibrium equations are satisfied only if the function p is independent of φ and z, and satisfies the equation

$$
\begin{aligned}
\frac{dp}{dr} =& \frac{1}{r}\{\big[r^2(D^2+C^2)-\frac{A^2}{r^2}\big]\Phi+r^2\big[I_1(D^2+C^2)-r^2(D^2+C^2)-\\
&-(FD+EC)^2\big]\Psi-\frac{A^2}{r^2}(I_1-\frac{A^2}{r^2})\Psi\}-A^2\frac{d}{dr}\big[\frac{1}{r}\Phi+\frac{1}{r^2}(I_1-\frac{A^2}{r^2})\Psi\big]\,.
\end{aligned}
$$

Thus, the relations (5.5.29) define a thermomechanical deformation for which there exist the functions T and p so that the equations (5.5.2) and (5.5.3) are satisfied for any $\hat{\psi}$ and B_s.

iv) We now consider the thermomechanical deformation

$$
\begin{aligned}
&r^2=A\varrho^2+B\,, &&\varphi=C\vartheta+DZ+G_1\,,\\
&z=R\vartheta+FZ+G_2\,, &&q_r=\frac{1}{r}K\,,\quad q_\varphi=q_z=0\,,
\end{aligned}
\tag{5.5.38}
$$

Where $(r,\,\varphi,\,z)$ and $(\varrho,\,\vartheta,\,Z)$ are defined by (5.5.13), (5.5.15), and A, B, C, D, G_1, G_2, R, F, K are constants which satisfy the condition (5.5.30).

The metric tensors G_{ij} and G^{ij} are given in (5.5.17). From (5.5.15), (5.5.30) and (5.5.38) we obtain

$$
\begin{aligned}
&g_{11} = \frac{r^2}{A^2\varrho^2}\,, \quad g_{22} = A^2(F^2\varrho^2 + E^2)\,, \quad g_{33} = A^2(D^2\varrho^2 + C^2)\,,\\
&g_{12} = g_{13} = 0\,, \quad g_{23} = -A^2(DF\varrho^2 + EC)\,, \quad g = r^2\,,\\
&g^{11} = \frac{A^2\varrho^2}{r^2}\,, \quad g^{22} = \frac{1}{\varrho^2}(D^2\varrho^2 + C^2)\,, \quad g^{33} = \frac{1}{\varrho^2}(F^2\varrho^2 + E^2)\,,\\
&g^{12} = g^{13} = 0\,, \quad g^{23} = \frac{1}{\varrho^2}(DF\varrho^2 + EC)\,.
\end{aligned}
\tag{5.5.39}
$$

Thus, we have

$$
\begin{aligned}
&I_1 = \frac{A^2\varrho^2}{r^2} + \frac{1}{\varrho^2}(C^2r^2 + E^2) + D^2r^2 + F^2\,,\\
&I_2 = \frac{r^2}{A^2\varrho^2} + \frac{A^2}{r^2}(F^2\varrho^2 + E^2) + A^2(D^2\varrho^2 + C^2)\,, \quad I_3 = 1\,,\\
&B^{11} = \frac{A^2}{r^2}[C^2r^2 + E^2 + \varrho^2(D^2r^2 + F^2)]\,,\\
&B^{22} = \frac{1}{A^2\varrho^2} + \frac{A^2}{r^2}(D^2\varrho^2 + C^2)\,, \quad B^{33} = \frac{A^2}{r^2}(F^2\varrho^2 + E^2) + \frac{r^2}{A^2\varrho^2}\,,\\
&B^{23} = \frac{A^2}{r^2}(DF\varrho^2 + EC)\,, \quad B^{12} = B^{13} = 0\,.
\end{aligned}
\tag{5.5.40}
$$

The nonzero components of the strain tensor γ^i_j are given by

$$
\begin{aligned}
&\gamma^1_1 = \frac{1}{2}\Big(\frac{A^2\varrho^2}{r^2} - 1\Big)\,, \qquad \gamma^2_2 = \frac{1}{2}\Big(D^2r^2 + \frac{r^2}{\varrho^2}C^2 - 1\Big)\,,\\
&\gamma^3_3 = \frac{1}{2}\Big(F^2 + \frac{1}{\varrho^2}E^2 - 1\Big)\,, \quad \gamma^2_3 = \frac{1}{r^2}\gamma^3_2 = \frac{1}{2\varrho^2}(DF\varrho^2 + EC)\,.
\end{aligned}
\tag{5.5.41}
$$

Clearly, the equation (5.5.3) is identically satisfied. The invariants J_4, J_5 and J_6 are given by

$$
J_4 = \frac{1}{r^2}K^2\,, \quad J_5 = \frac{K^2}{2r^2}\Big(\frac{A^2\varrho^2}{r^2} - 1\Big)^2\,, \quad J_6 = \frac{K^2}{4r^2}\Big(\frac{A^2\varrho^2}{r^2} - 1\Big)\,. \tag{5.5.42}
$$

From (5.5.27) we get

$$\begin{aligned}\frac{\partial T}{\partial r} &= [B_1 + \frac{1}{2}(\frac{A^2\varrho^2}{r^2} - 1)B_2 + \frac{1}{4}(\frac{A^2\varrho^2}{r^2} - 1)^2 B_3]\frac{K}{r}\,, \\ \frac{\partial T}{\partial \varphi} &= \frac{\partial T}{\partial z} = 0\,.\end{aligned} \tag{5.5.43}$$

The functions T and B_i are independent of φ and z. The function T can be determined from the equation $(5.5.43)_1$. The components of the stress tensor are given by

$$\begin{aligned}\tau^{11} &= \frac{A^2\varrho^2}{r^2}\Phi + \frac{A^2}{r^2}[C^2r^2 + E^2 + \varrho^2(D^2r^2 + F^2)]\Psi + p\,, \\ \tau^{22} &= \frac{1}{\varrho^2}(D^2\varrho^2 + C^2)\Phi + [\frac{A^2}{r^2}(D^2\varrho^2 + C^2) + \frac{1}{A^2\varrho^2}]\Psi + \frac{1}{r^2}p\,, \\ \tau^{33} &= \frac{1}{\varrho^2}(F^2\varrho^2 + E^2)\Phi + [\frac{A^2}{r^2}(F^2\varrho^2 + E^2) + \frac{r^2}{A^2\varrho^2}]\Psi + p\,, \\ \tau^{23} &= \frac{1}{\varrho^2}(DF\varrho^2 + EC)\Phi + \frac{A^2}{r^2}(DF\varrho^2 + EC)\Psi\,, \\ \tau^{13} &= \tau^{12} = 0\,.\end{aligned} \tag{5.5.44}$$

It follows from (5.5.44) that

$$\begin{aligned}r^2\tau^{22} &= \tau^{11} + [\frac{r^2}{\varrho^2}(D^2\varrho^2 + C^2) - \frac{A^2\varrho^2}{r^2}]\Phi + [\frac{r^2}{A^2\varrho^2} - \frac{A^2}{r^2}(E^2 + F^2\varrho^2)]\Psi\,, \\ \tau^{33} &= \tau^{11} + [\frac{1}{\varrho^2}(F^2\varrho^2 + E^2) - \frac{A^2\varrho^2}{r^2}]\Phi + [\frac{r^2}{A^2\varrho^2} - A^2(D^2\varrho^2 + C^2)]\Psi\,, \\ \tau^{23} &= \frac{1}{\varrho^2}(DF\varrho^2 + EC)(\Phi + \frac{A^2\varrho^2}{r^2}\Psi)\,, \\ \tau^{13} &= \tau^{12} = 0\,.\end{aligned} \tag{5.5.45}$$

Clearly, the functions Φ and Ψ depend only on the variable r. The equilibrium equations are given by (5.5.23). From (5.5.23) we deduce that p depends only on the variable r. In view of the fact that $\tau^{11} - r^2\tau^{22}$ does not depend on p, the first equation from (5.5.23) determines the function τ^{11} in the form

$$\tau^{11} = \int_{r_0}^{r} (r^2\tau^{22} - \tau^{11})r^{-1}\, dr + \tau_0^{11},$$

where $r^2\tau^{22} - \tau^{11}$ is given by $(5.5.45)_1$.

Other controllable states are presented in [162], [231], [298].

5.6 Bibliographical Notes

In the Sections 5.1 – 5.5 an account is given for some of the developments in nonlinear thermoelasticity. In the following, an attempt is made to give a brief review of other problems of nonlinear thermoelasticity investigated by various authors. The equations of the dynamic theory of nonlinear thermoelasticity have been intensively studied in recent years. The first results in one-dimensional nonlinear dynamic theory were presented by Slemrod [327] who established global existence, uniqueness and asymptotic stability of classical smooth solutions for bounded domains and special boundary conditions. The Cauchy problem was studied by Zheng and Shen [366] and Hrusa and Tarabek [170]. Global smooth solutions and asymptotic stability in one dimension have been investigated by Racke and Shibata [308] and Jiang [206]. The development of singularities for large data was studied by Dafermos and Hsiao [95] and Hrusa and Messaoudi [171]. The problem of periodic solutions in the one-dimensional theory has been considered by Feireisl [127].

The Cauchy problem in the three-dimensional theory of nonlinear thermoelasticity has been investigated by Racke [305], [306]. Jiang and Racke [208] and Chrzeszczyk [74] have established local existence theorems. The asymptotic behavior of solutions was investigated by Jiang [207], [209] both for one-dimensional and three-dimensional theory. Numerical methods in the one-dimensional theory of thermoelasticity have been presented by Jiang [205], and De Moura and Suaiden [258]. For an extensive review of the literature on mathematical aspects in the dynamic theory of nonlinear thermoelasticity the reader is referred to the work of Racke [309].

The method of successive approximation originated by Signorini [324] (see also Wang and Truesdell [361], p.476) was generalized in [174] to nonlinear thermoelasticity. In the context of nonlinear elastostatics, when the applied tractions and body forces both contain a multiplying parameter, Stopelli [337] has given a proof of existence and uniqueness of solution and has shown that the displacement can be expanded as an absolutely convergent power series with non-zero radius of convergence, provided the parameter is sufficiently small (see also Truesdell and Noll [354], p.138, Wang and Truesdell [361], p.494). In [274], Nistor has extended Stopelli's results to the nonlinear thermoelastostatics.

The nonlinear theory of thermoelastic shells, plates and rods has been studied in various papers (see, e.g. Naghdi [266], p.561, Antman [5], p.656 and the literature cited therein). A comprehensive bibliography on thermal stresses in shells has been presented by Keene and Hetnarski [218]. The reader interested in the problem of thermoelastic stability will find a full account in the monograph of Knops and Wilkes [223], pp.187, 249.

References

[1] ACHENBACH, J.D. *Wave Propagation in Elastic Solids.* North-Holland Publishing Co., Amsterdam, 1973.

[2] ADAMS, R.A. *Sobolev Spaces.* Press, New York, 1975.

[3] AMENDOLA, G. and MANACORDA, T. Piccole deformazioni termoelastiche sovraposte ad una deformazione termostatica finita. *Riv. Mat. Univ. Parma* (4), **5** (1979), 487-501.

[4] AMES, K. and STRAUGHAN, B. Continuous dependence results for initially prestressed thermoelastic bodies. *Int. J. Engng. Sci.*, **30** (1992), 7-13.

[5] ANTMAN, S.S. *The Theory of Rods.* In vol. VI a/2 of Handbuch der Physik (Edited by C. Truesdell), Springer-Verlag, Berlin-Heidelberg-New York, 1972.

[6] ANWAR, M.N. and SHERIF H.H. State space approach to generalized thermoelasticity. *J. Thermal Stresses*, **11** (1988), 353-365.

[7] ASLANYAN, E.G. and BONDAREVA, V.F. Calculation of the thermoelastic state of cylinders of opposed-piston engines. *Strenght Mater.*, **16** (1984), 405-410.

[8] AZARKHIN, A. and BARBER, J.R. Transient thermoelastic contact problem of two sliding half-planes. *Wear*, **12** (1985), 1-13.

[9] BAHAR, L.Y. and HETNARSKI, R.B. State space approach to thermoelasticity. *J. Thermal Stresses*, **1** (1978), 135-145.

[10] BAHAR, L.Y. and HETNARSKI, R.B. Direct approach to thermoelasticity. *J. Thermal Stresses*, 2 (1979), 135-147.

[11] BAHAR, L.Y. and HETNARSKI, R.B. Coupled thermoelasticity of a layered medium. *J. Thermal Stresses*, **3** (1980), 141-152.

[12] BALL, J.M. Convexity conditions and existence theorem in non-linear elasticity. *Arch. Rational Mech. Anal.*, **63** (1977), 337-403.

[13] BALL, J.M. and KNOWLES, G. Lyapunov functions for thermomechanics with spatially varying boundary temperatures. *Arch. Rational Mech. Anal.*, **92** (1986), 193-204.

[14] BARBER, J.R. Contact problems involving a cooled punch. *J. Elasticity*, **8** (1978), 409-423.

[15] BARBER, J.R. Some implications of Dundurs' theorem for thermoelastic contact and crack problems *J. Mech. Eng. Sci.*, **22** (1980), 229-232.

[16] BARBER, J.R., DUNDURS, J., COMNINOU, M. Stability considerations in thermoelastic contact. *J. Appl. Mech.*, **47** (1980), 871-874.

[17] BARBER, J.R. The distortion of a cylinder with non-uniform axial heat conduction. *J. Elasticity*, **13** (1983), 441-458.

[18] BARBER, J.R. and ZHANG, R. Transient behavior and stability for the thermoelastic contact of two rods of dissimilar materials. *Int. J. Mech. Sci.,* **30** (1988), 691-704.

[19] BARGMANN, H.W. Recent developments in the field of thermally induced waves and vibrations. *Nuclear Eng. Design*, **27** (1974), 372-385.

[20] BARGMANN, H.W. On inertia effects in thermoelasticity and thermal shock parameters. *J. Thermal Stresses*, **15** (1992), 229-240.

[21] BEJU, I. A uniqueness theorem in thermoelasticity with thermal relaxation. *Int. J. Engng. Sci.,* **30** (1992), 657-664.

[22] BENTHIEN, G. and GURTIN, M.E. A principle of minimum transformed energy in elastodynamics. *J. Appl. Mech.,* **31** (1970), 1147-1149.

[23] BIOT, M.A. Thermoelasticity and irreversible thermodynamics. *J. Appl. Mech.,* **27** (1956), 240-253.

[24] BIOT, M.A. *Mechanics of Incremental Deformations.* New York, Wiley, 1965.

[25] BLEVINS, R.D. Elastic thermal stresses in bimetallic pipe welds. *J. Pressure Vessel Tech.,* **102** (1980), 430-432.

[26] BLOCKI, J. Axisymmetric problems of thermoelasticity by the BEM. *Z. Angew. Math. Mech.,* **72** (1992), T 471-T 473.

[27] BLOOM, F. On continuous dependence for a class of dynamical problems in linear thermoelasticity. *ZAMP,* **26** (1975), 569-579.

[28] BOLEY, B.A. and WEINER, J.H. *Theory of Thermal Stresses.* John Wiley, New York, 1960.

[29] BOLEY, B.A. and TOLINS, I.S. Transient coupled thermoelastic boundary-value problems in the half space. *J. Appl. Mech.*, **29** (1962), 637-646.

[30] BOLEY, B.A. and HETNARSKY, R.B. Propagation of discontinuities in coupled thermoelastic problems. *J. Appl. Mech.*, ASME, **35** (1968), 489-494.

[31] BOLEY, B.A. Survey of recent developments in the field of heat conduction in solids and thermoelasticity. *Nucl. Eng. Des.,* **3** (1972), 377-399.

[32] BOLEY, B.A. Thermal stresses: A survey. In *"Thermal Stresses in Severe Environments"* (Edited by D.P.H. Hasselman and R.A. Heller), Plenum Publ. Corp., New York, (1980), 1-11.

[33] BORS, C.I. *Teoria elasticitatii corpurilor anizotrope.* Ed. Acad., Bucuresti, 1970.

[34] BOYD, D.E. and KISHORE, B.R. Thermal stresses in axially loaded orthotropic cylinders. *AIAA J,* **6** (1968), 980-982.

[35] BROWDER, F.E. Strongly elliptic system of differential equations. *Ann. Math. Studies,* **33** (1954), 15-51.

[36] BRULIN, O. Grade consistent micropolar theory of thermoelasticity. New Problems in Mechanics of Continua. *Proc. 3rd. Swed.-Pol. Symp. Mech., Jablonna, (1982), SM Stud. Ser.,* **17** (1983), 265-272.

[37] BRUN, L. Sur l'unicité en thermoélasticité dynamique et diverses expressions analogues à la formule de Clapeyron. C.R. *Acad. Sci. Paris,* **261** (1965), 2584-2587.

[38] BRUN, L. Méthodes énergétiques dans les systèmes évolutifs linéaires. Premier Partie: Séparation des énergies. Deuxième Partie: Thèorémes d'unicité. *J. Mécanique,* **8** (1969), 125-166.

[39] BRUN, L. L'onde simple thermoelastique linéaire. *J. Mécanique,* **14** (1975), 863-872.

[40] BRUN, L. et MOLINARI, A. Un algorithme de calcul en thermoélasticité dynamique. *ZAMP,* **28** (1977), 123-139.

[41] BRUN, L. and POTIER-FERRY, M. Mechanics of elastic solids. Thermodynamics and stability of the elastic solid. *C.R. Acad Sci. Paris,* **296** (1983), 401-404.

[42] BUI, H.D., EHRALCHER, A. and NGUYEN, Q.S. Propagation of a crack in dynamical thermoelasticity (in French). *J. Mécanique,* **19** (1980), 697-723.

[43] BURCHULADZE, T.V. Three-dimensional dynamical problems of non-classical thermoelasticity theory. *Tr. Tbilis. Mat. Inst. Razmadze,* **75** (1984), 3-21.

[44] CARBONARO, B. and IGNACZAK, J. Some theorems in temperature-rate dependent thermoelasticity for unbounded domain. *J. Thermal Stresses,* **10** (1987), 193-220.

[45] CARICATO, G. Il teorema di Menabrea per transformazioni non isoterme di un corpo elastico vincolato anisotropo e non omogeneo, con stress iniziale. *Atti Accad. Naz. Lincei, Rend., Cl. Sci. Fis. Mat. Natur. VIII,* **44** (1968), I: 191-200, II: 363-369.

[46] CARLSON, D.E. *Linear Thermoelasticity.* pp. 297-346 of Flügge's Handbuch der Physik, vol. VI a/2 (Edited by C. Truesdell), Springer-Verlag, Berlin-Heidelberg-New York, 1972.

[47] CARSLAW, H.S. and JAEGER, J.C. *Conduction of Heat in Solids,* Clarendon Press, Oxford, 1948.

[48] CAVIGLIA, G., MORRO, A. and STRAUGHAN, B. Thermoelasticity at cryogenic temperatures. *Int. J. Non-linear Mech.,* **27** (1992), 251-264.

[49] CEKIRGE, H.M. and SUHUBI, E.S. Propagation of a plane wave in an initially stressed thermoelastic medium. *Bull. Techn. Univ. Istanbul,* **27** (1974), 98-107.

[50] CHADWICK, P. *Thermoelasticity. The Dynamical Theory.* In *Progress in Solid Mechanics.* (Edited by I.N. Sneddon and R. Hill), vol. I, pp. 263-328, North Holland Pubbl. Co., Amsterdam, 1960.

[51] CHADWICK, P. and POWDRILL, B. Singular surfaces in linear thermoelasticity. *Int. J. Engng. Sci.,* **3** (1965), 561-595.

[52] CHADWICK, P. and SEET, L.T.C. Second-order thermoelasticity theory for isotropic and transversely isotropic materials. *Trends in Elasticity and Thermoelasticity.* Witold Nowacki Anniversary Volume, Wolters-Noordhoff Pubbl., Groningen, 1971, 29-57.

[53] CHADWICK, P. and OGDEN, R.W. On the definition of elastic moduli. *Arch. Rational Mech. Anal.,* **44** (1971), 41-53.

[54] CHADWICK, P. and CURRIE, P.K. The propagation and growth of acceleration waves in heat-conducting elastic materials. *Arch. Rational Mech. Anal.,* **49** (1972), 137-158.

[55] CHADWICK, P. Interchange of modal properties in the propagation of harmonic waves in heat-conducting elastic materials. *Bull. Austr. Math. Soc.,* **8** (1973), 75-92.

[56] CHADWICK, P. and CURRIE, P.K. On the propagation of generalized transverse waves in heat-conducting elastic materials. *J. Elasticity,* **4** (1974), 301-315.

[57] CHADWICK, P. and CREASY, C.F.M. Weak nonlinear waves in homogeneously deformed heat-conducting elastic materials. *Quart. J. Mech. Appl. Math.,* **32** (1979), 419-436.

[58] CHADWICK, P. Basic properties of plane harmonic waves in a prestressed heat-conducting elastic materials. *J. Therm. Stresses,* **2** (1979), 193-214.

[59] CHANDRASEKHARAIAH, D.S. and SRIKANTAIAH, K.R. Thermoelastic plane waves in a rotating solid. *Acta Mechanica,* **50** (1984), 211-220.

[60] CHANDRASEKHARAIAH, D.S. Thermoelasticity with second sound: A review. *Appl. Mech. Rev.,* **39** (1986), 355-376.

[61] CHANDRASEKHARAIAH, D.S. A uniqueness theorem in heat-flux dependent thermoelasticity. *J. Elasticity,* **18** (1987), 283-287.

[62] CHANDRASEKHARAIAH, D.S. and KESHAVAN, H.R. Axisymmetric thermoelastic interactions in an unbounded body with cylindrical cavity. *Acta Mechanica,* **92** (1992), 61-76.

[63] CHAUDHRY, H.R. A note on second-order effects in plane strain thermoelasticity. *Int. J. Engng. Sci.* **9** (1971), 673-679.

[64] CHEN, P.J. *Growth and Decay of Waves in Solids.* pp. 303-400 of Flügge's Handbuch der Physik, vol. VI a/3 (Edited by C. Truesdell) Springer-Verlag, Berlin-Heidelberg-New York, 1973.

[65] CHEN, D.D. and PETROSKI, H.J. Controllable states of elastic heat conductors obeying a Fourier law. *Int. J. Engng. Sci.* **13** (1975), 799-814.

[66] CHEN, T.C. and WENG, C.I. Coupled transient thermoelastic response in an axisymmetric circular cylinder by Laplace transform-finite element method. *Comput. Struct.,* **33** (1989), 533-542.

[67] CHIRITA, S. Thermal stresses in anisotropic cylinders. *Acta Mechanica,* **23** (1975), 301-306.

[68] CHIRITA, S. Uniqueness and continuous dependence results for the incremental thermoelasticity. *J. Therm. Stresses,* **5** (1982), 161-172.

[69] CHIRITA, S. Uniqueness and continuous data dependence in dynamical problems of nonlinear thermoelasticity. *J. Therm. Stresses,* **5** (1982), 331-346.

[70] CHIRITA, S. and RUSU, GH. On existence and uniqueness in thermoelasticity of non-simple materials. *Rev. Roum. Sci. Techn., Mec. Appl.,* **30** (1985), 21-30.

[71] CHIRIŢA, S. Hölder stability and uniqueness results for linear thermoelasticity on unbounded domains. *Bollettino U.M.I.,* (7) **1**-B (1987), 387-407.

[72] CHOU, S.I. and WANG, C.C. Estimates of error in finite element approximate solutions to problems in linear thermoelasticity. *Arch. Rational Mech. Anal.,* **77** (1981), 263-300.

[73] CHRISTENSEN, R.M. *Mechanics of Composite Materials.* J. Wiley, New York, Interscience, 1979.

[74] CHRZESZCZYK, A. Some existence results in dynamical thermoelasticity. I. Nonlinear case. *Arch. Mech.,* **39** (1987), 605-617.

[75] CHRZESZCZYK, A. Some existence results in dynamical thermoelasticity. Part II. Linear case. *Arch. Mech.,* **39** (1987), 619-632.

[76] CHUN, P.L. Thermal-stresses-induced birefringence in single-mode elliptical optical fibre. *Electron. Lett.,* **18** (1982), 45-46.

[77] CIARLETTA, M. and SCALIA, A. Thermoelasticity of nonsimple materials with voids. *Rev. Roum. Sci. Techn. Mec. Appl.,* **35** (1990), 115-125.

[78] CIARLETTA, M. and SCALIA, A. On the nonlinear theory of nonsimple thermoelastic materials with voids. *ZAMM.,* **73** (1993), 67-75.

[79] CIARLETTA, M. A solution of Galerkin type in the theory of thermoelastic materials with voids. *J. Thermal Stresses,* **14** (1991), 409-419.

[80] CLAUSING, A.M. Heat transfer at the interface between dissimilar metals. The influence of thermal strain. *Int. J. Heat Mass Transfer.* **9** (1966), 791-801.

[81] CLEMENTS, D.L. A thermoelastic problem for a crack between dissimilar anisotropic media. *Int. J. Solids Struct.,* **19** (1983), 121-130.

[82] CLEMENTS, D.L. and ROGERS, C. A boundary integral equation for the solution of a class of problems in anisotropic inhomogeneous thermostatics and elastostatics. *Quart. Appl. Math.,* **41** (1983), 99-106.

[83] COLEMAN, B.D. and NOLL, W. The thermodynamics of elastic materials with heat conduction and viscosity. *Arch. Rational Mech. Anal.* **13** (1963), 167-178.

[84] COLEMAN, B.D. and MIZEL, V.J. Existence of caloric equations of state in thermodynamics. *J. Chem. Phys.* **40** (1964), 1116-1125.

[85] COLEMAN, B.D. and GURTIN, M.E. Waves in materials with memory IV. Thermodynamics and the velocity of general acceleration waves. *Arch. Rational Mech. Anal.* **19** (1965), 317-338.

[86] COLEMAN, B.D. and DILL, E.H. Thermodynamic restrictions on the constitutive equations of electromagnetic theory. *ZAMP,* **22** (1971), 691-702.

[87] COMNINOU, M. and DUNDURS, J. On the Barber boundary conditions for thermoelastic contact. *J. Appl. Mech.* **46** (1979), 849-853.

[88] COMNINOU, M. and BARBER, J.R. The thermoelastic Hertz problem with pressure dependent contact resistance. *Int. J. Mech. Sci.,* **26** (1984), 549-554.

[89] CRISTESCU, N. and SULICIU, I. *Viscoplasticitate.* Ed. Tehnica, Bucuresti, 1976.

[90] ČUKIČ, R. Transversal vibrations of the thin shell of revolution produced by the thermal shock. *Bull. Acad. Polon. Sci. Ser. Sci. Techn.,* **20** (1972), 401-411.

[91] DAFERMOS, C.M. On the existence and the asymptotic stability of solutions to the equations of linear thermoelasticity. *Arch. Rational Mech. Anal.* **29** (1968), 241-271.

[92] DAFERMOS, C.M. The second law of thermodynamics and stability. *Arch. Rational Mech. Anal.* **70** (1979), 167-179.

[93] DAFERMOS, C.M. Stability of motions of thermoelastic fluids. *J. Thermal Stresses,* **2** (1979), 127-134.

[94] DAFERMOS, C.M. and HRUSA, W.J. Energy methods for quasilinear hyperbolic initial boundary value problems. Applications to elastodynamics. *Arch. Rational Mech. Anal.* **87** (1985), 267-292.

[95] DAFERMOS, C.M. and HSIAO, L. Development of singularities in solutions of the equations of nonlinear thermoelasticity. *Quart. Appl. Math.,* **44** (1986), 463-474.

[96] DARGUSH, G.F. and BANERJEE, P.K. Time dependent axisymmetric thermoelastic boundary element analysis. *Int. J. Numer. Meth. Eng.,* **33** (1992), 695-718.

[97] DANILOVSKAYA, V.Y. Thermal stresses in an elastic half space due to a sudden heating of its boundary (in Russian). *Prikl. Mat. Mekh.,* **14** (1950), 316-318.

[98] DAS, N.C., DAS, S.N. and DAS, B. Eigenvalue approach to thermoelasticity. *J. Thermal Stresses,* **6** (1983), 73-92.

[99] DASSIOS, G. and GRILLAKIS, M. Dissipation rates and partition of energy in thermoelasticity. *Arch. Rational Mech. Anal.* **87** (1984), 49-91.

[100] DAY, W.A. Extensions of a property of the heat equation to linear thermoelasticity and other theories. *Quart. Appl. Math.,* **40** (1982), 319-330.

[101] DAY, W.A. Initial sensitivity to the boundary in coupled thermoelasticity *Arch. Rational Mech. Anal.,* **87** (1985), 253-266.

[102] DAY, W.A. *Heat Conduction Within Linear Thermoelasticity.* Springer Tracts in Natural Philosophy. vol. 30, Springer-Verlag, New York, 1985.

[103] DAY, W.A. On nonclassical behaviour of the temperature in coupled and dynamic thermoelasticity. *J. Elasticity,* **17** (1987), 271-278.

[104] DAY, W.A. An asymptotic formula for the maximum work done by a thermoelastic body. *Quart. J. Mech. Appl. Math.,* **44** (1991), 357-368.

[105] DAY, W.A. A theorem about work in dynamic linear thermoelasticity. *J. Mech. Appl. Math.,* **44** (1991), 35-44.

[106] DERESIEWICZ, H. Solution of the equations of thermoelasticity. *Proc. 3rd. U.S. National Congr. Appl. Mech. Brown University,* (1958), 287-291.

[107] DHALIWAL, R.J. and SINGH, A. *Dynamic Coupled Thermoelasticity.* Hindustan Publ. Corp., Delhi, India, 1980.

[108] DORNINGER, K. and RAMMERSTORFER, F.G. A layered composite shell element for elastic and thermoelastic stress and stability analysis at large deformations. *Int. J. Numer. Meth. Eng.,* **30** (1990), 833-858.

[109] DOW, T.A. and BURTON, R.A. Thermoelastic instability of sliding contact in the absence of wear. *Wear,* **19** (1972), 315-329.

[110] DRAGOS, L. Fundamental solutions in thermoelasticity. *Acta Mechanica,* **32** (1979), 197-203.

[111] DRAGOS, L. *Principles of Mechanics of Continua.* (in Romanian). Ed. Tehnica, Bucuresti, 1983.

[112] DUNDURS, J. Distorsion of a body caused by free thermal expansion. *Mech. Res. Comm.,* **1** (1974), 121-124.

[113] DUNDURS, J. and COMNINOU, M. Green functions for planar thermoelastic contact problems. Exterior contact. *Mechanics,* **6** (1979), 309-316.

[114] DUNDURS, J. Green functions for planar thermoelastic contact problems. Interior contact. *Mechanics,* **6** (1979), 317.

[115] DUNWOODY, J. On weak shock waves in thermoelastic solids. *Quart. J. Mech. Appl. Math.,* **30** (1977), 203-208.

[116] DUVAUT, G. et LIONS, J.L. *Les Inéquations en Mécanique et Physique. Dunod, Paris,* 1972.

[117] DUVAUT, G. Free boundary problem connected with thermoelasticity and unilateral contact. In *"Free Boundary Problems."* vol. II, Pavia, 1979.

[118] EASON, G., NOBLE, B. and SNEDDON, I.N. On certain integrals of Lipschitz-Hankel involving products of Bessel functions. *Phil. Trans. Roy. Ser. A.,* **247** (1955), 529-551.

[119] EASON, G. Thermal stresses in anisotropic cylinders. *Proc. Edinburg Math. Soc.,* **13** (1962), 159-164.

[120] ENGLAND, A.H. and GREEN, A.E. Steady-state thermoelasticity for initially stressed bodies. *Phil. Trans. Roy. Soc. London, Ser. A* **253** (1961), 517-542.

[121] ERICKSEN, J.L. Deformations possible in every isotropic incompressible, perfectly elastic body. *ZAMP*, **5** (1954), 466-489.

[122] ERICKSEN, J.L. *Introduction to the Thermodynamics of Solids.* Chapman & Hall, London, 1991.

[123] ERINGEN, A.C. *Mechanics of Continua.* John Wiley and Sons, New York, 1967.

[124] ERINGEN, A.C. and SUHUBI, E.S. *Elastodynamics.* vol. I, New York, Academic Press, 1974.

[125] ERINGEN, A.C. Theory of thermo-microstretch elastic solids. *Int. J. Engng. Sci.*, **28** (1990), 1291-1301.

[126] FABRIZIO, M. and MORRO, A. *Mathematical Problems in Linear Viscoelasticity.* SIAM Studies in Applied Mathematics, vol. 12, SIAM, Philadelphia, 1992.

[127] FEIREISL, E. Forced vibrations in one-dimensional nonlinear thermoelasticity as a local coercive-like problem. *Comment. Math. Univ. Carolinae,* **31** (1990), 243-255.

[128] FICHERA, G. *Existence Theorems in Elasticity,* pp. 347-388 of Flügge's Handbuch der Physik, vol. VI a/2 (Edited by C. Truesdell) Springer-Verlag, Berlin-Heidelberg-New York, 1972.

[129] FICHERA, G. *Problemi Analitici Nuovi nella Fisica Matematica Classica.* Quaderni del Consiglio Nazionale delle Ricerche, Gruppo Nazionale di Fisica Matematica, Scuola Tipo-Lito "Istituto Anselmi", Napoli, 1985.

[130] FLAVIN, J.N. and GREEN, A.E. Plane thermoelastic waves in an initially stressed medium. *J. Mech. Phys. Solids,* **9** (1961), 179.

[131] FLAVIN, J.N. Thermoelastic Rayleigh waves in a prestressed medium. *Proc. Cambridge Phil. Soc.*, **58** (1962), 532-538.

[132] FLAVIN, J.N. Surface waves in prestressed Mooney material. *Quart. J. Mech. Appl. Math.*, **16** (1963), 441-449.

[133] FLORENCE, A.L. and GOODIER, J.N. The linear thermoelastic problem of uniform heat flow disturbed by a penny-shaped insulated crack. *Int. J. Engng. Sci.*, **1** (1963), 533-540.

[134] FRANCHI, F. Wave propagation in heat conducting dielectric solids with thermal relaxation and temperature dependent electric permittivity. *Riv. Mat. Univ. Parma, N. Ser.* **11** (1985), 443-461.

[135] FRANCIS, P.H. Thermo-mechanical effects in elastic wave propagation. A survey. *J. Sound Vibration,* **21** (1972), 181-192.

[136] FRESSENGEASC, C. and MOLINARI, A. Transient stress induced by thermal shock. *J. Thermal Stresses,* **3** (1980), 379-390.

[137] FRYDRYCHOWICZ, W. and SINGH, M.C. Analysis of a moving heated punch over the surface of a thermoelastic half-space. *J. Thermal Stresses,* **5** (1982), 13-51.

[138] FU, W.S. Thermal stresses in an elastic solid weakened by two coplanar circular cracks. *Int. J. Engng. Sci.,* **11** (1973), 317-330.

[139] GALDI, G. and RIONERO, S. *Weighted Energy Methods in Fluid Dynamics and Elasticity.* Lecture Notes in Mathematics, 1134, Springer, Berlin-Heidelberg-New York, 1985.

[140] GALKA, A. Singular solutions of thermoelasticity. *Bull. Acad. Polon. Sci. Sér. Sci. Techn.,* **13** (1965), 523-529.

[141] GAMBY, D. and BOURDILLON, H. Thermal stresses in asymmetrically heated thick cylinder shells. *J. Thermal Stresses,* **6** (1983), 189-196.

[142] GATEWOOD, B.E. *Thermal Stresses.* Mc Graw-Hill, New York, Toronto, London, 1957.

[143] GAWINECKI, J. $L^p - L^q$- time decay estimate for solution of the Cauchy problem for hyperbolic partial differential equations of linear thermoelasticity. *Ann. Pol. Math.* **2** (1991), 135-145.

[144] GERMAIN, P. *Cours de Mécanique des Milieux Continus.* Masson et Cie., Paris, 1973.

[145] GLADWELL, G.M.L., BARBER, J.R. and OLESIAK, Z. Thermal problems with radiation boundary conditions. *Quart. J. Mech. Appl. Math.,* **36** (1983), 387-402.

[146] GOLDSTEIN, J.A. *Semigroups of Linear Operators and Applications.* Oxford University Press, New York, 1985.

[147] GRAFFI, D. Sui teoremi di reciprocitá nei fenomeni non stazionari. *Atti Accad. Sc. Ist. Bologna,* **11** (1963), 33-40.

[148] GREEN, A.E. and ADKINS, J.E. *Large Elastic Deformations.* Oxford, Clarendon Press, 1960.

[149] GREEN, A.E. Thermoelastic stresses in initially stressed bodies. *Proc. Roy. Soc. London,* Ser. A, **266** (1962), 1-19.

[150] GREEN, A.E. and RIVLIN, R.S. Multipolar continuum mechanics. *Arch. Rational Mech. Anal.* **171** (1964), 113-147.

[151] GREEN, A.E. and RIVLIN, R.S. Multipolar continuum mechanics: functional theory. *Proc. Roy. Soc. London,* A **284** (1965), 303-324.

[152] GREEN, A.E. and ZERNA, W. *Theoretical Elasticity*, 2nd Edition, Clarendon Press, Oxford, 1968.

[153] GREEN, A.E. and NAGHDI, P.M. On undamped heat waves in an elastic solid. *J. Thermal Stresses,* **15** (1992), 253-264.

[154] GRIMADO, P.B. On the thermal stress induced in a semi-infinite body as a result of a suddenly applied step heat input. *J. Appl. Mech.* **37** (1970), 1149-1150.

[155] GRIMADO, P.B. Interlaminar thermoelastic stresses in layered beams. *J. Thermal Stresses,* **1** (1978), 75-86.

[156] GRINDEI, I. *Termoelasticitate.* Ed. Did. Pedag., Bucuresti, 1967.

[157] GURTIN, M.E. Variational principles for linear elastodynamics. *Arch. Rational Mech. Anal.* **16** (1964), 34-50.

[158] GURTIN, M.E. *The Linear Theory of Elasticity,* pp. 1-295 of Flügge's Handbuch der Physik, vol. VI a/2 (Edited by C. Truesdell), Springer-Verlag, Berlin-Heidelberg-New York, 1972.

[159] GURTIN, M.E. and PODIO-GUIDUGLI, P. The thermodynamics of constrained materials. *Arch. Rational Mech. Anal.* **51** (1973), 192-208.

[160] GURTIN, M.E. On a path-independent integral for thermoelasticity. *Int. J. Fracture,* **15** (1979), 169-170.

[161] HASSELMAN, D.P.H. and HELLER, R.A. (Eds.) *Thermal Stresses in Severe Environments.* Plenum Press, New York, 1980.

[162] HAYES, M., LAWS, N. and OSBORN, R.B. The paucity of universal motions in thermoelasticity. *Q. Appl. Math.,* **27** (1969), 416-420.

[163] HERMANN, G. On second-order thermoelastic effects. *ZAMP,* **15** (1964), 253-262.

[164] HETNARSKI, R.B. Coupled one-dimensional thermal shock problem for small times. *Arch. Mech. Stosow,* **13** (1961), 296-306.

[165] HETNARSKI, R.B. The fundamental solution of the coupled thermoelastic problem for small times. *Arch. Mech. Stosow,* **16** (1964), 23-31.

[166] HETNARSKI, R.B. Solution of the coupled problems of thermoelasticity in the form of a series of functions. *Arch. Mech. Stosow,* **16** (1964), 919-941.

[167] HETNARSKI, R.B. An algorithm for generating some inverse Laplace transforms of exponential form. *ZAMP,* **26** (1975), 249-253.

[168] HLAVÀČEK, I. and HLAVÀČEK, M. On the existence and uniqueness of solution and some variational principles in linear theories of elasticity with couple-stresses. *Aplikace Matematiky,* **14** (1969), I:387-410; II:411-426.

[169] HLAVÀČEK, I. and NEČAS, J. On inequalities of Korn's type. *Arch. Rational Mech. Anal.* **36** (1970), 305-311.

[170] HRUSA, W.J. and TARABEK, M.A. On the smooth solutions of the Cauchy problem in one-dimensional nonlinear thermoelasticity. *Quart. Appl. Math.,* **47** (1989), 631-644.

[171] HRUSA, W.J. and MESSAOUDI, S.A. On formation of singularities in one-dimensional nonlinear thermoelasticity. *Arch. Rational Mech. Anal.* **111** (1990), 135-152.

[172] HSU, Y.S., REDDY, J.N. and BERT, C.W. Thermoelasticity of circular cylindrical shells laminated of bimodulus composite materials. *J. Thermal Stresses,* **4** (1981), 155-177.

[173] IEŞAN, D. Principes variationnels dans la théorie de la thermoelasticité couplé. *An. Şt. Univ. Iaşi, Matematică,* **12** (1966), 439-456.

[174] IEŞAN, D. Asupra teoriei termoelasticitatii neliniare. *An. Şt. Univ. Iaşi, Matematică,* **13** (1967), 161-175.

[175] IEŞAN, D. On the thermoelastic finite plane strain. *An. Şt. Univ. Iaşi, Matematică,* **15** (1969), 195-207.

[176] IEŞAN, D. On the linear coupled thermoelasticity with two temperatures. *ZAMP,* **21** (1970), 583-591.

[177] IEŞAN, D. On the thermal stresses in beams. *J. Eng. Math.* **6** (1972), 155-163.

[178] IEŞAN, D. *On some reciprocity theorems and variational theorems in linear dynamic theories of continuum mechanics.* Memorie dell'Acad. Sci. Torino, Cl. Sci. Fis. Mat. Nat., Serie 4, n. 17, 1974

[179] IEŞAN, D. Thermal stresses in the generalized plane strain of anisotropic elastic solids. *Bull. Acad. Polon. Sci. Sér. Sci. Techn.,* **18** (1970), 227-230.

[180] IEŞAN, D. Saint-Venant's problem for inhomogeneous and anisotropic elastic bodies. *J. Elasticity,* **6** (1976), 277-294.

[181] IEŞAN, D. Thermal stresses in composite cylinders. *J. Thermal Stresses,* **3** (1980), 495-508.

[182] IEŞAN, D. Incremental equations in thermoelasticity. *J. Thermal Stresses,* **3** (1980), 41-56.

[183] IEŞAN, D. Thermoelastic stresses in initially stressed bodies with microstructure. *J. Thermal Stresses,* **4** (1981), 387-405.

[184] IEŞAN, D. A theory of thermoelastic materials with voids. *Acta Mechanica,* **60** (1986), 67-89.

[185] IEŞAN, D. On Saint-Venant's problem. *Arch. Rational Mech. Anal.* **91** (1986), 363-373.

[186] IEŞAN, D. *Saint-Venant's Problem.* Lecture Notes in Mathematics. Vol. 1279, Springer-Verlag, Berlin, Heidelberg, New York, London, Paris, Tokyo, 1987.

[187] IEŞAN, D. A theory of initially stressed thermoelastic materials with voids. *An. St. Univ. Iaşi, Matematică,* **33** (1987), 167-184.

[188] IEŞAN, D. On some theorems in thermoelastodynamics. *Rev. Roum. Sci. Techn. Méc. Appl.,* **34** (1989), 101-111.

[189] IEŞAN, D. Reciprocity, uniqueness and minimum principles in the dynamic theory of thermoelasticity. *J. Thermal Stresses,* **12** (1989), 465-482.

[190] IEŞAN, D. and QUINTANILLA, R. On the grade consistent theory of micropolar thermoelasticity. *J. Thermal Stresses,* **15** (1992), 293-318.

[191] IGNACZAK, J. A completeness problem for stress equations of motion in the linear elasticity theory. *Arch. Mech. Stosow.* **15** (1963), 225-235.

[192] IGNACZAK, J. and NOWACKI, W. Singular integral equations of thermoelasticity. *Int. J. Engng. Sci.,* **4** (1966), 53-68.

[193] IGNACZAK, J. Radiation conditions in asymmetric elasticity. *J. Elasticity,* **2** (1972), 307-321.

[194] IGNACZAK, J. Thermoelastic counterpart to Boggio's theorem of linear elastodynamics. *Bull. Acad. Polon. Sci. Sér. Sci. Techn.,* **24** (1976), 129-138.

[195] IGNACZAK, J. Thermoelasticity with finite wave speeds. A survey. In *"Thermal Stresses in Severe Environments".* (D.P.H. Hasselman and R. Heller, eds.), *Plenum Press,* (1980), 15-30.

[196] IGNACZAK, J. Thermoelastic polynomials. *J. Thermal Stresses,* **6** (1983), 45-55.

[197] IGNACZAK, J., CARBONARO, B. and RUSSO, R. Domain of influence theorem in thermoelasticity with one relaxation time. *J. Thermal Stresses,* **9** (1986), 79-92.

[198] INAN, E. Coupled theory of thermoelastic plates. *Acta Mech.,* **14** (1972), 1-29.

[199] INAN, E. and ERINGEN, A.C. Nonlocal theory of wave propagation in thermoelastic plates. *Int. J. Engng. Sci.,* **29** (1991), 831-844.

[200] IONESCU-CAZIMIR, V. Problem of linear coupled thermoelasticity. Theorems on reciprocity for the dynamic problem of coupled thermoelasticity. *Bull. Acad. Polon. Sci. Sér. Sci. Techn.,* **12** (1964), 473-480.

[201] IONESCU-CAZIMIR, V. Problem of linear coupled thermoelasticity. IV. Uniqueness theorem. *Bull. Acad. Polon. Sci. Sér. Sci. Techn.,* **12** (1964), 565-573.

[202] IVANOV, T.P. and RACHEV, A.J. Thermoelastic waves in a prestressed circular cylinder. *Theoret. Appl. Mech.* **6** (1975), 53-63.

[203] JAUNZEMIS, W. *Continuum Mechanics.* Macmillan Comp., New York, London, 1967.

[204] JENTSCH, L. Zur Thermoelastostatik stückweise homogener Körper. *Arch. Rational Mech. Anal.* **24** (1967), 141-172.

[205] JIANG, S. A finite element method for equations of one-dimensional nonlinear thermoelasticity. *J. Comp. Appl. Math.* **21** (1988), 1-16.

[206] JIANG, S. Global existence of smooth solutions in one-dimensional nonlinear thermoelasticity. *Proc. Roy. Soc. Edinburgh,* **115** (1990), 257-274.

[207] JIANG, S. Far field behavior of solutions to the equations of nonlinear 1-d-thermoelasticity. *Appl. Anal.,* **36** (1990), 25-35.

[208] JIANG, S. and RACKE, R. On some quasilinear hyperbolic-parabolic initial boundary value problems. *Math. Meth. Appl. Sci.,* **12** (1990), 315-339.

[209] JIANG, S. Rapidly decreasing behaviour of solutions in nonlinear 3-d-thermoelasticity. *Bull. Austral. Math. Soc.,* **43** (1991), 89-99.

[210] JOHN, F. Hyperbolic and parabolic equations. *Partial Differential Equations.* (Edited by L. Bers and M. Schecter), pp. 1-123, New York, Wiley, 1964.

[211] JOHNSON, A.F. *Some Second Order Effects in Thermoelasticity Theory.* Report DNAM 84, National Physical Laboratory, 1974.

[212] JOSEPH D.D. and PREZIOSI L. Heat waves. *Rev. Mod. Phys.* **61** (1989), 41-73 and addendum **62** (1990), 375-391.

[213] JOU, D., CASAS-VÁZQUEZ, J. and LEBON, G. Extended irreversible thermodynamics. *Rep. Prog. Phys.,* **51** (1988), 1105-1179.

[214] KALISKI, S. and NOWACKI, W. Thermal excitations in coupled fields. In *Progress in Thermoelasticity.* (Edited by W. Nowacki), Warsaw, 1969.

[215] KASTI, N.A., LUBLINER, J. and TAYLOR, R.L. A numerical solution for the coupled quasi-static linear thermoelastic problem. *J. Thermal Stresses,* **14** (1991), 333-350.

[216] KATO, T. Linear evolution equations of hyperbolic type, II. *J. Math. Soc. Japan,* **25** (1973), 648-666.

[217] KATTIS, M.A., Thermoelastic plane problems with curvilinear boundaries. *Acta Mech.,* **87** (1991), 93-104.

[218] KEENE, F.W. and HETNARSKI, R.B. Bibliography on thermal stresses in shells. *J. Thermal Stresses,* **13** (1990), 341-540.

[219] KHDEIR, A.A., RAJAB, M.D. and REDDY, J.N. Thermal effects on the response of cross-ply laminated shallow shells. *Int. J. Solids Struct.* **29** (1992), 653-667.

[220] KNOPS, R.J. and PAYNE, L.E. On uniqueness and continuous dependence in dynamical problems of linear thermoelasticity. *Int. J. Solids Struct.* **6** (1970), 1173-1184.

[221] KNOPS, R.J. and PAYNE, L.E. Growth estimates for solutions in Hilbert space with applications to elastodynamics. *Arch. Rational Mech. Anal.* **41** (1971), 363-398.

[222] KNOPS, R.J. and PAYNE, L.E. *Uniqueness Theorems in Linear Elasticity.* Springer Tracts in Natural Philosophy, vol. 19. Berlin-Heidelberg-New York, Springer, 1971.

[223] KNOPS, R.J. and WILKES, E.W. *Theory of Elastic Stability.* pp. 125-302 of Flügge's Handbuch der Physik, vol. VI a/3 (Edited by C. Truesdell) Springer-Verlag, Berlin-Heidelberg-New York, 1973.

[224] KNOWLES, J.K. On the spatial decay of solutions of the heat equations. *ZAMP,* **22** (1971), 1050-1056.

[225] KOLYANO, Y.M., PROTSYUK, B.V. and SINYUTA, V.M. The axisymmetric static problem of thermoelasticity for a multilayered cylinder. *PMM; J. Appl. Math. Mech.,* **55** (1991), 920-926.

[226] KOVALENKO, A.D. *Thermoelasticity. Basic Theory and Applications.* Wolters-Noordhoff Publ., Groningen, 1969.

[227] KUPRADZE, V. *Potential Methods in the Theory of Elasticity.* (Edited by I. Meroz), Israel Program for Scientific Traslations, Jerusalem, 1965.

[228] KUPRADZE, V.D., GEGELIA, T.G., BASHELISHVILI, M.O. and BURCHULADZE, T.V. *Three-Dimensional Problems of Mathematical Theory of Elasticity and Thermoelasticity.* North-Holland Publ., 1979.

[229] LAGNESE, J.E. and LIONS, J.L. *Modelling Analysis and Control of Thin Plates.* Collection RMA, vol. 6, Masson, Paris, 1988.

[230] LAGNESE, J.E. The reachability problem for thermoelastic plates. *Arch. Rational Mech. Anal.* **112** (1990), 223-267.

[231] LAWS, N. Controllable state in thermoelasticity. *Quart. Appl. Math.,* **28** (1970), 432-436.

[232] LAWS, N. The overall thermoelastic moduli of transversely isotropic composites according to the self-consistent method. *Int. J. Engng. Sci.,* **12** (1974), 79-87.

[233] LEBON, G. *Variational Principles in Thermomechanics.* In "Recent Developments in Thermomechanics of Solids" (Edited by G. Lebon and P. Perzyna), Wien-New York, Springer Verlag, 1980.

[234] LEIS, R. *Exterior initial boundary value problems in thermoelasticity.* Univ. Strathclyde Sem. Appl. Math. Anal. Vibration Theory (Edited by G.F. Roach), Shiva, Nantwich, Ches., 1982.

[235] LEKHNITSKII, S.G. *Theory of Elasticity of an Anisotropic Elastic Body.* Holden-Day, Inc., San Francisco, 1963.

[236] LESSEN, M. and DUKE, C.E. On the motion of an elastic thermally conducting solid. *Proc. First Midwest Conf. Solid Mech., Univ. Illinois,* (1953), 14-20.

[237] LESSEN, M. Note on the symmetrical property of the thermal conductivity tensor. *Quart. Appl. Math.,* **14** (1956), 208-209.

[238] LESSEN, M. Thermoelasticity and thermal shock. *J. Mech. Phys. Solids,* **5** (1956), 57-60.

[239] LEVINE, H.A. On a theorem of Knops and Payne in dynamical linear thermoelasticity. *Arch. Rational Mech. Anal.* **38** (1970), 290-307.

[240] LI, Y.Y. and CHEN, Y. An addendum to "A numerical method in solving a coupled thermoelasticity equation and some results." *J. Thermal Stresses,* **7** (1984), 175-182.

[241] LINDSAY, K.A. and STRAUGHAN, B. Propagation of mechanical and temperature acceleration waves in thermoelastic materials. *J. Appl. Math. Phys.,* **30** (1979), 477-490.

[242] LIU, W.K. and CHANG, H.G. A note of numerical analysis of dynamic coupled thermoelasticity. *J. Appl. Mech.,* **52** (1985), 483-484.

[243] LOCHETT, F.J. and SNEDDON, I.N. Propagation of thermal stresses in an infinite medium. *Proc. Edin. Math. Soc.,* **11** (1959), 237-245.

[244] LOVE, A.E.H. *A Treatise on the Mathematical Theory of Elasticity.* Fourth Edition, Cambridge University Press, 1934.

[245] MARSDEN, J.E. and HUGUES, T.J.R. *Mathematical Foundations of Elasticity.* Prentice-Hall Inc. Englewood Cliffs, New Jersey, 1983.

[246] MARTIN, S.E. and CARLSON, D.E. The behavior of elastic heat conductors with second-order response functions. *Z. Angew. Math. Phys.,* **28** (1977), 311-330.

[247] MASSALS, C.V., ANAGNOSTAKI, E. and KALPAKIDIS, V.K. Some considerations on the coupled thermoelastic problems. *Int. J. Engng. Sci.,* **23** (1985), 677-684.

[248] MAYSEL, V.M. A generalization of the Betti-Maxwell theorem to the case of thermal stresses and some of its applications (in Russian) *Dokl. Acad. Sci. URSS,* **30** (1941), 115-118.

[249] MC CARTHY, M.F. *Singular Surface and Waves.* In vol. II of the *Continuum Physics* (Edited by A.C. Eringen), New York, Academic Press, 1975.

[250] MC CARTHY, M.F. and O'LEARY, P.M.O. Wave propagation in linear thermoviscoelastic materials. *J. Thermal Stresses,* **5** (1982), 347-363.

[251] MC CONNEL, A.J. *Applications of Tensor Analysis.* Dover Publ. Inc. New York, 1957.

[252] MIKHLIN, S.G. *Multidimensional Singular Integrals and Integral Equations* (in Russian), Nayka, Moskow, 1962.

[253] MIKHLIN, S.G. *Numerical Realization of Variational Methods* (in Russian), Moscow, 1966.

[254] MISICU, M. *Mechanics of Deformable Media. Fundamentals of Structural Elasticity.* (in Romanian). Edit. Acad. R.S.R., Bucuresti, 1967.

[255] MISRA, J.C. and ACHARI, R.M. On axisymmetric thermal stresses in an anisotropic hollow cylinder. *J. Thermal Stresses,* **3** (1980), 509-520.

[256] MOLINARI, A. Thermoélasticité dynamique: un demi-espace soumis à un échelon de flux thermique. *J. Mécanique,* **18** (1979), 73-101.

[257] MOSHAIOV, A. Finite element formulation of coupled thermo-elastic plastic beam bending and axial response. *Comput. Struct.,* **41** (1991), 1023-1028.

[258] DE MOURA, C.A. and SUAIDEN, T. Termoelastodinamica nao-linear: experimentos numéricos em torno do modelo unidimensional de Slemrod. *Atas II Journados Latino-Americanas de Matemática Aplicada,* **1** (1983), 22-45.

[259] MRITHYUMJAYA RAO, K. and KESAVA RAO, B. Longitudinal vibrations of a prestressed thermoelastic transversely isotropic circular cylinder. *Proc. Natl. Acad. Sci., India, Sect. A,* **46** (1976), 263-271.

[260] MUKI, R. and BREUER, S. Coupling effects in a transient thermoelastic problem. *Österr. Ing. Arch.,* **16** (1962), 349-368.

[261] MÜLLER, I. The coldness, a universal function in thermoelastic bodies. *Arch. Rational Mech. Anal.,* **41** (1971), 319-332.

[262] MUÑOZ RIVERA, J.E. Energy decay rates in linear thermoelasticity. *Funkcial. Ekvac.,* **35** (1992), 19-30.

[263] MURPHY, W.K. and CARLSON, D.E. On the characterization of the mixed problem of coupled, dynamic, linear thermoelasticity in terms of the temperature. *J. Thermal Stresses,* **6** (1983), 139-152.

[264] MUSKHELISHVILI, N.I. *Some Basic Problems of the Mathematical Theory of Elasticity.* (in Russian). Nauka, Moscow, 1966.

[265] NABARRO, F.R.N. The calculation of thermal stresses in cylinders. *Int. J. Engng. Sci.,* **19** (1981), 1651-1656.

[266] NAGHDI, P.M. *The Theory of Shells and Plates.* In "Handbuch der Physik". vol. VI a/2. (Edited by C. Truesdell) Springer-Verlag, Berlin-Heidelberg-New York, 1972.

[267] NARUKAWA, K. Boundary value control of thermoelastic system. *Hiroshima Math. J.,* **13** (1983), 227-272.

[268] NAVARRO, C.B. and QUINTANILLA, R. On existence and uniqueness in incremental thermoelasticity. *ZAMP,* **35** (1984), 206-215.

[269] NEČAS, J. *Les Méthodes Directes en Théorie des Equations Elliptiques.* Academia, Prague, 1967.

[270] NEČAS, J. and HLAVAČEK, I. *Mathematical Theory of Elastic and Elasto-Plastic Bodies: An Introduction.* Elsevier Scientific Publ. Comp. Amsterdam, New York, 1981.

[271] NEMAT-NASSER, S., KEER, L.M. and PARIHAR, K.S. Unstable growth of thermally induced interacting cracks in brittle solids. *Int. J. Solids Struct.,* **14** (1978), 409-430.

[272] NICKELL, R.E. and SACKMAN, J.L. Variational principles for linear coupled thermoelasticity. *Quart. Appl. Math.,* **26** (1968), 11-26.

[273] NICKELL, R.E. and SACKMAN, J.L. Approximate solutions in linear coupled thermoelasticity. *J. Appl. Mech.,* **35** (1968), 255-266.

[274] NISTOR, I. A theorem of existence, uniqueness and analyticity of the solution of the equations of nonlinear thermoelasticity. *An. St. Univ. "Al. I. Cuza". Iasi, Matematica,* **19** (1973), 465-476.

[275] NODA, N. Transient thermoelastic contact problem in a long, circular cylinder. *J. Thermal Stresses,* **7** (1984), 135-147.

[276] NOLL, W. A mathematical theory of the mechanical behaviour of continuous media. *Arch. Rational Mech. Anal.,* **2** (1958), 197-218.

[277] NOWACKI, W. *Thermoelasticity.* Reading, Mass., Addison-Wesley, 1962.

[278] NOWACKI, W. Green functions for the thermoelastic medium II. *Bull. Acad. Polon. Sci. Sér. Sci. Techn.,* **12** (1964), 465-472.

[279] NOWACKI, W. *Dynamics Problems of Thermoelasticity.* PWN-Noordhoff, Warsaw, 1975.

[280] NOWACKI, W. *Theory of Asymmetric Elasticity.* Polish Scientific Publishers, Warszawa and Pergamon Press, Oxford, New York, Paris, Frankfurt, 1986.

[281] NOWINSKI, J.L. Thermal waves in an elastic highly stretched cylindrical bar. *Acta Mech.,* **7** (1969), 45-57.

[282] NOWINSKI, J.L. *Theory of Thermoelasticity with Applications.* Sijthoff & Noordhoff, Int. Publ. Alphen aan Den Rijn, Netherlands, 1978.

[283] NOWINSKI, J.L. Note on the solution of the equations controlling stability of nonlinear thermoelastic waves in a spinning disk. *J. Thermal Stresses,* **7** (1984), 75-78.

[284] OCHIAI, Y, ISHIDA, R. and SEKIYA, T. Unsteady thermal stress analysis in three-dimensional problems by means of the thermoelastic displacement potential and the boundary element method. *J. Strain Anal.,* **25** (1990), 9-14.

[285] ODEN, J.T. and POE, J. On the Numerical Solution of a Class of Problems in Dynamic Coupled Thermoelasticity. In *Proc. 5th SECTAM 1970* (D. Frederick, ed.), Pergamon Press, Oxford, 1970.

[286] ODEN, J.T. *Finite Elements of Non-Linear Continua.* Mc Graw Hill, New York, 1972.

[287] OLESIAK, Z. and SNEDDON, I.N. The distribution of thermal stress in an infinite elastic solid containing a penny-shaped crack. *Arch. Rational Mech. Anal.,* **3** (1960), 238-254.

[288] OLESIAK, Z. On a method of solution of mixed boundary-value problems of thermoelasticity. *J. Thermal Stresses,* **4** (1981), 501-508.

[289] ORTHWEIN, W.C. Thermal buckling of a bimetallic strip. *J. Vib. Acoust. Stress Reliab.,* **106** (1984), 538-542.

[290] PADOVAN, J. and LESTINGI, J. Thermoelasticity of anisotropic cylindrical shells. *J. Thermal Stresses,* **3** (1980), 261-276.

[291] PADOVAN, J. *Anisotropic Thermal Stresses Analysis.* In "Thermal Stresses" (Edited by R.B. Hetnarski), pp. 143-262, North-Holland, Amsterdam, 1986.

[292] PARKUS, H. *Thermoelasticity.* Blaisdell Publ. Co., Waltham, Mass, 1968.

[293] PAYNE, L.E. *Improperly Posed Problems in Partial Differential Equations.* Regional Conf. Ser. in Appl. Math. Vol. 22, SIAM, 1975.

[294] PAYNE, L.E. and STRAUGHAN, B. Error estimates for the temperature of a piece of cold ice, given data on only part of the boundary. *Nonlin. Anal. Theory Meth. Appl.,* **14** (1990), 443-452.

[295] PAZY, A. *Semigroups of Linear Operators and Applications to Partial Differential Equations.* Springer, New York, Berlin, Heidelberg, Tokyo, 1983.

[296] PERCIVAL, C.M. A quantitative measurement of thermally induced stress waves. *AIAA J.,* **9** (1971), 347-349.

[297] PETROSKI, H.J. and CARLSON, D.E. Controllable states of elastic heat conductors. *Arch. Rational Mech. Anal.,* **31** (1968), 127-150.

[298] PETROSKI, H.J. and CARLSON, D.E. Some exact solutions to the equations of nonlinear thermoelasticity. *J. Appl. Mech.,* **37** (1970), 1151-1154.

[299] PETROSKI, H.J. On the finite torsion and radial heating of the thermoelastic cylinders. *Int. J. Solids Struct.,* **11** (1979), 741-749.

[300] PHURKHAO, P. and KASSIR, M.K. Thermoelastic wave propagation in an elastic solid containing a finite crack. *Int. J. Fracture,* **47** (1991), 213-228.

[301] PIPKIN, A.C. and RIVLIN, R.S. *The Formulation of Constitutive Equations in Continuum Physics.* Div. Appl. Math., Brown University Repost, September, 1958.

[302] PODSTRIGACH, Ya.S. and SHVETS, R.N. *Thermoelasticity of Thin Shells* (in Russian) Naukova Dumka, Kiev, 1978.

[303] QUINTANILLA, R. A short note on incremental thermoelasticity. *Stochastica,* **7** (1983), 163-167.

[304] RACKE, R. Eigenfunction expansions in thermoelasticity. *J. Math. Anal. Appl.,* **120** (1986), 596-609.

[305] RACKE, R. On the Cauchy problem in nonlinear 3-d-thermoelasticity. *Math. Z.,* **203** (1990), 649-682.

[306] RACKE, R. Blow-up in nonlinear three-dimensional thermoelasticity. *Math. Meth. Appl. Sci.,* **12** (1990), 267-273.

[307] RACKE, R. $L^p - L^q$- estimates for solutions to the equations of linear thermoelasticity in exterior domains. *Asymptotic Analysis,* **3** (1990), 105-132.

[308] RACKE, R. and SHIBATA, Y. Global smooth solutions and asymptotic stability in one-dimensional nonlinear thermoelasticity. *Arch. Rational Mech. Anal.,* **116** (1991), 1-34.

[309] RACKE, R. *Mathematical Aspects in Nonlinear Thermoelasticity.* SFB 256 Preprint No. 25, Universität Bonn, 1992.

[310] REISS, R. and HAUG, E.J. Extremum principles for linear initial value problems of mathematical physics. *Int. J. Engng. Sci.*, **16** (1978), 231-251.

[311] RIONERO, S. and CHIRIŢA, S. The Lagrange identity method in linear thermoelasticity. *Int. J. Engng. Sci.*, **25** (1987), 935-947.

[312] RIVLIN, R.S. and TOPAKOGLU, C. A theorem in the theory of finite elastic deformations. *J. Rational Mech. Anal.*, **3** (1954), 581-589.

[313] RIVLIN, R.S. *Foundations of Continuum Thermodynamics.* (Edited by N. Nina and J. Whitelaw), McMillan, London, 1975.

[314] ROKNE, J.G., DHALIWAL, R.S. and SINGH, B.M. Thermal stresses near a penny-shaped crack in an elastic sphere embedded in an infinite elastic space. *Int. J. Engng. Sci.*, **18** (1980), 681-701.

[315] RUSSO, R. Classical coupled thermoelasticity in unbounded domains. *J. Elasticity,* **22** (1989), 1-24.

[316] RUSU, GH. On existence and uniqueness in thermoelasticity of materials with voids. *Bull. Acad. Polon. Sci. Sér. Sci. Techn.*, **35** (1987), 339-346.

[317] SCALIA, A. Teoremi di unicitá per le equazioni di Maxwell con condizioni al contorno dissipative. *Ann. Univ., Ferrara, Sez. VII, Sc. Mat.*, **28** (1982), 105-113.

[318] SCALIA, A. A grade consistent micropolar theory of thermoelastic materials with voids. *ZAMM,* **72** (1990), 133-140.

[319] SCALIA, A. On some theorems in the theory of micropolar thermoelasticity. *Int. J. Engng. Sci.*, **28** (1990), 181-189.

[320] SCALIA, A. On some theorems in the linear micropolar thermoviscoelasticity. *An. St. Univ. Iasi, Matematica,* **37** (1991), 249-260.

[321] SCOTT, N.H. A theorem in thermoelasticity and its application to linear stability. *Proc. R. Soc. Lond. A* **424** (1989), 143-153.

[322] SHANKER, M.U. and DHALIWAL, R.S. Singular integral equations in a-symmetric thermoelasticity. *J. Elasticity,* **2** (1972), 59-71.

[323] SHERMAN, D.I. On the problem of plane strain in nonhomogeneous media. In *Non-Homogeneity in Elasticity and Plasticity.* Pergamon Press, London, 1959.

[324] SIGNORINI, A. Transformazioni termoelastiche finite. Memoria 3ª, Solidi incomprimibili. *Ann. Mat. Pura Appl.*, **39** (1955), 147-201.

[325] SINGH, B.M., ROKNE, J., DHALIWAL, R.S. and HETNARSKI, R.B. Thermal stresses in a long elastic cylinder containing a penny-shaped crack and embedded in a thermally conductive elastic infinite medium. *J. Thermal Stresses,* **2** (1979), 449-473.

[326] SLADEK, V. and SLADEK, J. Boundary integral equation method in thermoelasticity. Part. 2: Crack Analysis, *Appl. Math. Model.*, **8** (1984), 27-36.

[327] SLEMROD, M. Global existence, uniqueness, and asymptotic stability of classical smooth solutions in one-dimensional nonlinear thermoelasticity. *Arch. Rational Mech. Anal.*, **76** (1981), 97-133.

[328] SNEDDON, I.N. *Fourier Transforms.* Mc Graw-Hill Book Co., New York, 1951.

[329] SNEDDON, I.N. Boundary value problems in thermoelasticity. In *Boundary Problems in Differential Equations,* Proc. of the 2nd Symp. on Diff. Eqs., University of Wisconsin Press, Madison, Wis., (1960), 231-240.

[330] SNEDDON, I.N. and LOCKETT, F.J. On the steady-state thermoelastic problem for the half-space and the thick plate. *Quart. Appl. Math.,* **18** (1960), 145-153.

[331] SNEDDON, I.N. and LOCKETT, F.J. The steady-state thermoelastic problem for an elastic layer resting on a rigid fondation. *Ann. Mat. Pura Appl. s. IV,* **50** (1960), 309-317.

[332] SNEDDON, I.N. Solutions of the equations of thermoelastic equilibrium. *Arch. Mech. Stosow.,* **14** (1962), 113-120.

[333] SOLOMON, L. *Élasticité Linéaire.* Masson, Paris, 1968.

[334] SOÓS, E. The Green's function (for short time) in the linear theory of the coupled thermoelasticity. *Arch. Mech. Stosow.,* **18** (1966), 101-109.

[335] SOÓS, E. *Discrete and Continuous Models of Solids* (in Romanian) Editura Stiintificá, Bucuresti, 1974.

[336] STENBERG, E. and MC DOWELL, On the steady-state thermoelastic problem for the half-space. *Quart. Appl. Math.,* **14** (1957), 381-398.

[337] STOPPELLI, F. Sull'esistenza di soluzioni delle equazioni dell'elastostatica isoterma nel caso di sollecitazioni dotate di assi di equilibrio. *Ricerche Mat.,* **6** (1957), 241-287; **7** (1958), 71-101; 138-152.

[338] STRAUGHAN, B. *Instability, Non-Existence and Weighted Energy Methods and Related Theories.* Pitman, Boston, 1982.

[339] STRAUGHAN, B. Stability problems in electrodynamics, ferrohydrodynamics and thermoelectric magnetohydrodynamics. In *"Mathematical Topics in Fluids."* (J.F. Rodrigues and A. Sequeira, eds.), pp. 163-192, Pitman, Longman, 1992.

[340] SUGANO, Y. A general treatise of plane thermoelastic problem in multiply-connected region exhibiting temperature-dependent properties. *Z. Angew. Math. Mech.,* **64** (1984), 163-172.

[341] ŞUHUBI, E.S. Propagation of a plane wave in an initially stressed thermoelastic medium. *Bull. Techn. Univ. Istanbul,* **24** (1972), 54-73.

[342] ŞUHUBI, E.S. *Thermoelastic Solids.* In vol. II of the *Continuum Physics* (Edited by A.C. Eringen), New York, Academic Press, 1975.

[343] SVE, C. Thermoelastic waves in a periodically laminated medium. *Int. J. Solids Struct.,* **7** (1971), 1363-1373.

[344] TAKEUTI, Y. On some three-dimensional thermal stress problems in industrial applications. In *Thermal Stresses I* (Edited by R.B. Hetnarski), Elsevier, Amsterdam, 1986.

[345] TAUCHERT, T.R. Thermoelastic analysis of laminated orthotropic slabs. *J. Thermal Stresses,* **3** (1980), 117-132.

[346] TAUCHERT, T.R. Thermal stresses in plates-statical problems. In *Thermal Stresses* (Edited by R.B. Hetnarski), pp. 23-142, North-Holland, Amsterdam, 1986.

[347] TAUCHERT, T.R. and HETNARSKI, R.B. Bibliography on thermal stresses. *J. Thermal Stresses,* **9** (1986), Supplement 1-128.

[348] TEODORESCU, P.P. *Dynamics of Linear Elastic Bodies.* Abacus Press, Tunbridge Wells. Kent, England, 1975.

[349] THOMAS, T.Y. *Plastic Flow and Fracture Solids,* Academic Press, New York, 1961.

[350] THOMAS, T.Y. *Concepts from Tensor Analysis and Differential Geometry,* Academic Press, New York, 1961.

[351] TOMAR, J.S. and JAIN, R. Thermal effect on frequencies of coupled vibrations of pretwisted rotating beams. *AIAA J.,* **23** (1985), 1293-1296.

[352] TRUESDELL, C. and TOUPIN, R. *The Classical Field Theories.* In vol. III/1 of the *Handbuch der Physik* (Edited by S. Flügge) Berlin-Heidelberg-New York, Springer-Verlag, 1960.

[353] TRUESDELL, C. General and exact theory of waves in finite elastic strain. *Arch. Rational Mech. Anal.,* **8** (1961), 263-296.

[354] TRUESDELL, C. and NOLL, W. *The Non-Linear Field Theories of Mechanics.* In vol. III/3 of the *Handbuch der Physik* (Edited by S. Flügge) Berlin-Heidelberg-New York, Springer-Verlag, 1965.

[355] TRUESDELL, C. Existence of longitudinal waves. *J. Acoust. Soc. Am.,* **40** (1966), 729-730.

[356] TRUESDELL, C. *History of Classical Mechanics.* Die Naturwissenschaften 63, Part. I: to 1800, 53-62; Part II: the 19th and 20th Centuries, 119-130. Springer Verlag, 1976.

[357] TRUESDELL, C. Some challenges offered to analysis by rational thermomechanics, pp. 495-603 of *Contemporary Developments in Continuum Mechanics and Partial Differential Equations* (Edited by G.M. de la Penha and L.A. Medeiros), North-Holland, 1978.

[358] TRUESDELL, C. *Rational Thermodynamics.* Springer-Verlag, New York, Berlin, Heidelberg, Tokyo, 1984.

[359] VAN der POL, B. and BREMMER, H. *Operational Calculus Based on the Two-Side Laplace Integral,* Cambridge, 1950.

[360] VIGAK, V.M., KOSTENKO, A.V. and ZASADNA, K.E. Optimum control of the heating of simple thermoelastic bodies by means of heat sources. *Sov. Appl. Mech.* (Eng. tr.), **25** (1989), 271-274.

[361] WANG, C.C. and TRUESDELL, C. *Introduction to Rational Elasticity.* Leyden, Noordhoff, 1973.

[362] WEINER, J.H. A uniqueness theorem for the coupled thermoelastic problem. *Quart. Appl. Math.,* **15** (1957), 102-105.

[363] WILKES, N.S. Continuous dependence and instability in linear thermoelasticity. *SIAM J. Math. Anal.,* **11** (1980), 292-299.

[364] WILLIAMS, H.E. Asymptotic analysis of the thermal stresses in a two-layer composite with an adhesive layer. *J. Thermal Stresses,* **8** (1985), 183-204.

[365] YOSIDA, K. *Functional Analysis.* Springer-Verlag, Berlin-Heidelberg-New York, 1968.

[366] ZHENG, S. and SHEN, W. Global solutions to the Cauchy problem of quasilinear hyperbolic parabolic coupled systems. *Sci. Sinica, Ser. A,* **30** (1987), 1133-1149.

[367] ZIBDEH, H.S. Reliability of thermally loaded cylinders. *J. Pressure Vessel Tech.,* **112** (1990), 303-308.

[368] ZIEGLER, H. Discussion of some objections to thermomechanical orthogonality. *Ing. Arch.,* **50** (1981), 149-164.

[369] ZIENKIEWICZ, O.C. and CHEUNG, Y.K. *The Finite Element Method in Structural and Continuum Mechanics.* New York, Mc Graw Hill, 1967.

[370] ZIENKIEWICZ, O.C. The finite element method; from intuition to generality. *Appl. Mech. Rev.,* **23** (1970), 249-256.

[371] ZORSKI, H. Singular solutions for thermoelastic media. *Bull. Acad. Polon. Sci. Sér. Sci. Techn.,* **6** (1958), 331-339.

Index

MIX
Papier aus verantwortungsvollen Quellen
Paper from responsible sources
FSC® C105338

If you have any concerns about our products,
you can contact us on
ProductSafety@springernature.com

In case Publisher is established outside the EU,
the EU authorized representative is:
Springer Nature Customer Service Center GmbH
Europaplatz 3, 69115 Heidelberg, Germany

Printed by Libri Plureos GmbH
in Hamburg, Germany